Majewski
MuPAD Pro Computing Essentials

Springer

Berlin
Heidelberg
New York
Hong Kong
London
Milan
Paris
Tokyo

Miroslaw Majewski

MuPAD Pro
Computing Essentials

Second Edition
Revised for MuPAD Pro 3.0

 Springer

Miroslaw Majewski
Zayed University
College of Information Systems
P.O. Box 4783
Abu Dhabi
United Arab Emirates
http://www.mupad.com/majewski/

Library of Congress Control Number: 2004104812

Mathematics Subject Classification (2000): 97U50, 97U70

ISBN 3-540-21943-9 Springer-Verlag Berlin Heidelberg New York
ISBN 3-540-43574-3 1st edition Springer-Verlag Berlin Heidelberg New York

Springer-Verlag is a part of Springer Science+Business Media

springeronline.com

© Springer-Verlag Berlin Heidelberg 2002, 2004
Printed in Germany

Typesetting by the author using Scientific Notebook from MacKichan Software, Inc.
Cover graphics by the author using Strata 3D®
Cover design: *design & production* GmbH, Heidelberg

Printed on acid-free paper   40/3142/LK - 5 4 3 2 1 0

PREFACE

The second edition of this wonderful book promises to have a significant impact on how college-level mathematics is taught and learned. While the first edition was mainly addressed to mathematics teachers, its audience turned out to be significantly wider than intended. In fact, user reaction to the first edition suggests that the book has already changed the way certain aspects of undergraduate mathematics are viewed and understood.

Apart from the breadth of topics covered in this book, three features make this book a must-read for most mathematics and many science students. The book shows how the use of a computer-algebra system can enhance, facilitate and accelerate the learning of mathematics. This is particularly true for MuPAD, given its flexible pedagogical strengths and focus. The book also shows how structural programming encourages, motivates, and justifies mathematical rigor. The third and perhaps most striking feature of the book is its emphasis on visualization. This book takes its readers gently by the hand and helps them explore the visual properties of a large collection of basic mathematical objects in a way that is simply superb. The conversational style adopted by the author builds confidence, creates excitement, and may even have an influence on the content of the undergraduate mathematics curriculum in the years to come.

The selection of topics making up mainstream mathematics has always been in a state of flux, depending on existing mathematical knowledge and discovery, our changing understanding and interpretation of basic mathematical theorems and concepts, newly-found solutions to important mathematical problems, the interests of young researchers, and the computational needs of users of mathematics. This book adds new dimensions to this dynamic by helping to shape the view of mathematics of a new generation and by stimulating their visual imagination. This book is one of the first to provide us with an exciting glimpse into the vast range of possibilities for rethinking what and how we teach in our mathematics courses.

As mentioned by the author, neither the first edition, nor this new addition of *MuPAD Pro Computing Essentials* pretends to be all things to all people. They represent a very personal account of a new perspective of how mathematics can be taught and studied with the help of computer algebra. The selection of topics in these books is broad enough to satisfy the needs of most college and undergraduate university mathematics majors programs. However, user feedback has already resulted in significant changes and improvements to the first addition. While reader influence is apparent in almost all chapters of this second edition, the author also takes full advantage of advances in the development of the MuPAD computer algebra system. This particularly apparent in Chapter 7, which is completely new, and precedes the descriptive exploration of curves and surfaces in Chapter 8 with a fascinating and manageable introduction the dynamic world of interactive graphics and computer animation.

Teachers of mathematics are currently still locked in vigorous debate about the virtues of computer-assisted teaching and learning. Opponents of the use of this technology argue that student fails to learn the basics. All they manage to acquire is a facility for pressing appropriate buttons to achieve mathematical output that they fail to understand. This is precisely why it is essential that the proponents of computer-assisted teaching and learning write good books that illustrate the pedagogical and mathematical benefits of technology. The present text is an excellent example of what is needed. It shows clearly the pedagogical value of a modest form of structural programming, and explains in motivational detail the basic steps and structure of many of the algorithms usually studied by mathematics undergraduates.

Let us briefly consider the range of topics covered in the text. It illustrates the comprehensive nature and extent of possible use of this book as a stand-alone textbook for college-level mathematics major programs.

The first five chapters still deal with the mechanics of using MuPAD, but in much greater detail and with more mathematical emphasis than the corresponding chapters in the first edition. They provide a careful introduction to basic principles of mathematical programming and algorithmic thinking. This is appropriate for several reasons. First of all, it is required reading for those interested in using MuPAD. But it is also indispensable for all mathematics students who hope to use their knowledge in the workplace. Today, and in the years to come, most mathematics graduates worth their salt are expected to be able to program in much the same way as they were expected to be able to use logarithm tables, slide rules and other gadgets in the past.

Chapters 9 to 13 provide an excursion into the more traditional topics of college mathematics: the language of sets, number systems, and some algebra, trigonometry, calculus and linear algebra. As such, *MuPAD Pro Computing Essentials* represents, in a real sense, a launch pad for the study of deeper mathematics with the help of MuPAD. The rapid development of specialized and advanced MuPAD libraries makes it possible to advance this project well beyond the practical limits set for this book.

I am looking forward to introducing my students to this new edition of *MuPAD Pro Computing Essentials*.

Prof. Fred E. Szabo

Concordia University

Montreal, Canada

Contents_____

Chapter 1 _____

Introduction

If you are like me, then you are definitely eager to know what is inside this book and how it can be useful to you. Let me take this opportunity to tell you in just a few words, for whom I wrote this book, and what my objective was.

1.1 For whom this book was written

Originally I wrote this book for mathematics teachers who want to explore new ways of teaching mathematics with a computer. However, when the first edition of this book came out into the daylight, I found that my readers were not only teachers, but also many students of various courses looking for new ways of solving mathematical problems. I was not surprised when I found that a number of mathematics courses in a few European universities had been built based on this book. In fact, I have made heavy use of large parts of my book for the *Computing Foundations* course at my own university. Teachers and university instructors can use this book as a starting point to any course where the computer can make a difference and then build the rest of the course around it.

1.2 The goal

This book should be considered as the first steps through mathematics with MuPAD. It is not a MuPAD reference book and, in fact, many MuPAD-related topics are not discussed here at all. Nor is it a text for a regular computer-assisted course of mathematics. Instead, it is an exciting excursion through different areas of mathematics assisted by MuPAD. I will show you the basic instructions that are useful for these specific areas. I will explore many topics and show you many examples. However, it may turn out that the particular topic you are interested in has been omitted. If this is the case, you have two possible choices—one, you can try to work it out on your own; and two, you can write to me, and I will try to add this topic in the next edition of the book. In fact some of the topics added in this edition were suggested by my readers.

I have tried to make this book as interesting and approachable as possible. As you have probably noticed by now, the style of this book is *me* talking to *you*. This is not only because I think it is easier to read, but also because this writing style is what comes naturally for me.

I like to stay in touch with my readers. For this purpose, a web site has been developed for this book. There you will find my current e-mail address, the source code for the MuPAD programs mentioned in this book, as well as bug fixes and some updates. You can find the web site for my book at www.mupad.com/majewski/.

For the second edition of this book, I have used MuPAD Pro version 3.0. Due to the changes between versions 1.x, version 2.0, and the later versions, some examples and constructions in this book might not work in earlier versions of MuPAD. However, in many cases you should easily be able to convert them to older versions. It is important to mention that MuPAD Pro version 3.0 introduces a completely new standard of mathematical graphics. Therefore, some of the graphical examples presented here cannot be translated to earlier versions at all.

1.3 Why we should care about MuPAD

We all know that the teaching of mathematics can benefit a lot from Computer Algebra Systems (CAS). By using CAS, we can visualize mathematical concepts, and especially various types of functions and equations; we can also solve complicated equations without tedious calculations or transform formulae without making difficult-to-find errors. There are several powerful CAS that can be used to teach mathematics. Each of them has its own good features and drawbacks. Sometimes, the drawback can be the price of the package; sometimes its difficult syntax; sometimes hardware requirements that are too high to meet in a school environment.

MuPAD is one of the youngest of these mathematical packages, and for this reason it is not as popular as Maple, Mathematica or Derive. MuPAD's development started in the early nineties. Until this time, the software market had been dominated by commercial and often very expensive packages. However, in the early nineties we saw a new trend in software development—the so-called open source software. It was at this time that such famous systems as Linux, MuPAD, POV-Ray, and many other free or inexpensive programs were developed.

MuPAD is neither freeware nor open source software, but shares similar beginnings with the latter. Its development started with a few students' master's theses at the University of Paderborn. Since then, over the last decade, many students and staff members of the university have contributed to its development. Currently, MuPAD is the most serious competitor for such powerful packages as Mathematica or Maple. For mathematics educators, its innovative features and the low price of the software are especially important. Indeed, if you are going to teach a course with MuPAD, you can find a number of inexpensive ways of getting MuPAD for your classroom—just check the web site `www.mupad.org/muptan.html` to find the best option for you.

1.4 What is inside

The first seven chapters of this book are focused on the basics of using MuPAD. We begin with the syntax of MuPAD commands and declarations (chapters 2 and 3), and programming control structures (chapter 4). In chapter 5, we move on to writing procedures and using MuPAD libraries. Chapters 6, 7 and 8 are devoted to graphics in MuPAD. I discuss there the syntax of plotting commands, showing how to plot curves and surfaces and how to develop MuPAD animation. At this point, we move on to my major point of interest — applications of MuPAD in mathematics. In chapter 9, I describe some uses of MuPAD graphics in calculus and geometry. Chapter 10 is devoted to different types of numbers. Finally, in the last four chapters, we move on to elementary algebra (chapter 11), logic and set theory (chapter 12), calculus (chapter 13) and linear algebra (chapter 14).

Each chapter of this book ends with a summary of the MuPAD elements that were introduced in the chapter. At the end of each chapter, I also enclose a set of programming exercises to be done by my readers. Most of these exercises are at a basic level; usually I ask my readers to develop a short MuPAD program or procedure with no more than 20 lines. This will help you to better understand the nature of MuPAD programming, and at the same time to build your confidence in using MuPAD in the classroom. I suppose the sets of exercises should really be much larger. However, I believe that almost any topic in high school or university mathematics can lead to a number of interesting programming activities.

1.5 Style conventions

Some items appearing in this book are given a special appearance to set them apart from the regular text. Here is how they look:

- `constant width text with a bullet`
  `or without a bullet but with an indent on the left side`

is used for MuPAD code listings.

Text using the Times New Roman font

is used for the output from MuPAD commands and programs.

Text with gray background

is used to highlight syntax of MuPAD statements.

Inside paragraphs, I use *Text in italics* to emphasize various terms, names and things that should be set apart from the rest of text, and `text with constant width` for MuPAD statements. Finally, the black square ■ is used to mark the end of each major example.

1.6 Writing "between"

The best word to describe this book is probably the word "between." This book was written between many things—different places, different versions of MuPAD, different times, different interests, and different people.

While writing this book I have moved from Far East Asia to the Middle East. The beginning of this book was written in the small city of Macau on the coast of the South China Sea, some of the middle chapters were written in Poland, and the last few chapters were written in Abu Dhabi, in the Persian Gulf. While writing this book, I have thus moved from the tranquility of East Asia to the atmosphere of war in the Middle East after September 11, 2001.

Writing this book, I continued to remain trapped between my two major interests — computer graphics and mathematics. In this book, you will notice a lot of influence of computer graphics. It is for this reason that the chapters about MuPAD graphics occupy about one third of the book. In the second edition of my book, these chapters have been completely rewritten in order to cover the new graphics standard introduced in MuPAD 3.0. I have also added here a large section about animation with MuPAD. However, I still have a feeling that too many things related to graphics are missing here. You will find them in my next book, which will be completely devoted to mathematical graphics with MuPAD.

I used a number of unofficial versions of MuPAD during the writing of the book. I started off with version 2.0; the later chapters however, were written with the alpha and beta tests of version 2.5. The second edition of my book was has been completely rewritten to cover version 3.0. I was thus able to capture some of the most significant changes in MuPAD. A few times, while writing this text, I got the feeling that one feature or another could work in a different way. It was quite surprising for me that some of my suggestions were implemented in MuPAD in a matter of hours or days. Working with the MuPAD team was and still is a real adventure. The ever-changing MuPAD was the biggest challenge for me. Many features and even concepts changed from version to version and definitely may also change in future versions. However, it was a great pleasure to witness such wonderful development.

Finally, I wrote this book between many people and each of them, knowingly or not, had some influence on my work. Let me introduce some of them here.

I shall start with Enrique Wintergerst and Barry MacKichan who, a few years ago, encouraged me to look into MuPAD. This was at the time when MacKichan Software Inc. had just decided to implement MuPAD as the computing engine in their Scientific Workplace® and Scientific Notebook® packages. In fact, thanks to MacKichan Software Inc. I use Scientific Notebook® to write all my texts, including this book, and prepare them for publishing. I cannot imagine how I could write any mathematical text without this wonderful tool.

A number of people from the MuPAD Research Group also had a great influence on my work. Allow me to mention some of them. Frank Postel gave me a lot of hints at the beginning of my MuPAD path, when I was trying to work out some puzzling features of MuPAD. Oliver Kluge, Ralf Hillebrand, Christopher Creutzig, Klaus Drescher, Stefan Wehmeier, Walter Oevel, Jürgen Billing, Torsten Metzner, Andreas Sorgatz and many others are those who were so patient in discussing with me the various features of MuPAD. Christopher and Stefan carefully read my book and suggested many improvements. When talking about the MuPAD team, I must also mention Prof. Benno Fuchssteiner, who invited me to Paderborn and encouraged me to work with his team.

I am very grateful to the many individuals who reviewed the first edition of the book, made valuable suggestions, and brought a large number of errors and omissions to my attention. They include: Dr. Wilhelm Forst from the University of Ulm in Germany, Dr. Friedrich Schwarz from University of Paderborn in Germany, and Dr. Fred Szabo from the Concordia University in Canada.

I need to also mention my son Jakub, who spent a significant amount of time searching for spelling and grammar errors in the manuscript. Furthermore, I could not omit here my wife and daughter, who both had a rather difficult time whenever I tried to concentrate on my work.

Finally, I would like to extend my gratitude to Georgios Dalaras, whose music gave me a lot of joy and inspiration while writing this book.

I thank all of you for your help, encouragement, or at least tolerating my passion for this work.

Zayed University, UAE, Spring 2004   *Miroslaw Majewski*

Chapter 2

A Quick Introduction to MuPAD

Before ordering or downloading MuPAD, you should make sure that you have hardware good enough to run this program.

In order to run MuPAD on your computer, you need an IBM compatible PC (Pentium III or higher) with 64 MB of RAM, a CD-ROM drive, and about 110 MB of available disk space. However, 256 MB of RAM and a Pentium III with an 800 MHz processor or better are recommended. More resources are especially needed for the new 3D graphics. Your system must be running Windows 98/ME/NT 4.0/2000/XP, Mac OS or the Linux operating system.

In order to take full advantage of the many resources provided by the MuPAD team, you may also need an Internet connection and access to e-mail.

The second thing you will need is, of course, the MuPAD software. MuPAD is distributed by SciFace Software GmbH—a software company in Germany. You can download a trial version of MuPAD, which will work for one month, from their web site at http://www.mupad.com. You can also check the MuPAD TAN server at http://www.mupad.org/muptan.html to find detailed licensing information, download MuPAD, buy it online and obtain the registration key.

2.1 The Very First Look

In order to start MuPAD on a Windows-based system, you only need to click on its icon from the Start menu, or find the shortcut to the MuPAD executable file on your desktop. It should look like this:

Now you are ready to start MuPAD. Double-click the icon on the desktop, and the MuPAD window will open up on your computer screen. At first, it will look similar to the one pictured below, but after a few days of using MuPAD you will almost certainly have introduce some modifications.

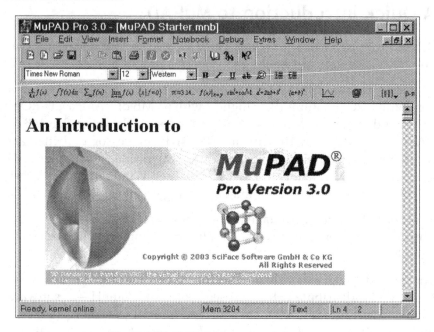

Fig. 2.1 MuPAD Pro version 3.0 screen

If this is your first contact with MuPAD, you may want to read the document on the screen. Alternatively, you can just start your exploration of MuPAD. In order to start working with MuPAD, choose the **File** menu and then the option **New Notebook**. This can also be done using the shortcut [Ctrl]+[N].

CONVENTION: Throughout this book I will frequently ask you to press a key or a combination of keys. It may look like this, [Ctrl] or, [Ctrl]+[N]. A single square bracket means that you have to press a single key with the given text or character; for example [Ctrl] means the Control key and [N] the key with letter "N". Two keys with "+" between them will mean that you have to press these two keys at the same time. For instance, [Ctrl]+[N] means that you should press the [Ctrl] key, and then while holding it, press the [N] key. Combinations of keys like the [Ctrl]+[N] combination mentioned above are called key sequences, and are very useful shortcuts to many operations that are available through menus. Note however that the key sequences in MuPAD may change from version to version.

MuPAD is a command-line program. So, in order to obtain any result, you first have to type in the appropriate command and press [Enter]. For example:

- 1.78 + 1.2 /* now press [Enter] */

  2.98

- 123 + 345

  468

The bullet on the left side means that you are ready to type in a command that will be evaluated by MuPAD. We will call the space to the right of the bullet the input region. In Unix, a different character will be used instead of a bullet. Below the input region is the output region. Every time you press [Enter], MuPAD tries to evaluate your command and to print the obtained result. After producing the output, a new input region will be created for you automatically, so that you can easily type in the next command.

Usually, the text in the input region is red, while the output text or a formula is blue. However, within this book, I have made both of them black—this will make the black-and-white printout more clear.

Notice that the key [Enter] is used in MuPAD to evaluate the input, while the key sequence [Shift]+[Enter] is used to break an input line. If you feel that this is inconvenient, you can easily redefine it. I will show you later how to do this. Now, let's practice some MuPAD commands to taste its flavor.

2.2 Getting Started

We will start with something simple, but not trivial. Otherwise you might get the impression that you are using a standard calculator. The first difference between a standard calculator and MuPAD is the possibility of typing in complicated formulae and getting the results at once. In a calculator, the same operation might take many steps. For instance, imagine that you need to calculate the expression $(123.897 + 23.987)/(67.987 - 763.98)$. In a calculator you would have to first work out $123.897 + 23.987$ then $67.987 - 763.98$ and finally divide the first result by the second. In MuPAD you can obtain the final result with a single command:

- `(123.897 + 23.987)/(67.987 - 763.98)`

    -0.2124791485

Now let's do something more difficult. For instance, let's check how MuPAD calculates large integers. For this purpose, a factorial of a three or four digit integer will be ideal—say, 456!

- `456!`

    15077739277771706590332856279829748293276484996630 1\
    31532490229569779798080299949204927547058084059358 2\
    70055615465499791246765367283619056736394453658144 4\
    39678603902841941715955316985293965273349948437443 2\
    64712140900271303471688527355766056829451423865130 4\
    20402642102621779712243747458104270667499750554877 4\
    52938755218526446930474587994433589633498013472757 6\
    77126247769970491381477880116497637996331651471303 2\
    78630508301684739445511160770117715636312520669764 2\
    49735244198904963740679910538715209329965485619444 6\
    88747483140592135972232472099655395620016540051906 9\
    67046884568611851786092655942132784522771298286524 2\
    89085201158791214855893492522925977886516475310237 1\
    91080161473206196510412973056159083940814744625294 8\
    84101178964170622576388723410067608455200549775376 4\
    49654638386469415990997949543246999330611024297348 6\
    33043279652233162891541853375858225215375329141289 7\
    34933536315430891192797224230480510976000000000000 0\
    00000000000000000000000000000000000000000000000000 0\
    00000000000000000000000000000000000000000000000000

You may perhaps be wondering how far you can go with large numbers in MuPAD. The best response is—just try. Calculate a few more examples like the one above and see what you can get. For example, see how much will be 9999! and how much time it will take to calculate it. Depending on how powerful your computer is, it may take from a few seconds to a few minutes to obtain a very impressive result—several pages of digits ended by about 40 lines of zeros. It seems that we've got quite an interesting toy.

Let's try something different. For example let's try to solve a few equations. In our equations, the powers of x, e.g. x^2 or x^4, will be represented by x^2 and x^4 respectively.

- solve(x^2 - 2*x - 5 = 0, x)

$$\left\{ \sqrt{6} + 1, 1 - \sqrt{6} \right\}$$

- solve(x^2 + 1 = x^4, x)

$$\left\{ -\frac{\sqrt{1-\sqrt{5}}\cdot\sqrt{2}}{2}, \frac{\sqrt{1-\sqrt{5}}\cdot\sqrt{2}}{2}, -\frac{\sqrt{\sqrt{5}+1}\cdot\sqrt{2}}{2}, \frac{\sqrt{\sqrt{5}+1}\cdot\sqrt{2}}{2} \right\}$$

Both equations are very simple, but imagine trying to solve the second equation by hand. It could take a while before you would be able to write the final solution.

You have seen that MuPAD easily solves equations. How about inequalities? These are usually painful problems for my students.

- solve(x^3 - 2*x^5 + x > 0, x)

$$(0, 1) \cup (-\infty, -1)$$

In our solution, we got two open intervals: $(-\infty, -1)$ and $(0,1)$ which is quite obvious if you try to plot the above inequality.

While working with mathematics, we use a lot of plots. Let's try to do a few plots with MuPAD. We will start with a very basic plot of the function $y = \sin 2x \cos 3x$.

- plotfunc2d(sin(2*x)*cos(3*x))

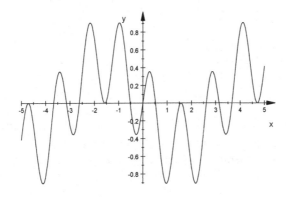

The obtained picture looks very interesting, but what about 3D plots? Let's try a similar function, but in 3D. Let $z = \sin 2x \cos 3y$.

- ```
  plotfunc3d(sin(2*x) * cos(3*y))
  ```

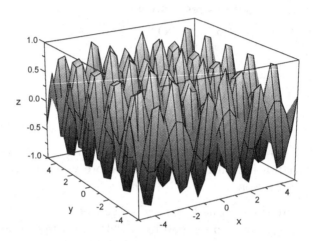

The obtained graph was plotted with MuPAD using its default parameters, but of course we can make it more accurate and choose only the part that we need. We will come to MuPAD plots later.

You may have perhaps noticed that up to now, all the operations were done in a way similar to how a calculator works—type in a single command and press [Enter]. However, in MuPAD you can go much further. You can write programs like in many programming languages. MuPAD programs are similar to Pascal programs. For instance, can you recognize what the program shown below does?

- ```
  n := 10:
  MyFact := 1:
  for i from 1 to n do
    MyFact := MyFact*i
  end
  ```
  3628800

Well, at least for some of us it was easy to guess that this program calculates the product of integers between 1 and 10. This means that the result is the factorial of 10. Let's try to check it in MuPAD:

- 10 !

  3628 800

You have already learned about a few interesting features in MuPAD. Now, it is time to explore its interface.

2.3 A Quick Look at MuPAD's Interface

You have already seen the MuPAD interface on your computer screen or in the picture enclosed at the start of this chapter. It doesn't look very complicated and it certainly reminds you of many other programs for MS Windows. You can easily identify the menu at the top of the window and three toolbars: the standard toolbar, format toolbar and command bar.

Most of the menu options are similar to what you know from MS Word and other Windows programs. However, there are a few things that are different and that are worth checking.

Using options from the **File** menu, you can open a new notebook—this is something you will be doing frequently; you can also save and print your notebook. The **Export** option is also worth noting, as it allows you to export your MuPAD notebook as an RTF document or TXT file. Saving a notebook as TXT is not particularly useful for us, as we will lose all the formatting. On the other hand, when you export your notebook as an RTF document, you will be able to open this document in any word processing program and all the formatting will still be there. This means that the fonts, colors and pictures will be the same as they were in the original MuPAD notebook.

The **Edit** menu looks exactly the same as in other Windows programs, so you do not need to worry about it now.

Some of the operations from the **File** and **Edit** menus are also available from the standard toolbar.

Fig. 2.2 Standard toolbar in MuPAD

You can add or remove toolbars from the screen using the **View** menu. The most important for us is the option **Options** in the **View** menu (Fig. 2.3). Let's take a look at it, as you will find a few things there that you may need in the future. Here you can change MuPAD's default behavior for the [Enter] key in the input regions. Using the **Evaluate On** option, you can choose **Shift+Enter** or **Enter only**.

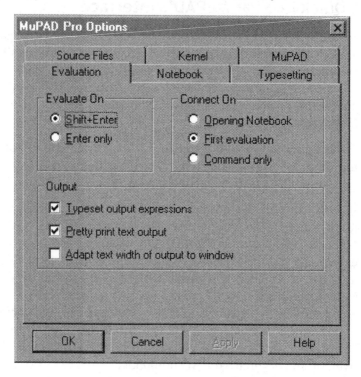

Fig. 2.3 MuPAD Pro options

The most important thing, for the rest of this chapter, will be the **Output** parameters. If you wish to improve the formatting of the output text and formulae on the screen and on the printout, you should check the two check boxes: **Typeset ...** and **Pretty print ...**

You will learn to use the **Insert** and **Format** menus later, while editing MuPAD notebooks. In the **Notebook** menu, you can find a very useful option—**Evaluate**. You can evaluate a single MuPAD input region, which works the same way as the [Enter] key, except that you can evaluate the whole notebook from the beginning to end. This can be very useful while working on large projects or a simulation. For example, you can place all the declarations in the beginning of the notebook, and then evaluate the notebook to get the final result; later, you may change these declarations and evaluate the notebook again to get the results for the new declarations, and again and again.

The **Debug** menu will be useful in more advanced applications, so you may leave it for your later investigations. Finally, the **Extras** menu is very useful. This is the place where you should look for shortcuts to MuPAD operations. Therefore, if you do not remember a MuPAD command or you do not wish to type in long commands, you can choose them from this menu. Let's try to use the **Extras** menu to draw a 3D plot. Just find the option: **Plot 3D function**. On the screen, you will get:

• plotfunc3d(%?)

Now replace %? by the equation of your function and press [Enter]. For instance:

• plotfunc3d(3*sin(x)*cos(y))

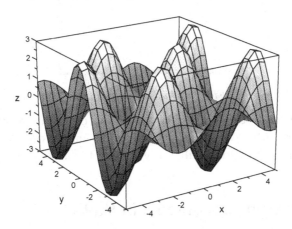

This is the easiest way to perform some less complicated MuPAD operations. However, you should remember that the collection of MuPAD commands and their parameters is so large that there is no way to put them all in a menu like this.

Some of the most frequently used MuPAD operations are also available from the command bar.

Fig. 2.4 MuPAD's command bar

There is another way to get quick access to MuPAD operations. Let's try it. Suppose that you wish to plot a graph of a function. In any empty input region, type in **plo** (the names of many plotting operations begin with "plo") and press [Ctrl]+[space]. You will get a pop-up helper box with the names of all the MuPAD operations that begin with "plo". It might look like the picture below:

Fig. 2.5 MuPAD helper box

Now you should scroll to the name of the operation you need and press [Enter]. MuPAD will complete the name of the command automatically and you will only need to add the appropriate parameters.

2.4 Formatting Documents in MuPAD

We are used to working with nice-looking documents where multiple fonts, styles and colors are applied. The editing functions in MuPAD are a bit limited in this regard. After all, MuPAD is not a word processing program or a typesetting tool. However, you do have enough tools provided to make your notebook look good.

You already know that there are two kinds of regions in MuPAD: the input and output regions. However, there is also a third type of regions, which is used for text. Thus, you can mix text and calculations in one document.

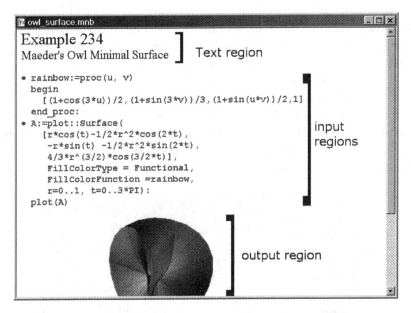

Fig. 2.6 MuPAD screen with text, input and output regions

Usually, MuPAD starts a new notebook with one input region on top of the screen. In order to insert a text area above or below this region choose the option **Text above** or **Text below** from the **Insert** menu. These two operations are available through the key sequences [Shift]+[Ctrl]+[T] and [Ctrl]+[T]. Now you may choose an appropriate font, its size, color and other parameters in the formatting toolbar. The picture below shows what the formatting toolbar offers.

Fig. 2.7 MuPAD's formatting toolbar

You can change the appearance of existing text by selecting it and then, while it is selected, choosing the appropriate parameters from the formatting toolbar.

There is also a way to define the editing parameters for the whole document. For example, you may wish to have all the input commands in the Letter Gothic font, 10 points in size and in the black color. Or you could have all the output messages in italics. In order to do this, choose the option **Fonts** in the **Format** menu. You will get the **Fonts** dialog box, where you can define all the font settings (see Fig. 2.8). Here you can choose your favorite font, color, size and attributes, and then apply these font settings to any region (text, input or output) in your notebook (see Fig. 2.6). Do not forget to check the option **Apply to whole document**. Changes made here will not affect formulae in the output region. However, you can change them too, by using the **Expressions** option from the **Format** menu.

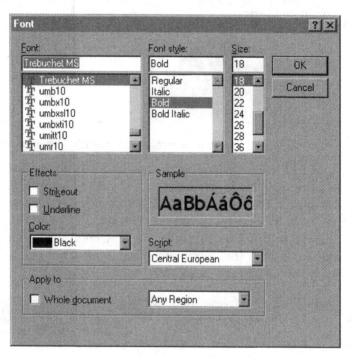

Fig. 2.8 Font setting dialog box

MuPAD doesn't allow us to delete or edit an output region. In order to remove it or edit, you must first change it to a text or input region using the **Notebook** menu; only then can you apply the appropriate changes.

In order to insert an input region above or below the current cursor position, choose the option **Input above** or **Input below** from the **Insert** menu. These two operations are available through the key sequences [Shift]+[Ctrl]+[I] and [Ctrl]+[I].

2.5 Getting Help

You may sometimes feel a bit lost while working with MuPAD. You might, for example, have forgotten the syntax or the exact name for a command. In such a situation, you may consult MuPAD about the problematic issue. For instance, let's see what MuPAD can tell us about the object **student**. You can ask about this in a few different ways—by issuing the commands **info**, **help** or just **?student**. Here is how each of them works.

The command info(student) produces the following output:

```
• info(student)
  Library 'student': the student package
  -- Interface:
  student::Kn,              student::equateMatrix,
  student::isFree,          student::plotRiemann,
  student::plotSimpson,     student::plotTrapezoid,
  student::riemann,         student::simpson,
  student::trapezoid
```

Meanwhile, the other two commands, as shown below, will display the MuPAD help browser box (see Fig. 2.9).

```
• help("student")
• ?student
```

After clicking on the [**Display**] button, you will see a very detailed hypertext help page with a number of examples showing how the command can be applied. Every term with a yellow background is linked to a page with further explanations. For example, by clicking on the word examples with a yellow background, you will go to a page with examples showing how to use the given command. You can copy each example to your notebook by clicking the >> arrows with a green background in front of the example.

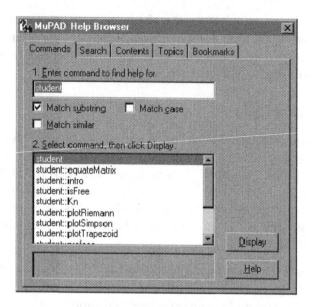

Fig. 2.9 MuPAD help browser

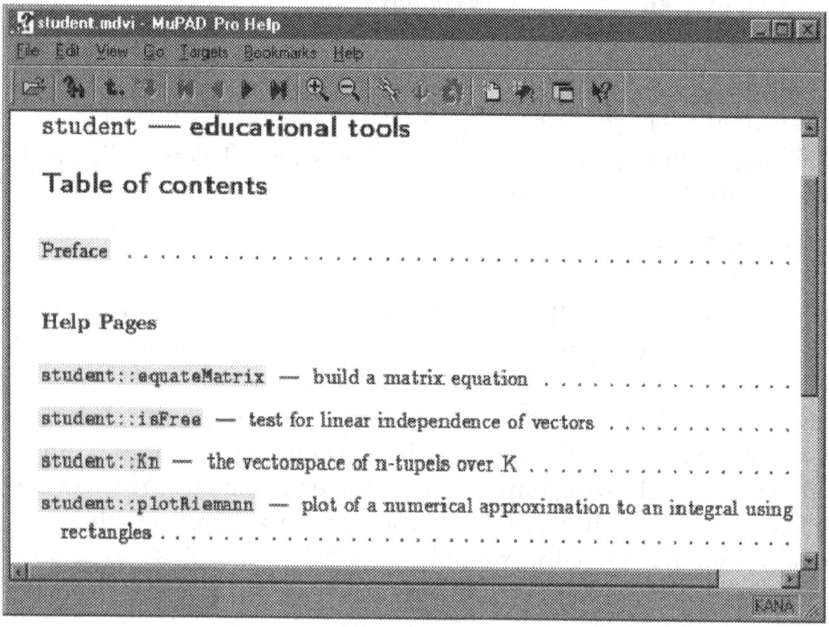

Fig 2.10 Help page for the student library

The MuPAD help browser can be very useful for learning about MuPAD features. You can search for a given word (which does not necessarily have to be a MuPAD command), or you can browse the MuPAD manual, tutorial and other files included with your copy of the program.

The MuPAD help files cannot be printed; however, if you are connected to the Internet, you can always access the online help and print out the pages that you are interested in.

2.6 Chapter Summary

Before we proceed to the next chapter, let's take a look at what we have learned in this chapter.

We have gotten accustomed to the MuPAD interface and the notebook's structure. We have also learned a few MuPAD shortcuts and instructions. Here are all of them.

2.6.1 Selected MuPAD Shortcuts

[Ctrl]+[N] - *open a new notebook*

[Enter] (*in the input region*) - *execute the command*

[Shift]+[Enter] (*in the input region*) - *add a new line in the current input region*

[Shift]+[Ctrl]+[T] - *insert text region above current line*

[Ctrl]+[T] - *insert text region below current line*

[Shift]+[Ctrl]+[I] - *insert input region above current line*

[Ctrl]+[I] - *insert input region below current line*

[Ctrl]+[Space] - *complete a MuPAD command name*

2.6.2 MuPAD Commands

n! - *produce the factorial of the integer n*

solve(*equation*) - *solve an equation with one variable*

plotfunc2d(*function*) - *plot a graph of a function with one variable*

plotfunc3d(*function*) - *plot a graph of a function with two variables*

info(*command*) - *display basic information for the given command*

help("*command*") - *display the help pages for the given command*

?*command* - *display the help pages for the given command*

Note that the **help** command requires that its parameter be given with double quotes (i.e. as a string).

2.7 Review Exercises

1. Open a new MuPAD notebook and save it as review_01.mnb. Above the existing input region, insert a text area. Place the cursor at the beginning of the text line and use the formatting toolbar to change the font settings to Arial, 18 points, Bold, color Maroon. Now type in, Review exercises, set 1. This is the title of your notebook. Below the title, insert a new line for text (just press [Enter]) and change its font settings to Times New Roman, italic, 16 points, color Navy. In this line type in, Calculations in MuPAD. This will be the title of the first computing section.

2. Place the cursor beam at the beginning of the input region, below the section title, and type in the command to calculate 234/345 + 34/789. Press [Enter] to obtain the result. In the same way, calculate 34/234 + 34/123, 33!, 333! and 3333!

3. Below the last output, create a new text area and change its font to Times New Roman, italic, 16 points, color Navy. In this line, type in MuPAD 2D plots. This will be the next section of your notebook. Below it, you should already have an empty input region. It was created after executing the last command.

4. In the last input region, type in the command to obtain the plot for the function $y = \sin(5 \cdot x) + \cos(3 \cdot x)$. In the same way, obtain the plots for functions $y = (1/x)\sin(x)$, $y = \sin(1/x)$, and $y = x \cdot \sin(1/x)$. Each command should be typed in as a separate input region. You will obtain four separate plots.

5. Now it's time to practice 3D plots. Below the last picture, insert a new line for text and again set its font to Times New Roman, italic, 16 points, color Navy. In this line, type in, MuPAD 3D plots. Below it, you will already have an empty input region. Like before, it was created after executing the last command.

6. Type in the commands to plot the 3D plots of the functions: $y = (1/x) \cdot \sin(x) \cdot \cos(y)$, $y = \sin(1/(x \cdot y))$, and $y = (x \cdot y) \cdot \sin(1/(x \cdot y))$. Again, you need to type each

command in a separate input region, and you will get three separate plots.

7. At the end of your review notebook, insert a new section entitled MuPAD help. Like the other headers before, set it in the Times New Roman font, italic, 16 points, color Navy.

8. Below the last header, insert a new input region. Here, type in the word in and then press [Ctrl]+[space]. From the pop-up helper box, choose the info command. Now use the info command to get information about plot and solve. Use the help and ?solve commands to get more information about the solve command. Reminder: help(solve) will not work, you need to type in help("solve").

9. Finally, go to the **Notebook** menu and click on **Delete Output**. This will remove all the outputs from your notebook. Now calculate all the commands in your notebook again—go to the **Notebook** menu and there, choose **Evaluate**, followed by **Any Input**.

10. Save your notebook.

Chapter 3 _____

Programming in MuPAD - Declarations

Many mathematicians consider programming as a skill that is completely irrelevant to mathematics. However, when looking at mathematics, we may notice that working in mathematics requires many activities that are not available in visual CAS. For instance, while solving mathematical problems, we have to explore algorithms, use recursive functions, or create complex constructions in 2D or 3D geometry. All these activities consist of many steps and their nature resembles writing a program. These kinds of activities can be practiced if we use a computer program where a programming language is available. This is the point where you will appreciate MuPAD.

Programming in MuPAD is structured very similarly to writing programs in Pascal. However, in MuPAD you can represent mathematical concepts in a more efficient and flexible way. Thus, you can follow the natural way of mathematical thinking with greater ease.

In this chapter you will start learning the basics of MuPAD programming. In the later chapters, we will expand this knowledge.

3.1 Writing Formulae in MuPAD

In the previous chapter, we had tried a few very simple formulae with MuPAD. The first thing that you had perhaps observed is that we cannot omit the multiplication symbol and brackets for functions arguments. For example, in real life we usually write $3x \sin x$ or $\sin x \cos y$, while in MuPAD we have to type in `3*x*sin(x)` or `sin(x)*cos(y)`. It is important to realize that, for example, 3x and 3*x mean completely different things in MuPAD. In fact, 3*x means the multiplication 3 times x, but 3x is just a nonsense expression that will produce an error message. This example suggests that we should consider for a moment how we can build formulae in MuPAD so they will mean what we want them to mean.

Up to now, you have seen only a very limited set of MuPAD arithmetical operations: +, −, *, /, and taking to the power ∧, for example, x∧3 meaning x^3. MuPAD knows a number of other operations and functions used in mathematics. You can use not only basic arithmetic operations but also functions like tan$x$, sin$x$, ln$x$ and many others. Creating formulae with these operations and functions can sometimes be confusing. For example, how do we write the formula x^{y^2} in MuPAD's notation, and how will x*y∧2 be evaluated? What will be calculated first, x*y or y∧2?

Let's learn some general rules.

1. **Brackets are the strongest.** Operations enclosed in brackets are calculated first. This means that in the expression 3*(3+9), MuPAD will first calculate 3+9 and then the multiplication. The same is true for 5*sin(x*0.01). First, x*0.01 will be calculated, then sin(x*0.01) and finally multiplication by 5.

2. **Next goes "∧".** For example, in the formula 2∧3*5 MuPAD will first calculate 2∧3 (i.e. 2^3) and then it will multiply it by 5.

3. **Operations * and / are stronger than + and −.** Thus, in the formula 5+3/2 or 3/2+5, the operation 3/2 will be calculated first and then 5 will be added.

4. **Binary operations not using brackets (like mod, div, etc.) are the weakest.** For example, 2+3 mod 5 really means (2+3) mod 5 and thus, 2+3 mod5 = 0. Note that "mod" denotes the binary operation modulo.

5. If you are not sure how your formula will work, use brackets.

Sometimes, long and complicated formulae may look very crowded, and it is very hard to read and understand them. For instance try to read a formula like this one: 3*x∧5-8*x∧4+9*x∧3-2*x∧2+x-5. It is difficult to understand, isn't it?

You should always write code that is easy to read and understand. Let's make an agreement that will help us to develop clear MuPAD code.

Rule 1 There are always spaces after keywords and surrounding binary operations.

Using this rule, you can convert the above example into:

    3 * x^5 - 8 * x^4 + 9 * x^3 - 2 * x^2 + x - 5

Some programmers prefer not to use a space around * and /, which produces more modular look of the formulae. For example:

    x*y*z + 2*x*y + 3*x*z + 5*y*z + 2*x/y - 3*y/z

In the future, you will learn a few other rules that are useful in producing easy to read, well-structured code.

3.2 Declaring Variables and Constants

While working in mathematics or writing mathematical papers, we used to say *"let $x = \pi$"*, or *"let L be a segment joining points A and B"*, or *"let $F_0 = 0$, $F_1 = 1$ and $F_n = F_{n-1} + F_{n-2}$"* or even *"let m denote the equation $x^2 + y^2 = 1$"*. In each case, we make a kind of label and we attach a mathematical object to it—a number, a line, a function, an equation, etc. Well, perhaps it's better to say that we stick these labels to some objects. The expressions x, π, L, F_0, F_1, F_n and m are such labels. We can use the same label for different objects at different times, and there is nothing wrong with this practice. Today x may mean a real number, tomorrow a plane in 3D and later something completely different.

However, among our labels there is one special label, π. We stick it to one specific number, that is, 3.1416. We call labels like x, L, F_0 or F_1 variables, and labels like π, e, etc. we call constants.

In MuPAD, you can deal with variables and constants in a similar way to mathematics. However, we will consider variables rather like boxes with labels. For instance, the declarations

- x := 1.7893
- y := 3.4567
- m := (2*v^2 + 3*u^2 = 1)

mean, let us drop 1.7893 into the box with the label x, 3.4567 into the box y and the equation $2v^2 + 3u^2 = 1$ into the box m. The symbol := used here is what is called an assignment operator.

Now we can follow the way how we work and think in mathematics:

- (3*x + 4*y)/2
    9.59735

- (x^2 + y^5)

  496.7257888

- solve(m, u)

$$\left\{ -\frac{\sqrt{3} \cdot \sqrt{1 - 2 \cdot v^2}}{3}, \frac{\sqrt{3} \cdot \sqrt{1 - 2 \cdot v^2}}{3} \right\}$$

- solve(m,v)

$$\left\{ -\frac{\sqrt{2} \cdot \sqrt{1 - 3 \cdot u^2}}{2}, \frac{\sqrt{2} \cdot \sqrt{1 - 3 \cdot u^2}}{2} \right\}$$

Variables that already have been assigned values can always be redefined. For example, you can assign different values to the same variables in different parts of your calculations, thus exchanging the values in the boxes for different ones. Let us analyze this example:

- x := 4.567

  4.567

- y := 2*x

  9.134

- x := 4.567 + 1

  5.567

- y := 2*x

  11.134

Notice that the second value of x was in fact obtained from its first value, by adding 1 to it. Thus, instead of using explicit numbers, this value can be expressed in the following form:

- x := x + 1

This operation will be understood as, let x be equal to its current value plus 1. This kind of assignment will be very useful in later examples. We will frequently use declarations: x:=x+a, or x:=x^2, or y:=x, etc., where, for example, y:=x means, let y have the same value as the current value of x. This means that the two boxes x and y will contain the same value. However, if we later replace the value assigned to x by another value, the value of y will still remain the same.

In our mathematical practice, we treat constants as something permanent. Our mathematical tradition does not let us assign to π or e constant values other than $\pi = 3.1416$ and $e = 2.7183$. Of course we can always change our habits, but doing this would be considered as a bad custom. In MuPAD, these values are also constants (PI and E) and they are protected. Let's see what will happen if we try to assign a different value to π.

- PI := 1

    Error: Invalid left-hand side [PI]

For your purposes, you can make your own constants and protect them before changing their values. There are two levels of such protection — weak protection, which only produces a warning, and strong protection, which produces an error and does not permit you to override the default value of the constant. These forms of protection work like this:

- MY_E := 2.71

2.71

- protect(MY_E):
- MY_E := 1

    Warning: protected variable MY_E overwritten

Now let's apply stronger protection:

- protect(MY_E, ProtectLevelError):
- MY_E := 1

    Error: Identifier 'MY_E' is protected [_assign]

As you can see, the declaration protect(MY_E) did not really protect the identifier MY_E; it issued a warning message, but still assigned the new value to it. On the other hand, protect(MY_E, ProtectLevelError) produced an error message, and MuPAD didn't accept the new value for MY_E.

By issuing the command anames(All) you can display the list of all the identifiers that have a value in the current MuPAD session:

- anames(All)

    {'*', '+', '-', '/', D, O, '**', '^', x, y, C_, Ci, Ei, Q_,

R_, Im, Re, Ax, Pi, Z_, Si, id, fp, ln, RGB, is, op, Cat,

Dom, SEED, _if, gcd, Sum, _in, abs, ode, adt, csc, arg,

rec, sec, has, lcm, erf, map, _or, log, tan, val, min,

cos, cot, max, lhs, fun, new, sin, MY_E, int, psi, exp,

rhs, zip, sum, My_E, LEVEL, ORDER, Pref, _and, igcd,

.....

You can try to check which of these identifiers are protected. Notice that the constant MY_E that was declared by us just a while ago is also included in this list.

All the protected objects can be unprotected by the `unprotect` command, like this:

• `unprotect(MY_E)`

Values stored in variables are very convenient for later use. You can use them anywhere in your notebook, provided that you have declared them before. There is also another way to refer to the result obtained before. For instance, suppose that you had calculated:

• `x^5 - x^4 + x^3 - x^2 + x - 1`
  $$x - x^2 + x^3 - x^4 + x^5 - 1$$

You can now evaluate the obtained expression for x=2.

• `subs(%, x=2)`
  21

The command `subs` (substitute) replaces x by 2 in our polynomial. MuPAD had simplified the obtained expression automatically. The operator ditto (%) which was applied here refers only to the last obtained output and is useful in some basic calculations. You shouldn't, however, use it in serious applications. Thus, it is worth remembering:

Rule 2 Always use a variable to store every MuPAD object that you are going to use later.

You have already seen a few different names of variables, constants and functions in MuPAD. Names in MuPAD can contain any alphanumeric character and underscores, but they cannot begin with digits. For instance you can use MyFun1, my_fun23, x23_98, etc., but not 2x, 32myf.

Writing programs in MuPAD does not differ too much from writing programs in other languages. There are some good programmers' rules regarding the creation of names in programs:

Rule 3 Constant names are UPPERCASE, with an occasional UNDER_SCORE, while variables names are always in lowercase.

3.3 Declaring Functions and Operations

In mathematics, we often define new functions. Sometimes a new function is just a shortcut for a formula, while in other cases a new function is defined in a less conventional way. Let us consider some simple examples here.

Suppose that we have a formula representing the function,

$$x \to \frac{x^5 - x^4 + x^3 - x^2 + x - 1}{x}$$

and in further calculations we wish to use a short notation for it, such as *myfun*(x). In MuPAD this can be done in the following way:

- myfun := x -> (x^5 - x^4 + x^3 - x^2 + x - 1)/x

$$x \to \frac{x^5 - x^4 + x^3 - x^2 + x - 1}{x}$$

You can now use this function just like other functions in mathematics. You can use *myfun*(x) in calculations, and you can even plot it:

- (myfun(2) + myfun(3))/myfun(5)

$$\frac{305}{2232}$$

- plotfunc2d(myfun(x), x=1..2)

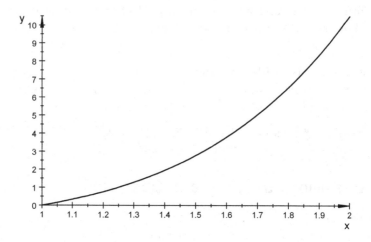

Sometimes you may feel that declaring functions this way is too formal, and you may want to take a shortcut and just assign a name to a formula. Note, however, that this will not work like you might expect. Let's look at an example. Suppose that we make this declaration:

- newfun := x*sin(x)

Now, let's execute the command,

- newfun(0)

  $x(0)\cdot\sin(x)(0)$

As you can see, MuPAD treats newfun as a label for the formula x*sin(x), but not as a new function.

In many applications, we have to use functions that are combined out of pieces of different functions. For example, suppose that our function should behave like $-x\sin 2x$ for $x \le 0$, then should equal 0 for $0 < x \le \pi$ and finally, equal $\sin x$ for $x > \pi$. Such a function can be defined with the use of the keyword piecewise, like this:

```
• f := x -> piecewise(
      [x <= 0,-x*sin(2*x)],
      [x >= 0 and x <= PI, 0],
      [x >= PI, sin(x)]
  ):
```

Here is the plot of our function:

- plotfunc2d(f)

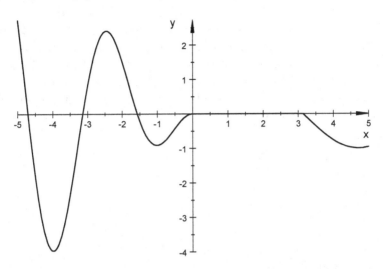

In mathematics we frequently use the term, operation. Usually we refer to a function that has two arguments and is expressed by an operation symbol that is placed between arguments, like for example, 4 + 5, 8/9, etc. In fact, we can talk about a more general term — operators, and we may consider operators with one, two or more arguments. However, for the purpose of this chapter, let us look at some very simple situations. Imagine that you wish to declare a new operation, for example an operation that for two integers will produce their greatest common divisor or lowest common multiplier. First, we have to choose a symbol for our new operation. Suppose that we decided to use the vertical bar "|". So, we expect that, for example 9 | 6 should produce the greatest common divisor of the numbers 9 and 6. In order to declare such an operation, we may use the operator command:

- operator("|", gcd):

Now we can check what we can obtain with this new command:

- 18 | 12

  6

- (18 | 12) + (24 | 72)

  30

Later, you will find that most of the symbols, or even combinations of some symbols, that you can type in from the keyboard are used in MuPAD for some other purposes. Of course you can use them as operators, but their original functionality will be lost for a while. However, you can create your operation symbols out of combinations of two or more characters. For example, here is how the two operations *gcd* and *lcm* can be declared as operations on integers with the use of the characters < and >.

- `operator("<>", lcm )`
- `operator("><", gcd)`
- `23 >< 34`

    1

- `23<>25`

    575

- `(18 >< 12) <> (28 >< 72)`

    12

Observe, in this example we redefined a very important MuPAD operator <>. We use it to express an inequality. Therefore, after finishing experiments with new operators we should reset the whole session or delete new operators.

The `operator` command can be used to create some nice examples for algebra. However, let us leave this task for a later time when we will truly be ready to use it.

3.4 Writing More than One Line of Code

Up to now, our examples have been just single MuPAD commands. Now you will start learning how to write programs with multiple commands.

In MuPAD, like in many other programming languages, you must separate each two consecutive commands that are in **the same input region**. For this purpose you can use two separators: ":" (colon) and ";" (semicolon). There is a slight difference between how these two separators work. The first one, ":" blocks the output of the preceding command to the screen; on the other hand, the separator ";" allows the printing the output. You can choose either of them depending on your preference. Let us analyze two examples showing the difference between the two separators.

You can easily see that the separator "**:**" doesn't produce any output from the first command in this example:

- ```
  z := 12:
  z^2 + 12
  ```

  156

Meanwhile, the separator "**;**" produces unwanted output from the first command:

- ```
  z := 12;
  z^2 + 12
  ```

  12 <– unwanted output from the first command

  156

This way, by using "**:**" and "**;**" you can limit MuPAD output to the elements that you really want to produce, i.e. output from the most important commands. All the unnecessary output can be hidden.

Note that if you forget to separate two commands in the same input region, MuPAD will display an error message:

  Error: Unexpected 'identifier' [line 2, col 2]

Some programmers write their code in a very crowded way, like this:

- ```
  x:=12:y:=x+2:u:=x/y*3:z:=x^2+y^(x+y+1)+1/u
  ```

It doesn't look good, does it? Who would be able to read it? Let's make another agreement that will help us to develop clear MuPAD code.

Rule 4 Do not put two MuPAD commands in the same line

According to this rule, as well as rule 1, the above example should be written like this:

- ```
  x := 12:
  y := x + 2:
  u := x/y*3:
  z := x^2 + y^(x + y + 1) + 1/u
  ```

Now it definitely looks better.

3.5 The Basics of the Domain Concept

Mathematicians deal with many different worlds. Usually, we work with real numbers. Which really means that while solving equations, for example, we look for solutions that are real numbers. However, quite often we have to extend the area of our interest into complex numbers, or narrow it down to integers or rational numbers. In each case we say that we are looking for solutions in the domain of real, complex, integer or rational numbers respectively. We deal with functions or equations where coefficients are also from a different domain than real numbers. The domain concept can be quite sophisticated, but for the purpose of this book let us use some basic examples.

MuPAD has a number of predefined domains, as well as tools that allow us to define our own domains. You do not need to always be so formal and refer to a specific domain. Quite often you can make assumptions without referring to any domain. Let's start with a simple example.

You certainly know that the equation

$$12 - 8x - 7x^2 + 2x^3 + x^4 = 0$$

has four roots: $-2, -3$, 2 and 1. You can easily confirm this by typing and executing the command:

- `solve(12 - 8*x - 7*x^2 + 2*x^3 + x^4 = 0, x)`
  $\{-3, -2, 1, 2\}$

However you may wish to obtain only positive solutions. You can do this by assuming that x is a positive number:

- `assume(x > 0):`
  `solve(12 - 8*x - 7*x^2 + 2*x^3 + x^4 = 0, x)`
  $\{1, 2\}$

By using multiple `assume` commands you can build more complex sets, that may not be domains in MuPAD sense. For example, you can restrict solving an equation to a specific interval, such as $(-5, 5)$.

- `assume(x < 5):`
  `assume(x > -5,_and)`

  $(-5,5)$

- `solve(cos(x)/sin(x) = 0,x)`

  $$\left\{-\frac{\pi}{2}, \frac{\pi}{2}, -\frac{3 \cdot \pi}{2}, \frac{3 \cdot \pi}{2}\right\}$$

Note how the interval was build. The first `assume` command restricted x to values that are less than 5. The second `assume` command then additionally restricted x to values that are bigger than –5. This was possible by adding the parameter _and in the second `assume` command. Note the underscore in front of the word and. This way, we obtained the set:

$$\{x \in R \mid x < 5 \text{ and } x > -5\} = \{x \in R \mid -5 < x < 5\}$$

You can also use the _or parameter, like in this example:

- `assume(x < -2):`
  `assume(x > 2,_or)`

  $(-\infty, -2) \vee (2, \infty)$

- `solve(cos(x)/sin(x) = 1, x)`

  $$((2,\infty) \cup (-\infty, -2)) \cap \left(\left\{\frac{\pi}{4} + \pi \cdot k \mid k \in \mathbb{Z}\right\} \setminus \{\pi \cdot k \mid k \in \mathbb{Z}\}\right)$$

Notice how our set was built this time:

$$\{x \in R \mid x < -2 \text{ or } x > 2\} = \{x \in R \mid x < -2\} \cup \{x \in R \mid x > 2\}$$

There is another way to restrict calculations to an interval. This can be done by using MuPAD's predefined types. Here is how it can be done using the interval type:

- `assume(y, Type::Interval(0,5))`

  $(0,5)$

- `solve(sin(y)/cos(y) = 1, y)`

  $$\left\{\frac{\pi}{4}, \frac{5 \cdot \pi}{4}\right\}$$

Using the `Type::Interval` option, you can define sets more complex than a single interval. For example, how about the set $(-4, -2) \cup (2, 4)$?

- `assume(x,Type::Interval(2,4)):`

```
assume(x,Type::Interval(-4,-2),_or);
```
$$(2,4) \vee (-4,-2)$$

However, solving equations in such a set is not very efficient. In many situations similar to the one above, MuPAD produces meaningless results that need further evaluation. It is much safer and more effective to solve the same equation for each interval separately.

In all the examples discussed in this section, the command assume has affected the entire notebook. This means that we could, for example, use assume(x>0) in the beginning of the notebook, and then perform many different computations that will all be influenced by our assumption. If you need to delete an assumption made earlier, you can use unassume command. For example:

- `assume(x>0):`
- `solve(x^2 - 1 = 0,x)`

   $\{1\}$

- `unassume(x)`
- `solve(x^2 - 1 = 0,x)`

   $\{-1, 1\}$

We can also use the command unassume() to remove all the assumptions from the whole notebook. Another way of removing the properties of a variable is to use the command delete. For example,

- `x := sqrt(3)`

   $\sqrt{3}$

- `delete x`
- `x`

   x

Finally, we can use assumptions for only local calculations. For example this statement will only produce positive roots:

- `solve(x^4-3*x^2+1=0,x) assuming x>0`

   $\left\{ \frac{\sqrt{5}}{2} - \frac{1}{2}, \frac{\sqrt{5}}{2} + \frac{1}{2} \right\}$

On the other hand, the statement below produces all roots, both positive and negative:

- solve(x^4 - 3*x^2 + 1 = 0,x)

$$\left\{ -\frac{\sqrt{5}}{2} - \frac{1}{2}, \frac{1}{2} - \frac{\sqrt{5}}{2}, \frac{\sqrt{5}}{2} - \frac{1}{2}, \frac{\sqrt{5}}{2} + \frac{1}{2} \right\}$$

In the above assumptions, we frequently used variable type checking. MuPAD contains a number of predefined data types, which we will call types for short, that can be used directly or with some parameters. Here is a list of the most useful types.

Type::Complex - *complex numbers*

Type::Even - *even integers*

Type::Function - *functions*

Type::Imaginary - *complex numbers with real part equal 0*

Type::Integer - *integer numbers*

Type::Interval - *intervals*

Type::NegInt - *negative integers*

Type::NegRat - *negative rational numbers*

Type::Negative - *negative real numbers*

Type::NonNegInt - *non-negative integers* (≥ 0)

Type::NonNegRat - *non-negative rational numbers* (≥ 0)

Type::NonNegative - *non-negative real numbers* (≥ 0)

Type::NonZero - *complex numbers without 0*

Type::Odd - *odd integers*

Type::PosInt - *positive integers* (> 0)

Type::PosRat - *positive rational numbers* (> 0)

Type::Positive - *positive real numbers* (> 0)

Type::Prime - *prime numbers*

Type::Rational - *rational numbers*

Type::Real - *real numbers*

Type::Zero - $\{0\}$

You can also use open and closed intervals with numbers from a given type:

Type::Interval(a, b, *given_type*) = $\{x \in given\_type \mid a < x < b\}$

Type::Interval($[a]$, b, *given_type*) = $\{x \in given\_type \mid a \leq x < b\}$

Type::Interval(a, $[b]$, *given_type*) = $\{x \in given\_type \mid a < x \leq b\}$

Type::Interval($[a]$, $[b]$, *given_type*) = $\{x \in given\_type \mid a \leq x \leq b\}$

Here is an example of a variable that represents any integer number greater than 2 and less than 100:

- `myvar:=Type::Interval(2,100,Type::Integer)`

  $(3,99) \cap \mathbb{Z}$

In order to obtain information about the types predefined in MuPAD, you can use such statement:

- `info(Type)`

Finally, sometimes you may need information about the type of a given variable. This can be obtained by `getprop(x)` command. For instance, suppose that we had declared:

- `assume(x,Type::Interval(2,4)):`

Then we will get:

- `getprop(x)`

  $(2,4)$

Information about the type of a MuPAD object can be very useful in some situations. It might be that in one notebook, you frequently change the types of variables by assigning them different values or solving equations with these variables. It may happen that, after some time you will start obtaining confusing results. In such a situation, you can check your variables' type, and if necessary, delete their current values using the `unassume` or `delete` commands.

If you happen to do several MuPAD exercises or experiments using the same notebook, it is a good idea to start each new exercise by executing the command:

- `reset()`

This command resets the current MuPAD session and clears all the data in MuPAD's memory. Your notebook will behave like a completely fresh notebook. You can prevent many unexpected errors this way. So, let's remember:

TIP   In a MuPAD notebook with multiple independent exercises, start each new exercise by executing the `reset()` command.

3.6 Chapter Summary

Well, this was a rather long chapter and you have learned a lot of new things about MuPAD. Let us try to summarize your new knowledge.

3.6.1 MuPAD Instructions and Operators

In this chapter, you have learned a number of new MuPAD instructions.

myvar := a - *assigns value a to variable myvar*

protect(*MY_CONST*) - *protects MY_CONST by warning message only*

protect(*MY_CONST*, ProtectLevelError) - *protects MY_CONST against overwriting*

unprotect(*MY_CONST*) - *unprotects MY_CONST*

anames(All) - *display all names defined in MuPAD in current notebook*

myfun := x -> f(x) - *declaration of a new function*

myfun := x -> piecewise([cond1, fun1],[cond2, fun2], ..,[cond_n, fun_n]) - *piecewise function declaration*

subs(%, x=a) - *substitutes x=a in last obtained output*

operator("*symbol*", *function*) - *defines a new binary operation*

assume(x▼a) - *makes assumption about* x

assume(x▼b,_or) - *makes additional assumption about* x *(logical* or*)*

assume(x▼c,_and) - *makes additional assumption about* x *(logical* and*)*

assuming - *declaration of a local assumption*

In the above commands, the symbol ▼ can be understood as one of relations <, <=, >, >=, =. Note that <= means ≤ and >= means ≥.

getprop(x) - *produces information about* x

assume(x, Type::Interval(a,b)) - *makes assumption that* a<x<b

unassume(x) - *deletes all assumptions about* x

delete(x) - *deletes current value of* x

reset() - *resets all variables in the current notebook*

You have also learned that the separator ":" doesn't allow a MuPAD command to display the output on the screen, while the separator ";" forces output to be displayed on the screen.

3.6.2 MuPAD Coding Guidelines

In order to write MuPAD code clear and easy to read, we should apply some general programming guidelines.

Rule 1 There are spaces after keywords and surrounding binary operations.

Rule 2 Always use a variable to store each MuPAD object that you are going to use later.

Rule 3 Constant names are UPPERCASE, with an occasional UNDER_SCORE, while variables names are always in lowercase.

Rule 4 Do not put two MuPAD commands in the same line.

We had also one important tip.

TIP In a MuPAD notebook with multiple independent exercises, start each new exercise by executing the reset() command.

3.6.3 Selected MuPAD Types

MuPAD has a number of predefined types. Here are the most basic ones.

Type::Complex - *complex numbers*

Type::Even - *even integers*

Type::Function - *functions*

Type::Imaginary - *complex numbers with real part equal* 0

Type::Integer - *integer numbers*

Type::Interval - *intervals*

Type::Interval$(a, b, Type) = \{x \in Type \mid a < x < b\}$

Type::Interval$([a], b, Type) = \{x \in Type \mid a \leq x < b\}$

Type::Interval$(a, [b], Type) = \{x \in Type \mid a < x \leq b\}$

Type::Interval$([a], [b], Type) = \{x \in Type \mid a \leq x \leq b\}$

Type::NegInt - *negative integers*

Type::NegRat - *negative rational numbers*

Type::Negative - *negative real numbers*

Type::NonNegInt - *non-negative integers* (≥ 0)

Type::NonNegRat - *non-negative rational numbers* (≥ 0)

Type::NonNegative - *non-negative real numbers* (≥ 0)

Type::NonZero - *complex numbers without* 0

Type::Odd - *odd integers*

Type::PosInt - *positive integers* (>0)

Type::PosRat - *positive rational numbers* (>0)

Type::Positive - *positive real numbers* (>0)

Type::Prime - *prime numbers*

Type::Rational - *rational numbers*

Type::Real - *real numbers*

Type::Zero - $\{0\}$

3.7 Review Exercises

1. Use the `anames(All)` command to produce a complete list of names that have values in the current session of MuPAD. Analyze this list. Can you identify the meaning of at least 20 names on this list? Which of them are binary operations, which are constants and which are functions? Which of these names are protected?

2. Declare three new variables equation1, equation2, and equation3, and assign to them the equations $x^2 - 4 = 0$, $2x^2 + x - 1 = 0$ and $x^3 - 13x + 12 = 0$. Use these variables in the `solve` command to obtain the solutions of these equations.

3. Declare two new constants that are approximations of the number π with 5 and 10 decimals ($\pi = 3.141592654$). Protect these constants with warning and error options. Check how MuPAD will react if you try to assign new values to these constants. Unprotect these constants and check again if you can assign new values to them.

4. Define the functions $\sin x(\sin 3x + \cos 5x)$ and $x^2(\sin 3x + \cos 5x)$. Use the names of these functions to obtain their plots. Do not bother about the range and accuracy of the plot. You will learn about these things later.

5. Define the identifiers `newfun1:=sin(x)` and `newfun2(x):=cos(x)`. Can you calculate `newfun1(0)`, `newfun2(0)` or `subs(%,x=0)` for both declarations? Explain

what is wrong with these declarations?

6. Type the formulae listed below in MuPAD. Compare each MuPAD output with the original formula. Correct your typing if the MuPAD output is different from the original formula. Use the commands `subs(y, x = a)` and `subs(%, x = 0)` to evaluate each formula for a given x, say, $x = 0$ or $x = 1$. Note that $\sqrt{formula}$ shall be written as `sqrt(formula)`.

  a. $y = \dfrac{8a^3}{x^2 + 4a^2}$

  b. $y = (x^2 + x^3 + a^2)^2 - 4a^2x^2 - b^2$

  c. $y = \cos^2 x - \sin^2 x$

  d. $y = 2\cos^2 x - 1$

  e. $y = \dfrac{\sin x}{1 + \cos x}$

  f. $y = \sin x \cos x - \sqrt{\dfrac{1 - \cos x}{1 + \cos x}}$

7. Define the following domains using the relations $<, <=, >, >=$, and the `assume` command:

  a. interval $A = [0, 1)$ of real numbers,

  b. interval $B = [-10, 10]$ of rational numbers,

  c. set $C = (-\infty, -1] \cup [1, \infty)$ of integer numbers,

  d. set $D = (-\infty, 0)$ of prime numbers.

8. Define the following domains using the `Type::Interval` option:

  a. interval $A = [0, 1)$ of real numbers,

  b. interval $B = [-10, 10]$ of rational numbers,

  c. set $C = (-\infty, -1] \cup [1, \infty)$ of integer numbers,

  d. set $D = (-\infty, 0)$ of prime numbers.

Chapter 4 _____

Programming in MuPAD - Control Structures

In the previous chapter we started talking about programming in MuPAD. However, so far we haven't written anything that would deserve to be called a program. Well, perhaps these four lines can be called a program:

- ```
  x := 12:
  y := x + 2:
  u := x/y*3:
  z := x^2 + y^(x + y + 1) + 1/u
  ```

Of course, we cannot do too much with just the assignment operator := and selected arithmetical operations. In this chapter, I will show you the most important MuPAD control structures. You will learn how to implement decisions in MuPAD programs, how to program iterations and finally how to program situations with multiple choices. This is our goal for the next few pages. Let us begin with decisions.

4.1 Decisions

In mathematics we are used to dealing with sentences like "if something then something else". For instance, while plotting a graph of the function $y = \sqrt{1 - x^2}$ we always have to suppose that $1 - x^2 \geq 0$ and then we plot the graph only inside the interval $-1 \leq x \leq 1$. This makes perfect sense, since the square root $\sqrt{1 - x^2}$ only exists as a real number if $1 - x^2 \geq 0$.

Let's write a short MuPAD program to calculate the value of $y = \sqrt{1 - x^2}$ for a given x. First, let's formulate it in the same way that we would in every day. We say:

1. Choose a value for x; this can be $x = \frac{1}{2}$ or any other real number.

2. if $1 - x^2 \geq 0$ then calculate $\sqrt{1 - x^2}$. Otherwise, do not do anything.

In MuPAD notation, the above two lines can be formulated in the form of the following program:
  ```

- x := 1/2:
  if 1 - x^2 >= 0 then y := sqrt(1 - x^2) end_if

$$\frac{\sqrt{3}}{2}$$

Now you can run this program a few times for different values of x. You will see that for some values of x, you will get a real number while for some others you will get the result *NIL*, which really means that for this x the value of y is not a real number. In the above program, the calculations were performed according to the condition 1 - x^2 >= 0. If the condition was true, than the value of y was calculated; otherwise, the output was a *NIL* value. We will call the programming construction presented here an if..then statement, or an if statement for short. Note that the whole construction was ended by the keyword end_if, which will help you to recognize that we have just finished an if..then construction. However, when writing short and simple programs, you can be less formal and use end instead of end_if.

Observe that we didn't bother about what would happen if the condition would be false. We simply supposed that MuPAD will find its way to deal with such a situation, and it worked perfectly. However, in some situations you will need to also program the other possibility. Here is such a situation. The function absolute value can be defined as:

$$|x| = \begin{cases} x & \text{if } x \geq 0 \\ -x & \text{if } x < 0 \end{cases}$$

Let's write a MuPAD program to calculate the values of this function. This problem sounds a bit similar to the previous example:

- x := -5:
  if x >= 0 then y := x else y := -x end_if;

Now you can run this program for different values of x and see if it works as you expected.

The conditions, and their resulting operations in an if statement can be quite elaborate. Writing all of them in one line is sometimes impossible and the readability of such a program can be quite poor. For this purpose, it is more convenient to write the above if statement in the following form:

```
if x >= 0 then
    y := x
else
    y := -x
end_if;
```

Here you can easily see that the commands y := x and y := -x are nested inside the if statement.

Let's summarize our current knowledge about the if statement.

4.1.1 MuPAD Syntax: the "if" Statement

```
if condition then
    statement
end_if
```

or

```
if condition then
    statement 1
else
    statement 2
end_if
```

Please note that at some point you may wish to perform more than one statement as a consequence of a given condition. This can be done in MuPAD by placing multiple statements separated by : or ;. The resulting program may look like this:

```
if condition then
    statement 1;
    statement 2;
    .....
    statement n;
end_if
```

or:

```
if condition then
    statement 1;
    statement 2;
```

```
     .....
     statement n
else
     another statement 1;
     another statement 2;
     .....
     another statement n
end_if
```

4.1.2 Indentation in MuPAD

In all the examples used in this chapter, we had indented some lines of the MuPAD code. This can improve the readability of the program, as well as showing the different nesting levels. Do you think that this is important? Well, let's check it. Imagine that you have the following program:

```
if x>0 then
    if y>0 then
        z := x*y
    else
        z := -x*y
    end_if
else
    if y>0 then
        z := -x*y
    else
        z := x*y
    end_if
end_if
```

Here, you can properly see the structure of the program. You can easily find out which statement is a consequence of which condition. Now, imagine that somebody wrote the same program like this:

```
if x>0 then
if y>0 then
z:=x*y
else
z:=-x*y
end
else
if y>0 then
z:=-x*y
else
```

```
z:=x*y
end
end
```

or even like this:

```
if x>0 then if y>0 then z:=x*y else z:=-x*y end else
if y>0 then z:=-x*y else z:=x*y end end
```

Do you like it? Can you read and understand such program? I guess not many people will be patient enough to read even a few lines of such code. This is the reason why we use indents in programming. Some programmers prefer to use indents equal to 2 spaces, while some others use 3 or even more spaces. You can easily see that 2 spaces for indents are not enough, however, while 4 spaces are too much. So, within this book, I have used indents equal to 3 spaces. Here we are coming up to another well known programers' rule.

Rule 5 Use indents to emphasize the nesting level in your programs

4.1.3 The Quadratic Equation Example

The if statement is ideal for many high school examples. One such example is the quadratic equation. While solving it, we need to pass a few conditions and this makes the whole solving process quite interesting. Let's see how we would do it using the paper and pencil technique.

Suppose that we are dealing with the equation $ax^2 + bx + c = 0$, where the coefficients a, b, and c are real numbers. The very first step while solving this equation is to check if $a = 0$ or not. If $a = 0$ then we are only dealing with a linear equation. Thus, we have the first if statement:

```
if a<>0 then
```

      BLOCK 1: statements to solve the equation using the quadratic
                formula

```
else
```

      BLOCK 2: statements to solve a linear equation

```
end_if
```

Now we can think about filling in the two missing blocks of statements. The second block is pretty obvious. We already know that $a = 0$ and our equation has the form $bx + c = 0$. Depending on the value of *b* we can produce a result or not, like this.

```
if b<>0 then
    x1 := -c/b
else
    print("No solutions")
end_if
```

Now we can concentrate on the first block, which is a bit more difficult. We may expect two real roots, different or not, or no real roots at all. This is determined by these two formulae:

$$x_1 = \frac{-b + \sqrt{b^2 - 4ac}}{2a}$$

$$x_2 = \frac{-b - \sqrt{b^2 - 4ac}}{2a}$$

In order to simplify the formulae in the program, let us declare an additional variable:

```
delta := b^2 - 4*a*c
```

Now, depending on the value of delta, we can either produce some results or not:

```
if delta >= 0 then
    x1 := (-b + sqrt(delta))/(2*a);
    x2 := (-b - sqrt(delta))/(2*a);
    print(x1, x2)
else
    print("No real solutions")
end_if
```

Finally we can combine all of these declarations into one MuPAD program:

```
if a<>0 then
    delta := b^2-4*a*c;
    if delta >= 0 then
        x1 := (-b + sqrt(delta))/(2*a);
        x2 := (-b - sqrt(delta))/(2*a);
        print(x1, x2)
    else
        print("No real solutions")
    end_if
else
    if b<>0 then
        x1 := -c/b;
        print(x1)
    else
```

```
            print("No solutions")
        end_if
    end_if
```

Now our program to calculate the roots of a quadratic equation is almost ready. We just need to declare the coefficients a, b, c and test it.

Let's add following three declarations at the start of the program:

```
a := 1:
b := -21:
c := 3:
```

Now we can run the whole program.

```
• a := 1:
  b := -21:
  c := 3:
  if a<>0 then
      delta := b^2 - 4*a*c;
      if delta >= 0 then
          x1 := (-b + sqrt(delta))/(2*a);
          x2 := (-b - sqrt(delta))/(2*a);
          print(Typeset, x1, x2)
      else
          print("No real solutions")
      end_if
  else
      if b<>0 then
          x1 := -c/b
      else
          print("No solutions")
      end_if
  end_if
```

$$\frac{\sqrt{429}}{2} + \frac{21}{2}, -\frac{\sqrt{429}}{2} + \frac{21}{2}$$

You can change the declarations for a, b, and c and run the program again. You will see how the output differs for different input values.■

There are many ways we could improve our program. You could change it to produce complex roots of the equation, to produce more sophisticated messages on the computer screen, and so on. There is one more thing worth mentioning here. Suppose that you need to calculate roots of different quadratic equations in two or more places in your notebook. Do you need to type in separate instances of the program for each of these equations and run each of them separately?

Of course this can be done much more efficiently. Later, you will learn how to develop MuPAD procedures. This will improve the usability of our programs. For now let's talk a bit about other things.

4.1.4 Formatting Output and Input

In three different places of the program to solve quadratic equations, we had used the `print` command to display a message on the computer screen. The command `print` can be used to print text or data on the screen. Text to be printed must be given in quotes, for example "No real solutions". You can print more than one message with one `print` statement:

- `print("Hello", "John & Marry")`

  "Hello", "John & Marry"

If you do not like the quotes in the output, you can force MuPAD to not display quotes by using the parameter `Unquoted`:

- `print(Unquoted, "Hello", "John & Marry")`

  Hello, John & Marry

We have removed the quotes. However, as you can see, there is still one problem with the above output. When printing two messages from one `print` command, MuPAD places a comma between them. In order to remove the comma, replace it by a dot, "." and add one space following the word `Hello`. You will get:

- `print(Unquoted, "Hello "."John & Marry")`

  Hello John & Marry

Now, let me explain the dot story. A text enclosed in quotes is called a string. Any word or sequence of words can be used as a string. In order to distinguish between strings and named variables or constants we use quotes. Thus "PI" means only the word *PI*, while at the same time `PI` (without quotes) means the number $\pi = 3.141592654$. The dot is an operator for strings, connecting two strings in one string. In MuPAD, the dot operator is very useful and we will return to it later.

You can also display computed data using the same `print` command. Let's see how it works.

- ```
  r := 3:
  pi := 3.14:
  area := pi*r^2:
  print(Unquoted,"The area of a circle with radius",r,
      "is equal",area)
  ```

  The area of a circle with radius, 3, is equal, 28.26

Now, again we have to fix the comma problem. In this example we cannot just replace them by a dot operator because area is not a string. First we need to convert it to a string. This can be done using the command expr2text. Thus, after replacing the commas with the dot operator, adding some additional spaces and using expr2text, we get quite pleasing results:

- ```
  r := 3:
  pi := 3.14:
  area := pi*r^2:
  print(Unquoted,"The area of a circle with radius "
          .expr2text(r)." is equal ".expr2text(area))
  ```

  The area of a circle with radius 3 is equal 28.26

The main goal of the print command is to output on the screen multiple numerical data separated by commas. Therefore we have to use some tricks in order to change its standard functionality. One of such situations is when we use the print command to output a formula. The standard output doesn't look good even with checked "Pretty Print Text" in **Format** menu. Let see what we may get.

Open the menu **Format** and check options **Typeset Expressions** and **Pretty Print Text**. Now type in and execute:

- ```
  print(sqrt(1/(1+x)))
          /   1   \1/2
          | ----- |
          \ x + 1 /
  ```

The 2D shape of the formula is the result of the **Pretty Print Text** option. Without this option we get even worse looking result.

- ```
  print(sqrt(1/(1+x)))
  ```

  $(1/(x + 1))^{\wedge}(1/2)$

Since version 3.0 we have in MuPAD a new option Typeset, that forces the print command to produce a nice looking result.

- `print(Typeset,sqrt(1/(1+x)))`

$$\sqrt{\frac{1}{1+x}}$$

For now, I guess we have solved the problem of getting different printouts on the screen.

Another aspect of our programs that isn't important but is still worth improving is the input of data into the program. We have used the assignment operator to input data into the program in all the examples so far. We used to begin each program with declarations like `x:=1/2` or `r:=3`, etc. In case of longer programs, this means scrolling back and forth through a few screens to input data and then see the result. It is more convenient to use the `input` command. For example, if you execute the command,

- `input("Input radius of the circle", r)`

MuPAD will display a dialog box like the one below:

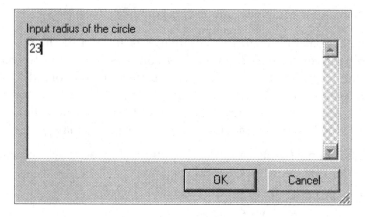

Here we can type in our input and then proceed with further calculations by pressing the [OK] button. The input box seems to be too large for our data. However, later you will be able to use it to input larger objects, like matrices, sequences of numbers, etc.

The `input` command can also be used in a slightly different way:

- `r := input("Input radius of the circle")`

In both cases the input value will be assigned to the variable r.

Now we can return again to MuPAD control structures.

4.1.5 "elif" a Useful Shortcut

Perhaps you have noticed that too many nested if statements make a program crowded and difficult to control. In a few situations, you can simplify the multiple nested if structures and make your program more flat. Imagine for a moment that you are dealing with a structure like this code:

```
• if x<0 then
     y := -1
  else
     if x<1 then
        y=0
     else
        if x<2 then
           y=1
        else
           if x<3 then
              y=2
           else
              y=3
           end_if
        end_if
     end_if
  end_if
```

Using the word elif, which combines the word else and the word if that follows it into a single word, you can obtain a simpler and easier to understand structure. Note, using elif we can reduce also the number of end_if keywords in our programs.

```
• if x<0 then
     y := -1
  elif x<1 then
     y=0
  elif x<2 then
     y=1
  elif x<3 then
     y=2
  else
     y=3
  end_if
```

You can run both programs in MuPAD for different values of x and check if you get the same results.

In undergraduate mathematics there are many interesting examples where you can apply the if statement. You will find a few of them at the end of this chapter in the programming exercises section.

4.2 Programming Iterations in MuPAD

Do you remember how often you had to calculate long sums of numbers? Sometimes it was obvious how many terms you have to add. However, in some situations it was not easy to determine how many terms you need to add. In this section, we will concentrate on those situations where the number of operations to be performed is known.

4.2.1 The "for" Loop

Let's consider the sum $1 + \frac{1}{2} + \frac{1}{3} + \frac{1}{4} + ... + \frac{1}{100}$. Adding these numbers by hand is very time consuming. In MuPAD, you could add all these numbers with a single command, but let's practice our programming skills instead. The above sum can be calculated with the use of the for loop.

- ```
  mysum := 0:
  for i from 1 to 100 do
     mysum: = mysum + 1/i:
  end_for
  ```
  $$\frac{14466636279520351160221518043104131447711}{2788815009188499086581352357412492142272}$$

In the for loop, you can perform operations from the smallest value of the index upwards, or from the largest value of the index down, like here:

- ```
  mysum := 0:
  for i from 100 downto 1 do
     mysum := mysum + 1/i:
  end_for
  ```

You can also produce a sum of every second term, every third term, etc. Here is how you would add every fifth term:

- ```
  mysum := 0:
  for i from 1 to 100 step 5 do
     mysum := mysum + 1/i:
  end_for
  ```
  ```

$$\frac{7546139536701184555}{4562486820030563424}$$

4.2.2 MuPAD Syntax: the "for" Loop

for *var* from *starting_val* to *ending_val* [step *step_val*] do

   statement 1;

   statement 2;

   ....

   statement n

end_for

or

for *var* **from** *starting_val* **downto** *ending_val* **[step** *step_val*] **do**

   statement 1;

   statement 2;

   ....

   statement n

end_for

Like in the case of the if control structure, you may use the word end instead of end_for.

It may happen that in some situations we wish to stop further execution of the loop and exit from it. For example, suppose that we are processing a list with sequences of digits of the same length n. It could be that by mistake in some of these sequences a character different than a digit may occur. We would wish to stop processing the sequence or even the whole list if such an error occurs. In order to do this we need to place another instruction inside the for loop:

```
if ... then break end_if;
```

The resulting program may look like this example:

```
a := 1:
for i from 1 to 100 do
  a := a*i;
  if a > 1500 then print(i,a); break end_if
end_for;
```

I guess now you might like to practice using the for loop. You can take a break in the reading of this text, and do a few of the exercises enclosed at the end of this chapter.

4.2.3 The "while" and "repeat" Loops

In a situation where the number of iterations depends on logical condition rather than the number of steps to be performed, we can apply two different kinds of loops — the while loop and the repeat loop. Both are very simple.

Here is an example showing how to calculate the sum of numbers $\frac{1}{n^2}$, adding only the terms larger than a certain small value.

- ```
  mysum := 0:
  n := 1:
  small := 1/1000:
  while (1/(n*n) > small) do
      mysum := mysum + 1/(n*n):
      n := n+1:
  end_while;
  print(mysum);
  ```

32

$$\frac{840971882932111177603170460}{5213096522073683233230240000}$$

The whole construction depends on the condition (1/(n*n)>small). I placed this condition in brackets in order to distinguish it from the rest of the program, but you can skip the brackets and the program will still work exactly the same way as it did before. The set of instructions between the do and end_while keywords will be performed as long as the condition is true. This means that you have to be careful how the condition is constructed, and make sure that there really is a situation when the condition will become false. For instance, imagine that in our example we forgot to add the instruction n:=n+1. The loop would have been performed up to infinity, because the condition 1/(n*n)>small never becomes false, and you will always have 1 > 1/1000.

The placement of the instruction n:=n+1 is also very important. If you put it before mysum:=mysum+1/(n*n), then you will skip the first term of the sequence. Try and see how it works for large values of small.

If you need to use more sophisticated conditions, you can always create them using the logical operations and, or, not. For instance:
  ```

```
x>0 and x<1
(n>0 and m<5) or (n<0 and m>5)
(x>2 or x<-2) and (x<3 or x>-3)
```

Determining whether a complicated logical condition is true or false can be sometimes tricky. We will talk more about logic and logical operations later.

Did you notice that at the end of the last example I used the command print(mysum)? Perhaps you are wondering why we needed this? MuPAD, like many other mathematical programs, will always display the last calculated result for a given control structure. In our program, the number 32 was the last calculated value of n obtained while checking the loop entry condition. So, we need to force MuPAD to display the real final result.

The while loop is very similar to the repeat loop. The difference is where we place the condition. In the while loop the condition was placed on entry to the loop and it was checked before doing any operations placed inside the loop. In the repeat loop, we place the condition at the end. Thus, the condition is checked after performing all operations inside of the loop at least once. This is slightly riskier, but you can handle it as long as you are careful. Here is an example showing how the repeat loop works.

```
• mysum := 0:
  n := 1:
  small := 1/1000:
  repeat
     mysum := mysum + 1/(n*n):
     n := n+1:
  until 1/(n*n) < small end_repeat:
  print(mysum);
```

$$\frac{8409718829321111776031704609}{5213096522073683233230240000}$$

4.2.4 MuPAD Syntax: the "while" and "repeat" Loops

while *condition* do

   statement 1;

   statement 2;

   ...

   statement n;

```
end_while
```

and

```
repeat
    statement 1;
    statement 2;
    ...
    statement n;
until condition end_repeat
```

4.3 What is Wrong with my Loops and Conditions

It may happen sometimes that your program looks perfectly, its logical structure is correct, but the results produced are wrong. Finding such errors can be very difficult. In many cases we may even think that the fault is in MuPAD itself and MuPAD produces the wrong results. Before jumping to such conclusions, let us examine a few examples. Then we will try to find explanations for them along with possible solutions.

Let us start with a very simple if statement,

- `reset():`
- ```
  if 1 - 1.0 = 0 then
      print("OK")
  else
      print("something is wrong")
  end_if
  ```
  "something is wrong"

As you can see, the output is not this what we really expected. According to our mathematical knowledge, 1 and 1.0 are exactly the same numbers, so we would expect the result "OK". Let us check how MuPAD performs operations with floating point numbers. First, let us reset our current MuPAD session so we are sure that we are starting completely fresh calculations.

- `reset():`
- `a := 1.0 + 10^(-11)`
  1.0
  ```

- a

  1.0

It seems that MuPAD completely ignored the fraction 10^{-11}. Now, let us ask MuPAD to use more digits in floating point numbers. This can be done with the so-called environmental variable DIGITS. Let us set DIGITS to 15.

- `DIGITS := 15:`
- a

  1.00000000001

This is the result that we expected. However, let's make DIGITS even larger, for example 50 or even more, and see what we get this time.

- `DIGITS := 50:`
- a

  1.0000000000099999999600419720025001879548653960228

As we see in this latest output, the number 10^{-11}, when added to a decimal number, was converted to something that is only its approximate value.

Let us repeat some parts of this investigation without using decimal numbers. Therefore, we will use 1 instead of 1.0. Do you think that such a slight change will matter? Let us check it.

- `reset()`
- `a := 1 + 10^(-11)`

  $$\frac{100000000001}{100000000000}$$

This time the value of a was produced as a fraction. We can easily convert it to a decimal fraction using the `float` command. Here is what we now get for a very high value of DIGITS:

- `DIGITS := 500:`
- `float(a)`

  1.00000000001

Let us check another example. Like last time, we will start by resetting our current session.

```
• reset():
  i := 0.1:
  repeat
      i := i+0.1:
      if i>2 then print(i) end_if;
      if i>5/2 then break end_if;
  until i=2.0 end_repeat
```

2.1

2.2

2.3

2.4

2.5

2.6

This time the repeat loop was not stopped in the right place. If we remove the second if statement inside the repeat statement then our program will start producing thousands of numbers until we stop it by pressing the red button. You may already suspect that we got wrong results because we used decimal numbers in the program. This is the right conclusion. Replace all the decimal numbers used in the program with their exact values – this means 0.1 by 1/10 and 2.0 by 2. Now the program will behave exactly like we expected.

The explanation of these problems is very simple. MuPAD, like many other computer programs, uses binary numbers, i.e. numbers of form 100010.10011100, in all internal calculations. This means that a decimal number first has to be converted to a binary number, and only then will the appropriate operations be performed. The obtained result will then be converted back to decimal form. However, it happens that a decimal fraction with a finite number of digits has infinitely many digits in its binary form. Due to limited size of computer memory, this long binary number will be truncated to a finite binary fraction. This is the point where errors can be introduced. I guess you might be a bit confused by this explanation. If this is the case, just take a look at a very similar example from school mathematics. In decimal form, the number 1/3 should be written as 0.333333333333333..., where the three dots should actually be replaced by infinitely many digits 3.

What do we normally do? We cut this long fraction and leave 10 or even less digits. You certainly agree that 1/3 is not equal to 0.3333333333, but we don't have any other choice if we want to use a decimal representation of 1/3. The same is the case when converting decimal fractions to binary.

Therefore, we can conclude—it is much safer using exact values of numbers in MuPAD instead of decimal fractions.

Before we proceed to the next section of this chapter, let me show you one thing that you may need when using exact values.

In some situations, MuPAD does not immediately evaluate the value of an expression. Is such cases, we have to force MuPAD to calculate an expression and use its value. Look at the following program:

- ```
  number:=sqrt(5)-sqrt(3) // here number is a label only
  ```
  $$\sqrt{5} - \sqrt{3}$$

- ```
  if number > 0 then
      print("Is OK")
  else
      print("Doesn't work")
  end_if
  ```
  Error: Can't evaluate to boolean [_less]

However, if we use the command is(number>0), which forces the evaluation of the inequality, we can make our program work properly.

- ```
  if is(number > 0) then
      print("Is OK")
  else
      print("Doesn't work")
  end_if
  ```
  "Is OK"

In our example evaluating is(number>0) was not a problem. However, we can find many situations where MuPAD will not be able to evaluate the condition and it will return the value UNKNOWN. In such cases it can be much safer to use is(*condition*)=TRUE or sometimes is(*condition*)<>FALSE.

I think that you now know enough to make your loops and conditions work properly. So, let us move on to another control structure.
  ```

4.4 Selections

In some applications, we need to perform calculations depending on the value of certain parameters. For instance, we may wish to calculate the values of a function that will have a different formula for various intervals, like this:

$$
f(x) = \begin{cases}
-1 & x < -1/2 \\
-1/2 & -1/2 \le x < 0 \\
1/2 & 0 \le x < 1/2 \\
1 & 1/2 < x
\end{cases}
$$

You have already learned about a control structure that allows you to program such selections. Let me remind you about it. It was the `if` structure with `elif` keyword. Let's see how it works and if this is this what we need.

```
• x := 1/3:
  if x < -1/2 then
      y := -1
  elif x = 0 then
      y := -1/2
  elif x < 1/2 then
      y := 1/2
  else
      y := 1
  end
```

Running this program for different values of x, you will easily see that this construction works how we expected. However, if you wish to use floating point numbers for x then perhaps it would be wiser to protect the evaluation of conditions in the program. This may look like here,

```
• x := 0.01:
  if is(x < -1/2) then
      y := -1
  elif is(x = 0) then
      y := -1/2
  elif is(x < 1/2) then
      y := 1/2
  else
      y := 1
  end
```

There is also another programming construction in MuPAD that you will find useful in some situations where we have to choose between different cases. This is the case control structure. You can apply it like this:

```
• mark := 7:
  case mark
    of 0  do
    of 1  do
    of 2  do
    of 3  do
    of 4  do
    of 5  do
       print(mark, "Fail the course"); break;
    of 6  do
       print(mark, "Pass the course"); break;
    of 7  do
    of 8  do
       print(mark, "Well done"); break;
    of 9  do
    of 10 do
       print(mark, "Outstanding result"); break;
    otherwise print("Such a mark does not exist")
  end_case;
  7, "Well done"
```

In the above example, we used the case structure to produce a message informing the students if they passed the course and how good was their result. In this example, our grades are the integer numbers 0, 1, 2, .., 10. Let's analyze the above example to learn more of its secrets.

First, note that the values of the variable used in the case structure must be single objects. We cannot use intervals, relations or multiple values here. For example, we cannot write:

```
• mark := 2:
  case mark
    of mark <6  do
       print(mark, "Fail the course"); break
    of 6  do
       print(mark, "Pass the course"); break;
    of 7,8  do
       print(mark, "Well done"); break;
    of 9,10  do
       print(mark, "Outstanding result"); break;
```

```
    otherwise print("Such mark does not exist")
end_case;
```

That is, we can write such a construction, and MuPAD will not complain about the wrong use of the case statement, but the result will be a nonsense. You can try it out with the variable mark equal to 3, 6, 7 or 12.

In the case statement where you wish to produce the same result for multiple values of the input variable, you will need to use a block of multiple of..do statements followed by a command or commands that will be executed for the entire block. For instance, in our first example we had used this structure:

```
    of 0   do
    of 1   do
    of 2   do
    of 3   do
    of 4   do
    of 5   do
        print(mark, "Fail the course"); break
```

When executing this structure, MuPAD will produce the same message for the integers 0, 1, 2, 3, 4, 5.

There is another special feature of the case control structure. In order to avoid it, we had to place the keyword break at the end of the instructions for each group of cases. Let's take a look at how such a structure would work without the break command.

```
• mark := 6:
  case mark
      of 0   do
      of 1   do
      of 2   do
      of 3   do
      of 4   do
      of 5   do
          print(mark, "Fail the course");
      of 6   do
          print(mark, "Pass the course");
      of 7   do
      of 8   do
          print(mark, "Well done");
      of 9   do
      of 10  do
```

```
      print(mark, "Outstanding result")
   otherwise print("Such a mark does not exist")
end_case;
```

6, "Pass the course"

6, "Well done"

6, "Outstanding result"

"Such a mark does not exist"

Now we can see what happened. MuPAD found the line where the value of mark was equal to 6, and then performed all the commands not only for this case, but also for all the following cases. Thus, if you are going to use the case control structure, you should never forget the break keyword.

4.4.1 MuPAD Syntax: the "case" Structure

case *variable*
    of *case_1* do *statements* ... ; break
    of *case_2* do *statements* ... ; break
    ...
    of *case_k* do *statements* ... ; break
otherwise *statements* ... end_case

4.5 Chapter Summary

This was another chapter full of MuPAD-specific information. There will be one more chapter like this, and then you will spend more time with graphics and mathematics. Now let us summarize what you learned in this chapter.

The whole chapter was mostly devoted to MuPAD programming constructions, so-called control structures. You learned how to apply the if conditional structure, loops for, while, and repeat, as well as the case selection structure. Here is a summary of their syntax.

4.5.1 MuPAD Syntax: the "if" Statement

```
if condition then
    statements
end_if
```

or

```
if condition then
    statements_1
else
    statements_2
end_if
```

or with elif keyword

```
if condition_1 then
    statements_1
elif condition_2 then
    statements_2
elif condition_3 then
    statements_3
.......
elif condition_n then
    statements_n
else
    statements_n+1
end_if
```

4.5.2 MuPAD Syntax: the "for" Loop

```
for var from starting_val to ending_val [step step_val] do
    statements
end_for
```

or

```
for var from starting_val downto ending_val [step step_val] do
    statements
end_for
```

4.5.3 MuPAD Syntax: the "while" and "repeat" Loops

while *condition* do
   statements
end_while

and

repeat
   statements
until *condition* end_repeat

4.5.4 MuPAD Syntax: the "case" Structure

case *variable*
   of *case_1* do *statements* ... ; break
   of *case_2* do *statements* ... ; break
   ...
   of *case_k* do *statements* ... ; break
otherwise *statements* ... end_case

4.5.5 MuPAD Command Summary

In this chapter we learned very few useful MuPAD commands.

float(x) - *produces decimal value of x*
is(*condition*) - *checks if the condition is true*
break - *breaks calculations inside the loop and quits*

The break command was also very useful in improving the performance of the case control structure.

4.6 Programming Exercises

4.6.1 The "if" Statement Exercises

1. Write a program to input two real numbers and output the largest one.

2. Write a program that will read two real numbers and output them in order of increasing value.

3. Write a program that will read three real numbers and output the smallest one.

4. Write a program that will read three real numbers and output them in order of decreasing value.

5. Write a program that will produce the value of $\sqrt{1-x^2}$ for a given x if such a value exists; otherwise, an error message should be printed.

6. Write a program that will produce the value of $\frac{1}{1-x^2}$ for a given x if such a value exists; otherwise, an error message should be printed.

7. Write a program that will produce the value of $\sqrt{x^2-1} + \sqrt{4-x^2}$ for a given x if such a value exists; otherwise, an error message should be printed.

8. Write a program that will produce the value of $\sqrt{x^2-1} \cdot \sqrt{4-x^2}$ for a given x if such a value exists; otherwise, an error message should be printed.

9. Write a program that will produce the value of $\frac{1}{\sqrt{x^2-1}}$ for a given x if such a value exists; otherwise, an error message should be printed.

10. The radius of a circle inscribed in a triangle of sides a, b, and c is $r = \sqrt{\frac{(s-a)(s-b)(s-c)}{s}}$, where $s = \frac{1}{2}(a+b+c)$. Write a program that will calculate the radius for a triangle with given sides.

11. The radius of a circle circumscribed about a triangle of sides a, b, and c is $R = \dfrac{abc}{4 \cdot \sqrt{s(s-a)(s-b)(s-c)}}$, where $s = \frac{1}{2}(a+b+c)$. Write a program that will calculate the radius for a triangle with given sides.

12. The eccentricity of the ellipse $\dfrac{(x-h)^2}{a^2} + \dfrac{(y-k)^2}{b^2} = 1$ is given by the formula
$$e = \begin{cases} \frac{\sqrt{a^2-b^2}}{a} & \text{if } a > b \\ \frac{\sqrt{b^2-a^2}}{b} & \text{if } b > a \end{cases}$$

Write a program to calculate the eccentricity of the ellipse with given a and b.

4.6.2 Exercises for Loops

1. Write a program that will produce the sum of all integers from 1 to a given positive n.

2. Write a program that will produce the sum of the first n positive even integers.

3. Write a program that will produce the sum of the first n positive odd integers.

4. Write a program that will produce the sum of the first n positive integers that are divisible by 7.

5. Write a program that will produce $\sqrt{\dfrac{1+2+\ldots+n}{1 \cdot 2 \cdot \ldots \cdot n}}$ for a given positive integer n.

6. Write a program that will produce the product $n \cdot (n+1) \cdot (n+2) \cdot \ldots \cdot m$ for the two given positive integers n and m, where $n < m$.

7. An arithmetic progression has the form
$$a, a+d, a+2d, a+3d, \ldots$$
The n-th term is $a_n = a + (n-1)d$, and the sum of the first n terms is
$$S_n = \sum_{k=0}^{n-1}(a+kd) = \frac{n}{2}(a+a_n)$$
Write a program that will calculate the sum of the first n terms for given n, a, and d. Check if you get the same result using the formula on the right side.

8. The arithmetic mean of a and b is $\frac{a+b}{2}$. The arithmetic mean of n numbers $x_1, x_2, \ldots, x_n$ is the sum of the numbers divided by n:
$$\frac{x_1 + x_2 + \ldots + x_n}{n}$$
Write a program that will calculate the arithmetic mean of all the numbers $x_k = (\frac{1}{2})^k$ for $k \le 20$.

9. A geometric progression has the form

$$a, ar, ar^2, ar^3, \ldots$$

The sum of the first n terms is

$$S_n = \sum_{k=0}^{n-1} ar^k = \frac{ar^n - a}{r - 1}$$

Write a program that will calculate the sum of the first n terms for given n, a, and r. Check if you get the same result using the formula on the right side.

10. A harmonic progression is a sequence of numbers whose reciprocals form an arithmetic progression. It has the form

$$\frac{1}{a}, \frac{1}{a+d}, \frac{1}{a+2d}, \frac{1}{a+3d}, \ldots$$

The harmonic mean of a and b is $\frac{2ab}{a+b}$. The harmonic mean of n positive numbers $x_1, x_2, \ldots, x_n$ is the reciprocal of the arithmetic mean of the reciprocals.

$$\frac{n}{\sum_{i=1}^{n} \frac{1}{x_i}}$$

Write a program that will calculate the harmonic mean of the first n terms of the arithmetic progression with n, a, and d given.

11. It is well known, that if $0 < a < 1$ then $a^n \to 0$. Write a program that will display all the numbers a^n such that $0 < b < a^n < 1$ for a given $b < a$.

12. It is also well known that the number e can be calculated as a limit of the sequence

$$e_n = 1 + \frac{1}{1!} + \frac{1}{2!} + \frac{1}{3!} + \ldots + \frac{1}{n!}$$

Write a program that will calculate an approximate value of the number e, i.e. e_n such that $e_{n+1} - e_n < theta$, where $theta$ is a small number, like 1/1000. The program should also display the number of terms that were used in the sum.

13. In one of the standard methods of constructing fractals the iteration rule involves successive removals. To construct the shape known as the *Sierpinski triangle*, we begin with an equilateral triangle as a seed and then use the iteration rule: "remove from the middle of this triangle a smaller triangle with side lengths one-half of the original so that three congruent triangles remain." When this

rule is iterated infinitely often, we reach a limiting shape known as the *Sierpinski triangle*.

While repeating this iteration rule we produce a sequence of images that we may call *Sierpinski triangles* of a given order n, where n represents number of iterations. Here I show the original triangle and three images for iterations 1, 2 and 3.

a. Write a program to calculate the area of the *Sierpinski triangle* for a given number of iterations n.

b. Write a program to calculate the total length of sides of all small triangles in the Sierpinski triangle of order n.

14. There are more geometric objects constructed in the same manner like the *Sierpinski triangle* from the previous example. Here is another geometric construction that results in a fractal. Start with a square whose side has length 1. The iteration rule is to first divide the square into nine squares whose sides have length 1/3. Then remove the four corners as shown here,

 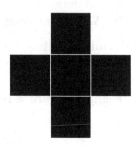

When this rule is iterated infinitely often, we obtain a cross like figure that we call the *cross fractal*. Here are shown resulting figures for two and three iterations,

Write a program to calculate the area of the *cross fractal* for a given number of iterations *n*.

15. The *Menger sponge* is a cube-like object that was created in the following way. Begin with a cube and divide it into 27 cubes of equal size. Remove seven of the smaller cubes: one from the middle of each face, and one from the center. This leaves 20 smaller cubes, each having 1/27 of the volume of the original cube. Now, repeat this process for each smaller cube, etc. This way, you will obtain the 1st, 2nd, ..., *n*-th approximation of a *Menger sponge*, or say, a *Menger sponge* of the order 1, 2, ..., *n*.

   a. Write a program to calculate the volume of the *Menger sponge* of the order *n* for a given number *n*.

   b. Write a program to calculate the total area of the *Menger sponge* of the order *n* for a given number *n*.

Chapter 5

Programming in MuPAD - Procedures

When working with MuPAD, we often deal with situations where the same piece of code will be reused in many parts of our notebook. For example, the code used to solve quadratic equations in the previous chapter could be used to solve many different equations. The major question that arises, then, is whether we will have to type in or copy and paste the quadratic equation code every time we need to use it. Imagine a student having to solve ten or more quadratic equations. Will he need to put the same code in ten different places in his notebook? Certainly, that would look silly. There are much more efficient methods to do it. Do you remember how we declared functions in MuPAD? We had to develop a single declaration for a function and then, every time we needed it, we simply called its name. Like this:

- ```
  mysin := x -> abs(sin(x));
  mysin(5*PI/3);
  ```
  $$\frac{\sqrt{3}}{2}$$

Now you will be able to apply the mysin function to many different problems. So you could, for example, use the commands,

- ```
  mysin(200*PI/201);
  ```
- ```
  mysin(20000*PI/201);
  ```
- ```
  (mysin(mysin(PI*1.2)))^2;
  ```

You will certainly agree, doing things this way makes more sense. Now, let's return to the quadratic equation example. Here is its original code:

- ```
  a := 1:
  b := -21:
  c := 3:
  if a<>0 then
      delta := b^2-4*a*c;
      if delta >=0 then
          x1 := (-b + sqrt(delta))/(2*a);
          x2 := (-b - sqrt(delta))/(2*a);
          print(x1, x2)
  ```
  ```

```
      else
         print("No real solutions")
      end
   else
      if b<>0 then
         x1 := -c/b;
         print(x1)
      else
         print("No solutions")
      end
   end
end
```

We would definitely be happy if we could transform this code into something similar to the function mysin. Fortunately, this is both possible and quite easy. We will call this transformed code a procedure. First, we need to choose a name for the procedure — this is how we will call this code later on. In fact, this would be the right time to think for a moment about how you will name various objects in MuPAD. Just imagine that your code to solve quadratic equations is called mycode or abc or xyz123. Who could guess what it is and how it works? Certainly, you will be the only person knowing this, and only for short time. Something like solvequadraticequation would be much better, but no one likes typing in such long commands. Let's make another agreement that will help us in further programming experiments.

Rule 6   The names of MuPAD functions, procedures and variables should be descriptive and not unreasonably long.

There is another issue related to names that's worth discussing right now. Have you noticed that most MuPAD procedures are usually in lower case? For example, solve, gcd, limit, taylor, random, and so on. It is good idea to agree on another rule.

Rule 7   Variable and procedure names are in lowercase, with occasional upperCase characters in the middle.

Now, after accepting rules 6 and 7, we may consider the words quadratic, quad, or quadequation as potential names for our code. Perhaps the last choice is the most descriptive but the two words sealed together are a bit difficult to read. If you wish to separate them, apply rule 7 and capitalize the second word — quadEquation.

In the next step of our transformation, you need to remove these three lines declaring a, b and c. Instead of them, you need to use this:

```
quadEquation := proc(a,b,c)
```

Finally, after this declaration you have to add these MuPAD keywords:

```
begin

end_proc
```

The rest of the program to calculate the roots of the quadratic equation is then inserted between these keywords. Here is the complete construction of your first MuPAD procedure.

- ```
  quadEquation := proc(a,b,c)
  begin
      if a<>0 then
          delta := b^2-4*a*c;
          if delta >= 0 then
              x1 := (-b + sqrt(delta))/(2*a);
              x2 := (-b - sqrt(delta))/(2*a);
              return(x1, x2)
          else
              print("No real solutions")
          end
      else
          if b<>0 then
              x1 := -c/b;
              return(x1)
          else
              print("No solutions")
          end
      end
  end_proc:
  ```

Note, the print statement prints data on the screen but it does not produce a real output from the procedure. This is just only a text on the screen. Therefore, in places where our procedure produces an output I used the return command that is the proper way of getting data from a procedure. Now, you can try using it. For example:

- ```
  quadEquation(1000,-2,-1)
  ```

$$\frac{\sqrt{1001}}{1000} + \frac{1}{1000}, \frac{1}{1000} - \frac{\sqrt{1001}}{1000}$$

- `quadEquation(23,1,-100)`

$$\frac{\sqrt{9201}}{46} - \frac{1}{46}, \quad -\frac{\sqrt{9201}}{46} - \frac{1}{46}$$

Now that you know how to convert your programs into procedures, you could try revising all your work from the previous chapters and add procedure declarations in as many places as necessary. The next sections of this chapter will help you make this work more efficient and show you a few secrets of MuPAD procedures.

5.1 What is a Procedure

The previous example gave us a basic idea of what a procedure is and how it works. Now we need to deepen and systematize our knowledge of procedures. We will start with the general syntax of a procedure.

proc_name := proc(*input parameters*)
local *local_variables*;
begin
    here is the body of the procedure, i.e. your code
end_proc

The procedure ending keyword `end_proc` can also be used in the simplified form `end`.

For now, this definition will be enough. However, later on we will add some other components. Right now, let us take a look at some examples of MuPAD procedures in order to gain some more experience. We will mainly use examples where the procedure calculates some numbers. This is because for such procedures, the knowledge introduced in the previous chapters will be sufficient. Later, we will also introduce some MuPAD data structures that can make our experiments more interesting.

5.2 Examples of Procedures

The simplest example of a MuPAD procedure is mentioned in the chapter 3 declaration of a function. You should already remember it. Here is an example of such a function.

- f := x -> sin(x)/x

The syntax of the function is very simple and you probably will not even realize that you are dealing with a procedure. However, you may notice that f is, in fact, just a shortcut for the procedure name, x is the input parameter, and sin(x)/x is the body of the procedure. Procedures declared this way do not have local variables.

You can also define functions with more than one variable. You do it like this:

- h := (x,m,n,p) -> ((x^m + x)^n +x)^p

Such declaration tells MuPAD that $h(x,m,n,p) = ((x^m + x)^n + x)^p$.

The above notation suggests that this function should be used for real values of x, and integer values of m, n, p. However, you may use all kinds of numbers. If you wish to restrict it to integer values of m, n, and p, then you will need to work on this later.

You may also have functions where one function uses another:

- s := (a,b,c) ->(a+b+c)/2
- heron := (a,b,c)->
  sqrt(s(a,b,c)*(s(a,b,c)-a)*(s(a,b,c)-b)*(s(a,b,c)-c))

These examples are probably not especially interesting for you. You have already developed a lot of them while solving the problems at the end of chapter 3. So, we can move on to more interesting procedures.

Example 5.1 Heron formula

The very simple Heron's formula to calculate the area of a triangle given the lengths of all three sides deserves a bit more of your attention. Here you can experiment a bit and see what you need to watch out for while writing procedures. We will start with a very basic structure.

- heron1 := proc(a,b,c)
  begin
    s := (a + b + c)/2;
    sqrt(s*(s - a)*(s - b)*(s - c))
  end:
- heron1(3,4,5)

  6

You will certainly agree that this procedure works for all examples where a, b, c are the sides of a real triangle. However, it also produces results when the sides of the triangle are negative numbers. This should not happen.

- `heron1(-4,2,3)`

$$\frac{3}{4}\sqrt{15}$$

It can also produce a result for an impossible triangle where one side is longer than the length of the other two sides combined, like this:

- `heron1(6,1,1)`

$$6 \cdot i \cdot \sqrt{2}$$

Of course, in the last case, the result is a complex number, but you agree that our triangle is also a kind of imaginary object. Therefore, everything seems to be correct. Nonetheless, you may want to restrict this procedure to real triangles only. Well, let's add some basic protection.

-
```
heron2 := proc(a,b,c)
  begin
     if a<0 or b<0 or c<0 then
        print("This is not a real triangle")
     else
        s := (a + b + c)/2;
        sqrt(s*(s - a)*(s - b)*(s - c))
     end
  end:
```

Now, if one of the input numbers is negative, the procedure `heron2` will display an error message. However, the `heron2` procedure still produces some output for non-existent triangles. You will have to correct this. Note that if one of the sides of the "triangle" is longer than the sum of the other two, then one of the numbers $(a-b-c)$, $(b-a-c)$ or $(c-a-b)$ will be positive and the other two negative. For a real triangle, all three numbers would be negative. So, you could ask whether $(a-b-c) \cdot (b-a-c) \cdot (c-a-b) > 0$. If this condition is true, then you are dealing with an impossible triangle. Here is the improved code.

-
```
heron3 := proc(a,b,c)
  begin
     if a<0 or b<0 or c<0 or (a-b-c)*(b-a-c)*(c-a-b)>0
     then
```

```
                print("This is not a real triangle.")
        else
            s := (a + b + c)/2;
            sqrt(s*(s - a)*(s - b)*(s - c))
        end
    end:
```

Now you can give it a try, to see how it works.

- `heron3(-1,1,1)`

   "This is not a real triangle"

- `heron3(6,1,2)`

   "This is not a real triangle"

- `heron3(1,1,1)`

   $$\frac{\sqrt{3}}{4}$$

There are a few more things that are worth changing in our procedure, but this can be done later. For now, let's fix only the two most important issues.

You perhaps remember that, while writing the syntax of the procedure, I mentioned local variables. In our procedure, there is definitely one variable that is local to the procedure. This is the variable s. So, we will declare it as a local variable. We will discuss why local variables are important later on. For now, just remember that all the variables that you introduce locally inside a procedure should be declared as local, as you may get into trouble otherwise.

Another thing worth remembering is how to obtain output from a procedure. In our heron, heron1, heron2 and heron3 procedures, we got the output from the statement sqrt(s*(s-a)*(s-b)*(s-c)). This will work as long as this is the last command executed inside the procedure. It will not work, however, if you put in anything after sqrt(s*(s-a)*(s-b)*(s-c)). For example, if you add the line a=a there, you will get the result 1=1. Which is of course the output from the procedure, but not what you expected. In order not to obtain such unwanted output, you have to force your procedures to return the desired result. The best way to do so is to use the return command, like here:

```
• heron := proc(a,b,c)
     local s;
  begin
     if a<0 or b<0 or c<0 or (a-b-c)*(b-a-c)*(c-a-b)>0
     then
        print("This is not a real triangle")
     else
        s := (a + b + c)/2;
        return(sqrt(s*(s - a)*(s - b)*(s - c)));
     end
  end:
```

Now, you can be sure that your heron procedure will give you the correct result, at least in most cases. Well, be careful — try to analyze even the strangest cases. For example, what will your procedure return when one of the input numbers is a complex number? Good question, give it a try. ∎

In the later sections of this chapter, we will explore more systematically the various aspects of MuPAD procedures. The goal of the examples in this section is to give you some experience in developing MuPAD procedures. So, let us try another example.

Example 5.2 The Euclid algorithm to find GCD of two integers

The Euclid algorithm to find the greatest common divisor of two numbers is considered to be one of the most classical examples of high school algorithms. Of course, it depends on the country and the school that you are in. However, the idea is quite simple. If you do not remember, just take a look at the procedure enclosed below and try to figure out how it works.

```
• Euclid := proc(a,b)
  begin
     while a <> b do
        if a > b then
           a := a - b
        else
           b := b - a
        end
     end;
     return(a);
  end:
```

It could be very useful to develop a table of values produced by the procedure. This would help us trace all the consecutive steps that are performed inside the procedure.

Suppose that we put in a=215 and b=125. We will then get:

step	a<>b	a>b	a	b	output	comment
0, input			215	125		
1	TRUE	TRUE	90	125		a=215-125=90
2	TRUE	FALSE	90	35		b=125-90=35
3	TRUE	TRUE	55	35		a=90-35=55
4	TRUE	TRUE	20	35		a=55-35=20
5	TRUE	FALSE	20	15		b=35-20=15
6	TRUE	FALSE	5	15		a=20-15=5
7	TRUE	FALSE	5	10		b=15-5=10
8	TRUE	FALSE	5	5		b=10-5=5
9	FALSE	FALSE				
10, output					5	

By analyzing this table, you can trace all the values from the input of a and b until the output of the final result.

The procedure Euclid produces proper results when both numbers are positive integers. However when one of the numbers is negative or a non-integer number, the results, if any, are completely wrong. We have to find a way to check the type of the input data. As you perhaps remember, we have various domains in MuPAD. One of them is the domain of integer numbers. Therefore, we can use the command testtype(a,Type::Integer) to check if the number a is an integer number.

- testtype(12,Type::Integer)

  TRUE

- testtype(12.23,Type::Integer)

  FALSE

Well, it seems that one problem has been solved. We know how to check if a given number is an integer. There is also another problem — what to do with negative integers. We do want the procedure to generate the GCD for negative integers, too. This can be done by introducing two additional variables, u:=abs(a) and v:=abs(b), where abs(x)=$|x|$. This will replace a and b by two positive integers having the same GCD. So now, we can write down the final version of our procedure. In order to distinguish it from the previous version, we will use a lowercase name.

```
• euclid := proc(a,b)
  local u,v;
  begin
     if testtype(a,Type::Integer) and
        testtype(b,Type::Integer)
     then
        u := abs(a);
        v := abs(b);
        while u<>v do
           if u>v then
              u := u - v
           else
              v := v - u
           end
        end;
        return(u);
     else
        error("use integer numbers")
     end;
  end:
```

Now we can try testing it and see what we get:

```
• euclid(-12,234)
```
6

```
• euclid(1.3,789870)
```
Error: use integer numbers [euclid]

```
• euclid(12355,23450)
```
35

Note that in this procedure, we used another form of displaying the message if the input is wrong. I used the command error, which displays an error message as well as the error popup box.

Well, you can certainly see that the procedure works just like we expected. However, you may still find a few things to do. For example, what result would be produced if we were to put in three or more integers, or what would happen if we were to use just one number or no input at all? We will solve these problems later. Right now, it is time to move on to the next section.   ■

5.3 Getting Output from Procedures

A procedure is like a black box. You drop some values into it, and you get some sort of output. The question now is, what do you wish to do with this output? In some situations, it will be enough if a procedure would print your results on the screen. In other situations however, you may need to use the obtained values in further calculations. So, you will have to store them as variables for future use.

The first case is easy to arrange. You merely need to use the command print to display the results on the screen. For example, this procedure will display information if the three given numbers are the sides of a rectangular triangle:

```
• rectangular := proc(a,b,c)
  begin
     if (a^2+b^2=c^2) or (a^2+c^2=b^2) or (b^2+c^2=a^2)
     then
        print("This is a rectangular triangle")
     else
        print("This is not a rectangular triangle")
     end
  end;
```

Now, if you try it, this is what you will get:

```
• rectangular(3,4,5)
```
   "This is a rectangular triangle"

- `rectangular(1,2,3)`

    "This is not a rectangular triangle"

You will agree that for many school or university examples, this output is sufficient. However, if you wish to keep the obtained information for further use, you will need to be careful about what kind of output you produce and where you put it. Certainly, a piece of text is not the type of information that you would be able to use later. So, the procedure should produce values that we can assign to variables. For example, the `rectangular` procedure can be modified to this form:

- ```
  rect := proc(a,b,c)
  begin
      bool((a^2+b^2=c^2) or (a^2+c^2=b^2) or (b^2+c^2=a^2))
  end
  ```

Now you can assign the obtained value to a variable and check what you get.

- `v:=rect(1,2,3)`

    FALSE

In the above example we used the `bool` function, which produces a logical value for a given expression. This value can be either TRUE or FALSE. The words "or", "and" and "not" are logical connectors that you have certainly already used in mathematics, not to mention real life. We will talk about working with logical expressions in MuPAD in one of the later chapters. For now, let us concentrate on MuPAD procedures.

Values produced by MuPAD procedures can be applied in more complex calculations. For example, the `rect` procedure can be used in the `if`, `while` or `repeat` constructions, like here:

- `if not rect(a,b,c) then a:=a+0.1 end_if;`

MuPAD procedures automatically return the last calculated value. This is the simplest, but not the best way of getting output from MuPAD procedures. If you do not force a MuPAD procedure to return the right values, you may get incorrect or incomplete results. Just take a look at the simplified quadratic equation procedure:
  ```

```
• simpleQuadEquation := proc(a,b,c)
  begin
     if a<>0 then
        delta := b^2 - 4*a*c;
        if delta>=0 then
           x1 := (-b + sqrt(delta))/(2*a);
           x2 := (-b - sqrt(delta))/(2*a);
        else
           print("No real solutions")
        end;
     end;
  end:
```

• simpleQuadEquation(1,6,-3)

$$-2\sqrt{3} - 3$$

Now, you are certainly wondering what happened to the other root. For sure, it exists. MuPAD returned the last calculated value, which is x2. The value of x1 was completely ignored. So, you need to use the return command to force the simpleQuadEquation to output the correct result. But first, there is something else here that's worth pointing out. Every time the return command is executed inside a procedure, MuPAD returns the requested value and stops all further calculations inside the procedure. If you use return twice or more inside of a procedure, only one of these commands will actually work. In the case of simpleQuadEquation, we want to return a pair of numbers. Have a look at what you can do to achieve this with simpleQuadEquation procedure:

```
• simpleQuadEquation := proc(a,b,c)
  begin
     if a<>0 then
        delta := b^2 - 4*a*c;
        if delta >= 0 then
           x1 := (-b + sqrt(delta))/(2*a);
           x2 := (-b - sqrt(delta))/(2*a);
           return([x1,x2])
        else
           print("No real solutions")
        end;
     end;
  end:
```

- `simpleQuadEquation(1,6,-3)`

  $[2 \cdot \sqrt{3} - 3, -2 \cdot \sqrt{3} - 3]$

Well, we have obtained what we wanted. By the way, the square bracket that we used here is not just a bracket. It is also an operator that makes a list of objects. For example, here $[2\sqrt{3} - 3, -2\sqrt{3} - 3]$ is a list of two real numbers. Of course, you can create a list of 0 numbers or of a few thousand. You can create lists of other objects, too. Every object in such a list can then be easily accessed. Later we will talk more about lists and other data structures that you can use in MuPAD.

5.4 Returning Unevaluated or Symbolic Expressions

It frequently happens that MuPAD is not able to complete a calculation. This is most often the case when a procedure was used with symbolic input data. For example, the command

- `1-2*sin(a)*cos(a)`

cannot be executed, as MuPAD does not know the current value of a. So, the result produced looks like this:

$1 - 2 \cdot \sin(a) \cdot \cos(a)$

There is nothing wrong with this output. This is not an error message. This is something that you still can use in further calculations. For example,

- `simplify(%)`

  $1 - \sin(2 \cdot a)$

Thus, we were able to transform the above output to simplified form, without knowing anything about a. I would like to remind you here, the ditto operator "%" denotes the result obtained in the previous line. But what if we decided to assign a value to the variable a? For example, `a:=PI/3`. Can we still produce the exact value of the first expression? Let us try it.

- `delete a`
- `U := 1-2*sin(a)*cos(a)`

  $1 - 2 \cdot \cos(a) \cdot \sin(a)$

- a := PI/3

$$\frac{\pi}{3}$$

- U

$$1 - \frac{\sqrt{3}}{2}$$

You are perhaps wondering why we are revising these old topics at this point. Well, just think about your procedures – it would be nice if you could write your procedures in such a way that they will behave the same way as sin(a) and cos(a) did in our examples. First of all, you must remember that MuPAD will not put you down even if you use the wrong data as input in your procedure. Let us see what might happen. We will declare a very simple procedure to calculate the geometric mean of two numbers.

- ```
  geomAv := proc(n,m)
    local av;
    begin
       av := sqrt(n*m);
       return(av);
    end:
  ```

Now, let us try to use this procedure to calculate the geometric mean of two numbers.

- geomAv(7.1, 5.6)

  6.305553108

- geomAv(-p,q)

$$\sqrt{-p \cdot q}$$

As you see, the last command produced a result that would be acceptable in many situations, especially if you wished to use it in further calculations. However, if you wished to obtain an evaluated result only when it is possible to get its final value, and in all other cases to get the function name, you would need to slightly modify your procedure. This can be done like this:

- ```
  geomAv := proc(n,m)
    local av;
    begin
       if testtype(n,Type::Real) and
          testtype(m,Type::Real)
       then
  ```

```
        av := sqrt(n*m);
        return(av);
     else
        procname(args())
     end
  end:
```
- geomAv(-p,q)

  geomAv(-p, q)

- geomAv(-7.1, 5.6)

  $6.305553108 \cdot i$

In our procedure, the command procname(args()) produces the unevaluated expression geomAv(-p,q). This can still be evaluated later provided that you assigned some real values to p and q. The function args() makes a copy of the sequence of all the input data. We will learn more about this interesting function in one of the later sections.

There are still many things that can be improved in the geomAv procedure. For example, it should be possible to produce results when the input data are complex numbers or anything else that can have a square root. I will leave all these improvements to you. Just remember, all these improvements can be done in a very simple way or a very complicated one. Try to make them as simple as possible. This way, you are saving your time, and the time of the other people who wish to follow your steps.

5.5 Shall We use Global or Local Variables?

Until now, we didn't really bother about what sort of variables we used and why we needed to declare internal variables inside procedures. However, there is much to think about here. First of all, note that we have only developed very simple examples so far. We have never developed a complex notebook with many operations and procedures where the dependency of each part of the notebook on other parts is more serious. From now on, we must start to care about the consistency of variables and procedure names in the whole notebook.

First, observe that all variables that are used inside of a procedure can be internal variables or variables used elsewhere in your notebook (global variables). The consequences of using global variables inside a procedure can be very serious. We will go through a series of examples to show what can go wrong, and how it can be fixed.

Case 1: A global variable can be modified accidentally by a procedure

Let us start by analyzing a very simple example.

- myVar := 123;

  123

- work := proc(x)
  ```
  begin
      myVar := myVar + x
  end:
  ```
- work(1)

  124

- myVar

  124

In line 1, we declared the global variable myVar and assigned it the value 123. Then we made the procedure work, where we used, accidentally or not, the same variable myVar. After the first use of the procedure, the global variable myVar received a new value. Therefore, the next time we use myVar, it may produce the wrong result. Of course, whether the result is wrong or not depends on what we wanted to achieve. However, we generally expect that global variables will indeed be global and that they will not be modified by accident. This means that no matter what operations were performed inside a procedure, the global variable should still have the value that was assigned to it before executing the procedure.

Case 2: Using local variables in procedures to protect global variables

First, let us declare myVar inside the procedure as a local variable

- myVar := 123

  123

- work2 := proc(x)
  ```
  local myVar;
  ```

```
begin
   myVar := myVar + x
end:
```
• work2(2)

Warning: Uninitialized variable 'myVar' used;
during evaluation of 'work2'
Error: Illegal operand [_plus];
during evaluation of 'work2'

In the above example we have two declarations of the variable myVar. The first is a global one, and the second is a local one inside of the procedure work2. While trying to execute the command work2(2), we obtained an error message that shows something important. Because we have declared myVar as an internal variable, the procedure no longer recognizes it as a global variable and doesn't want to use the current value of the global variable. We have to initialize this variable again inside the procedure. Look at the enclosed code.

• myVar := 123

  123

• ```
  work2 := proc(x)
  local myVar;
  begin
     myVar := 321;
     myVar := myVar+x
  end:
  ```
• work2(2)

  323

• myVar

  123

Notice what happened this time—the command work2(2) used the value of myVar that was assigned to it inside of the procedure work2, but the global value of myVar remains untouched. This is what we usually expect when we work on large projects.

It is also worth remembering that the values of local variables declared and used inside a procedure are not available elsewhere. So, if you produced anything inside a procedure and you did not use it as a return value, you will not be able to use it later on.
  ```

Good programmers, and I hope that you are going to be a good MuPAD programmer, do not use names of global variables inside their procedures at all. Many of them even use different names for global variables and local variables.

5.6 Introduction to Data Types and Domains

In most of our procedures so far, we have dealt with integers or real numbers. In one case, we did develop a procedure that produced a list of two real numbers, but this is all. We have never used any more complex objects in our procedures. However, in mathematics you may find a number of examples where the objects we use have a more complex structure than just numbers. We call such objects MuPAD data types.

This book is not a programmer's guide — this is a book for teachers and perhaps students who want to use MuPAD in teaching and learning mathematics. Thus, going into detail about MuPAD data types in the pages of this book may not be a good idea. However, it is worthwhile to learn some basics and see how you can use them in your work. We will start by exploring data types that are available in MuPAD.

In one of the previous chapters, I had introduced the basics of the domain concept; I have also frequently used type declarations for a given object. For example,

```
assume(x,Type::Interval(2,4));
myvar := Type::Interval(2,100, Type::Prime);
```

These are two things that can easily be confused in MuPAD — domains and types. Let me explain them briefly.

Domains should be considered as mathematical domains. We have the domains of real numbers, integer numbers, rational numbers, etc. You can find a number of predefined domains in MuPAD, and you can also build your own domains. Generally, a domain can be understood as a set of objects, which are not necessarily standard mathematical objects. Quite often, we require that a domain be a closed set for a particular operation. This means that the results produced by the operation on objects from this domain are once again objects in the same domain. For example, in the domain of integer numbers, multiplication of integers will still result in an integer number.

At the same time, we may also deal with objects that have a specific type. For instance, you could have an imaginary part of a complex number, relation, equation or arithmetical expression. You will easily see that there are many types that are related to an appropriate domain. However, many types can be applied to various objects from different domains. For example, you can talk about equations for real numbers, integers or complex numbers. You can also produce equations with polynomials and matrices.

If you wish to know more about which domains and types are defined in your version of MuPAD, try using the help commands `info(Dom)` and `info(Type)`. You should keep in mind that the development of MuPAD is a very dynamic process, and you will find slightly different collections of types and domains in different versions of MuPAD. Here is what I was able to get while writing this text:

- `info(Dom)`

  Library 'Dom': basic domain constructors

  – Interface:
  Dom::AlgebraicExtension,
  Dom::ArithmeticalExpression,
  Dom::BaseDomain,
  Dom::Complex,
  Dom::DenseMatrix,
  Dom::DihedralGroup,
  Dom::DistributedPolynomial,
  Dom::Expression,
  Dom::ExpressionField,
  Dom::Float,
  Dom::FloatIV,
  Dom::Fraction,
  Dom::FreeModule,
  Dom::FreeModuleList,
  Dom::FreeModulePoly,
  Dom::FreeModuleTable,
  Dom::GaloisField,
  Dom::Ideal,
  Dom::ImageSet,
  Dom::Integer,

Dom::IntegerMod,
Dom::Interval,
Dom::LinearOrdinaryDifferentialOperator,
Dom::Matrix,
Dom::MatrixGroup,
Dom::MonomOrdering,
Dom::Multiset,
Dom::MultivariatePolynomial,
Dom::Numerical,
Dom::PermutationGroup,
Dom::Polynomial,
Dom::Product,
Dom::Quaternion,
Dom::Rational,
Dom::Real,
Dom::SparseMatrixF2,
Dom::SquareMatrix,
Dom::SymmetricGroup,
Dom::UnivariatePolynomial

Meanwhile, the list of types available in my MuPAD looked like this:

- `info(Type)`

  Library 'Type': type expressions and properties

  – Interface:

```
Type::AlgebraicConstant,   Type::AnyType,
Type::Arithmetical,        Type::Boolean,
Type::Complex,             Type::Constant,
Type::ConstantIdents,      Type::Equation,
Type::Even,                Type::Function,
Type::Imaginary,           Type::IndepOf,
Type::Indeterminate,       Type::Integer,
Type::Intersection,        Type::Interval,
Type::ListOf,              Type::ListProduct,
Type::NegInt,              Type::NegRat,
Type::Negative,            Type::NonNegInt,
Type::NonNegRat,           Type::NonNegative,
Type::NonZero,             Type::Numeric,
Type::Odd,                 Type::PolyExpr,
Type::PolyOf,              Type::PosInt,
Type::PosRat,              Type::Positive,
Type::Predicate,           Type::Prime,
```

```
Type::Product,              Type::Property,
Type::RatExpr,              Type::Rational,
Type::Real,                 Type::Relation,
Type::Residue,              Type::SequenceOf,
Type::Series,               Type::Set,
Type::SetOf,                Type::Singleton,
Type::TableOf,              Type::TableOfEntry,
Type::TableOfIndex,         Type::Union,
Type::Unknown,              Type::Zero
```

I left the output of both commands almost untouched, without adding or removing anything. So, you may read the name of each domain and each type and try to figure out what they mean. How many types are related to a specific domain?

5.7 Using MuPAD Types and Domains

You might be a bit confused seeing these two apparently very similar things — domains and types. For example, what is the difference between `Dom::Real` and `Type::Real`? In this section we will try to capture the nature of domains and types and major differences between them.

Here is the meaning for each of them; there are some major differences between them.

Domains, like in mathematical investigations, can be considered as sets of objects. For example, a set of complex numbers, a set of integer numbers or a set of real numbers, a set of all the polynomials, etc. On the other hand, type is used to mark an object as having a given mathematical property. For example, `Type::Arithmetical`. Of course, there is no mathematical domain called Arithmetical. For some types, there are also domains of objects described by that given type, but this is not always the case.

You can define objects from different domains, in the same way that you can declare the other properties for objects. For example,

- `a := Dom::Quaternion([3,5,8,7]);`
- `b := Dom::Quaternion(2 + 7*I + 9*J -6*K);`

$3 + 5 \cdot i + 8 \cdot j + 7 \cdot k$

$2 + 7 \cdot i + 9 \cdot j - 6 \cdot k$

- a/b

$$\frac{71}{170} + \frac{10 \cdot i}{17} - \frac{9 \cdot j}{17} + \frac{43 \cdot k}{170}$$

In the example above, Dom::Quaternion was used to declare two numbers that are quaternions. However, with some domains you can go much further. Here is another example, where the domain product was used to create the new domain Cartesian product $C5 = \mathbb{R} \times \mathbb{R} \times \mathbb{R} \times \mathbb{R} \times \mathbb{R} = \mathbb{R}^5$, and then the elements of this new domain.

- C5 := Dom::Product(Dom::Real, 5)

  Dom::Product(Dom::Real, 5)

- a := C5([1, 2/3, 0, 3, 4])

  $[1, \frac{2}{3}, 0, 3, 4]$

- b := C5(2, 3, 4, 1, 2)

  $[2, 3, 4, 1, 2]$

- a + b, a*b, 2*a

  $[3, \frac{11}{3}, 4, 4, 6], [2, 2, 0, 3, 8], [2, \frac{4}{3}, 0, 6, 8]$

Here is yet another example. We first create a domain of rational polynomials of the two variables x and y, and then create two polynomials from this domain.

- PXY := Dom::MultivariatePolynomial(
      [x,y],Dom::Rational
  ):
  pol1 := PXY(2*x + 3*x*y + 5*y);
  pol2 := PXY(3*x*x + 7*x*y + 9*y*y)

  $3 \cdot x \cdot y + 2 \cdot x + 5 \cdot y$

  $3 \cdot x^2 + 7 \cdot x \cdot y + 9 \cdot y^2$

- pol := pol1*pol2

  $9 \cdot x^3 \cdot y + 6 \cdot x^3 + 21 \cdot x^2 \cdot y^2 + 29 \cdot x^2 \cdot y + 27 \cdot x \cdot y^3 + 53 \cdot x \cdot y^2 + 45 \cdot y^3$

You can always check an existing object's domain using the domtype command. Thus, in our case we have,

- domtype(pol)

  Dom::MultivariatePolynomial([x, y], Dom::Rational, LexOrder)

Types are used for a different purpose. For example, you would typically assume that a variable is of a given type, and then expect that MuPAD will produce results according to the type of the variable. Here is an example showing how this works:

- `assume(x, Type::Complex):`
  `solve(x^3 - 4*x - 15=0,x)`

$$\left\{ 3, -\frac{i}{2}\sqrt{11} - \frac{3}{2}, \frac{i}{2}\sqrt{11} - \frac{3}{2} \right\}$$

- `assume(x, Type::Real):`
  `solve(x^3 - 4*x - 15=0,x)`

$$\{3\}$$

In the above example, x is just a variable and cannot permanently belong to the type of real or complex numbers. It is just a symbol with properties assigned to it. Indeed, due to its dual nature—being a symbol means that it is just a character, but at the same time it is also an object carrying a specific property—we may need to check it from two different points of view. Let us again declare x as a complex number

- `assume(x, Type::Complex):`

There are two procedures that will be particularly useful at this point—`getprop` and `is`. The first one can be used to obtain the current properties of the variable x, while the second will produce TRUE or FALSE depending on how we formulate the command. Here you have an example of both procedures in use:

- `getprop(x)`

  $\mathbb{C}$

- `is(x, Type::Complex)`

  TRUE

- `is(x, Type::Real)`

  UNKNOWN

Note that with the second command, the result was TRUE, because we had declared x as a complex number. However, because the value of x is not known, and not every complex number is a real number, the third command produced the result unknown.

As we have noted before, x is a symbolic expression and we can investigate its formal nature (syntax). For this purpose, we can use the two commands `type` and `testtype`. The first command will produce information about the structure of x (syntactical form), while the second will allow us to check if x has a given syntax. Take a look at these commands, applied to the same x:

- `reset():`
- `type(x)`

  DOM_IDENT

- `testtype(x, Type::Complex)`

  FALSE

As you can see, the first command told us that x is an identifier; it is an object of DOM_IDENT, an internal data type of the MuPAD kernel. This simply denotes the fact that x is nothing more than a symbol. You can apply the `type` procedure to another objects, and in each case, you will get information about the syntax of that specific object. For example,

- `type(sin(x))`

  "sin"

- `type(sin(x)*exp(2))`

  "_mult"

The results obtained above mean that our objects are formed with the use of the function sine or a multiplication of two other objects.

Note that the way MuPAD prints results will differ depending on whether you have the option typesetting turned on or off. Therefore, in some situations you can obtain the output C_, PI, I, J, K if typesetting is off, and $\mathbb{C}$, π, i, j, k if typesetting is on.

Let us finish this section by analyzing one more case. We often declare input parameters in procedures, and we put in some domain and type-checking statements. For example,

```
proc(n:Dom::Real, m:Dom::Real)
proc(n:Type::Real, m:Type::Real)
```

This makes sense as long as we know how both declarations work, and what the difference is between them. Here are two examples showing what we can expect.

Let us take the procedure geomAv and apply domain and type-checking commands to it. In the first example, we will use Dom::Real to test the input values.

- ```
  geomAv := proc(n:Dom::Real, m:Dom::Real)
    local av;
    begin
       av := sqrt(n*m);
       return(float(av));
    end:
  ```
- geomAv(3.1,2.34)

  2.693325082

- geomAv(exp(12),2.34)

  617.1273871

And here is what I got using Type::Real to check the input parameters:

- ```
  geomAverage := proc(n:Type::Real,m:Type::Real)
    local av;
    begin
       av := sqrt(n * m);
       return(av);
    end:
  ```
- geomAverage(3.1,2.34)

  2.693325082

- geomAverage(exp(12), 2.34)

  Error: Wrong type of 1. argument (type 'Type::Real' expected, got argument 'exp(12)'); during evaluation of 'geomAverage'

The first example shows that after declaring an input parameter as the object of a given domain, MuPAD will check if the given object belongs to this domain. On the other hand, after declaring an input parameter as an object of a given type, MuPAD will check if the syntax of the object classifies it to the given type or not. For example, the numbers 3.1 and 2.34 will be considered as real numbers, but exp(x) will be treated as a formula, not a real number. You may check this with type(exp(x)) or testtype(exp(x),Type::Real).

Here is a short summary of the MuPAD procedures used in this section.

Domain Related Procedures

Dom::domain_name(*object*) - *declares an object from a domain domain_name*

domtype(*object*) - *returns the domain of a given object*

Type Related Procedures

assume(x, Type::*type_name*) - *declares the type of the variable* x

Type semantical checking (by value)

getprop(*x*) - *returns the mathematical properties of x (checks value)*

is(*x*, Type::*type_name*) - *checks if x has a given type (checks value)*

testtype(*x*, Dom::*domain_name*) - *checks if the value of x belongs to the given domain (checks value)*

Type syntactical checking (by form)

type(*x*) - *returns the syntactical type of x*

testtype(*x*, Type::*type_name*) - *checks if x looks like objects of type Type::type_name*

5.8 Using Procedures to Produce Data Structures

In the previous sections of this chapter, you have learned how to check the type of input parameters for a procedure and how to use different types of numbers. Now let us think a for a moment about obtaining more complex data through MuPAD procedures. Until now, we have been using procedures to produce simple data types, like numbers of different types. Every time we wanted to get multiple output from a procedure, we had to invent some trick. Look, for example, at the procedure we made to solve quadratic equations. There, we used a return(x1,x2) statement with two parameters. This way we produced a sequence of two numbers. Every time when we wish to get multiple outputs from a procedure, we will have to organize it into some kind of structure and then output the whole structure. The goal of this section is to show you how this can be done.

In high school mathematics, and especially at universities, we certainly do deal with data that are more complex than single numbers. For instance, in calculus we produce sequences of numbers, points or other objects. In geometry, we deal with lists of ordered pairs or triplets of real numbers to describe the coordinates of points on a plane or in 3D. In set theory, we deal with sets, and in linear algebra with matrices and vectors. These data types are different from real, integer or complex numbers. They have a specific structure. Therefore we will call them data structures. It is important to mention that the term "data structures" may have two meanings. One of them is a popular meaning derived from computer science, where by data structures we mean data structures used for programming – arrays, records, lists, trees, etc. The other meaning of data structures are the structures we use in mathematics that were mentioned above. In this book, we will concentrate on the mathematical data structures. However, you can easily find out that both meanings in practice refer to almost the same objects.

5.8.1 Sequences

There is no doubt that sequences are at the heart of calculus. We also use sequences to describe many other mathematical objects and properties. In order to define a short sequence in MuPAD, it is enough to declare it like this:

- `mySequence := 1, 2, 3, 4, 5, 6;`

However, this is not the best method do define a sequence with more than several terms. Imagine typing in such a declaration with a few hundred terms. Sometimes, you have to declare a sequence where the terms need to be calculated, and these tedious calculations should be left to MuPAD. An example of this would be if we wanted to declare the first one hundred terms of the sequence $a_n = (1 + \frac{1}{n})^n$. In MuPAD, there is a very useful operator $ to define long sequences like this. We will call it the sequence generator. The syntax for this operator can look like either of the following two lines:

$f(n) \$ n=n1 .. n2$

$f(n) \$ n$ in *set of values*

The expression on the left side is the formula of the sequence. On the right side, we give the range for the variable that will be used as an index for the sequence terms. Finally, the $ is the operator that will produce an appropriate number of terms in the sequence. The second declaration uses the construction n in *set_of_values*. This simply means using, for example, a set of numbers as indexes for the sequence elements. We will take a look at an example to give you a better idea about the $ operator.

Example 5.3 To produce a sequence

Suppose that you need to declare the sequence of the first 20 terms of $a_n = \left(1 + \frac{1}{n}\right)^n$, you would use:

- (1 + 1/n)^n $ n=1..20

This will produce 20 fractions. If you wish to get your sequence in floating point notation, you may use this command instead:

- float((1 + 1/n)^n) $ n=1..20

  2.0, 2.25, 2.37037037, 2.44140625, 2.48832, 2.521626372, 2.546499697, 2.565784514, 2.581174792, 2.59374246, 2.604199012, 2.61303529, 2.620600888, 2.627151556, 2.632878718, 2.637928497, 2.642414375, 2.646425821, 2.650034327, 2.653297705

Hence, a procedure extracting a subsequence from a_n can be written as follows:

- extract := proc(n1:Type::Integer, n2:Type::Integer)
  local sequence, n;
  begin
      sequence := float((1 + 1/n)^n) $ n=n1..n2;
      return(sequence)
  end:
- DIGITS := 20:
- mySequence := extract(5000,5001)

  2.7180100501018540468, 2.71801010443669972794

In this example, the procedure extract selects elements between n1 and n2. You can modify it to produce a subsequence starting from a given index and having a given number of elements. This way you would easily find out that a_n is convergent to something equal to *2.718281828*. This is the well-known Euler constant *e*. ∎

You can then access each element of a sequence by using its index. For example,

- MySequence := sqrt(3^n) $ n=2..23:
- MySequence[21]

    177147

You will find out more about sequences in the later chapters of this book. Right now, we will move on to another useful data structure in MuPAD—lists.

5.8.2 Lists

Lists are data structures very similar to sequences. We declare them the same way, but with square brackets, like here:

- myList := [1,2,3,[2,3,a],[a,b]]

    $[1,2,3,[2,3,a],[a,b]]$

You can then access each element of the list by using its index.

- myList[4]

    $[2,3,a]$

At first glance, it seems that there is not much difference between sequences and lists. However, the types of these two objects are different. To see the difference, check type(myList) and then type(mySequence). Another important difference is that lists can be nested but sequences not.

Furthermore, there is a special library listlib for lists, with a number of important operations that can be performed on lists—merge, insert, etc. All of these operations are very useful for working with lists, testing algorithms on lists, searching, and sorting lists. This library is very useful when teaching discrete mathematics.

Example 5.4 To produce a long list of numbers

You can produce long lists of objects in a similar way to sequences. You only need to add the square brackets. However, it is important to consider the placing of the brackets.

- elist := proc(n1: Type::Integer, n2:Type::Integer)
  local mylist, n;
  begin
      mylist := [float((1+1/n)^n) $ n=n1..n2];
      return(mylist)

```
   end:
• list1to5:=elist(1,5);
```
   [2.0, 2.25, 2.37037037, 2.44140625, 2.48832]

The above example produced the list with the same elements as did the sequence that we had discussed in the previous section. However, if you place the bracket like this,

```
   mylist := [ float((1+1/n)^n) ] $ n=n1..n2 ;
```
then you will obtain a sequence of one-element lists.   ■

5.8.3 Sets

From a mathematical point of view, sets can be considered as unordered sequences. This means that there is no first, no second element, and so on. We produce sets in a similar way to normal sequences or lists with one important change — we use the curly brackets.

```
• myset := {1, 3, 3, 4, 4, u, PI, a, b, 3}
```
   $\{1,3,4,\pi,a,b,u\}$

Internally, sets in MuPAD have a specific order of elements, and you can easily access each element in a set by using its internal index that (be careful) might be different than the order in which we placed the elements inside the curly brackets. For instance, take a look at the above example and compare the order of elements in the input line with the obtained output. There is more — this order is completely out of our control and it may change when you perform an operation on the set. Therefore, think of sets as unordered collections of objects.

You can use well-known set operations on MuPAD sets — union, intersection, difference (minus). There is also an empty set that can be represented by empty curly brackets {}. We will talk more about sets in one of the later chapters, when we investigate the applications of MuPAD in set theory.

5.8.4 Strings

Strings are sequences of characters. We declare them with the quotation operator.

```
• mstr := "Hi I am writing this book for you"
```

You can access each character of a string by its index: `mstr[1]`, `mstr[12]`, etc. Note, however, that in MuPAD 3.0, the very first element of a string has the index 1, not 0 as was the case in the previous versions. Thus,

- `mstr[1]  // in MuPAD 3.0`
  "H"

- `mstr[0]  // in MuPAD 2.0 and 2.5`
  "H"

As you may have noticed, a single character is also considered a one-element string.

In MuPAD you can join two or more strings using the dot operator "`.`", sometimes called the concatenation operator.

There is one trick that is worth exploring while talking about strings. Imagine that you need to produce a large collection of objects, assign each of them to a separate variable, and then use all of them in further calculations or for plotting. In such a situation, you could use the dot operator to generate indexed names of these variables: $x1, x2, \ldots, xn$.

Here is just a quick example.

- ```
  functions := (x.i := x^(1/i)) $ i=1..20:
  plotfunc2d(x1, x3, x5, x=0..2)
  ```

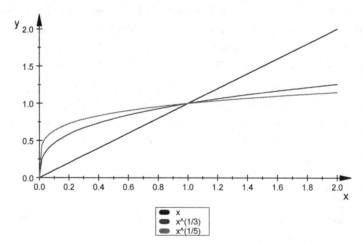
  ```

In this example, the construction $(x.i:=x^{(1/i)}) \$ i=1..20$ has produced a sequence of variables x1, x2, x3, ... From this moment on, you could use each of these variables in further calculations. Of course, if you needed it, you could also plot all these functions by using the name of the sequence that you have created.

- `plotfunc2d(functions, x=0..1)`

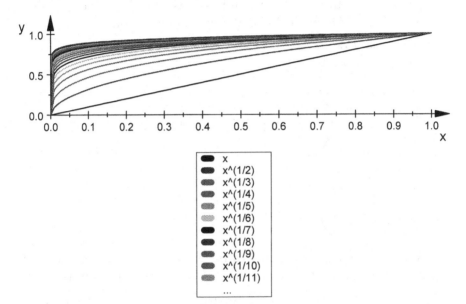

Comment: Both pictures presented here were produced with MuPAD using different colors for different lines. For printing purposes, I had to change all the colors to black and white and adjust some other parameters.

The dot operator is quite fascinating and it gives you a very unconventional way of declaring variables. Try to find as many applications as possible for it. However, it is now time to move on to something else that is also very important.

5.8.5 Vectors and Matrices

There is no analytic geometry without vectors, and no linear algebra without matrices. These two types of objects also play a very important role in many other areas of mathematics. The simplest way to declare a matrix is to use the command `matrix`. For example,

- A := matrix([[a,123],[345,b]])
- B := matrix([[1,7,31],[3,2,6],[4,6,7],[0,9,8]])

The first command produces a matrix with two rows and two columns and assigns it to the variable A. The second produces a matrix with four rows and three columns, and assigns it to the variable B. Let us see how MuPAD will deal with these two declarations.

- A := matrix([[a,123],[345,b]])

$$\begin{pmatrix} a & 123 \\ 345 & b \end{pmatrix}$$

- B := matrix([[1,7,31],[3,2,6],[4,6,7],[0,9,8]])

$$\begin{pmatrix} 1 & 7 & 31 \\ 3 & 2 & 6 \\ 4 & 6 & 7 \\ 0 & 9 & 8 \end{pmatrix}$$

If you have never used matrices before, you can easily see from the example above what I meant talking about rows and columns.

Accessing elements inside a matrix or changing them is very straightforward. The position of each element is determined by its two indexes, where the first represents the row number, and the second the column number. For example,

- A[1,2]

  123

- B[3,2]

  6

In a similar way, you can insert values into appropriate places in the matrix. Just try this.

- B[3,2] := c:
  B[3,3] := 1000:
  B[4,3] := 1/213:

• B

$$\begin{pmatrix} 1 & 7 & 31 \\ 3 & 2 & 6 \\ 4 & c & 1000 \\ 0 & 9 & \frac{1}{213} \end{pmatrix}$$

Declaring large matrices could be a very tedious task. It is best to use a procedure that will do it for you.

Example 5.5 To produce a matrix of any dimension

Suppose that we wish to produce a square matrix of given dimensions $n \times n$ where all the elements on the diagonal are equal to 1, elements above the diagonal are equal to 0 and elements under the diagonal are equal to $\frac{1}{kl}$, where k and l are row and column numbers respectively. Developing such a procedure is very straightforward. First, analyze the enclosed example, and later I will explain it.

```
• diagMatrix := proc(n)
    local row, column;
  begin
     A := matrix(n,n);
     for row from 1 to n do
        for column from 1 to n do
           if row=column then
              A[row,column] := 1
           end;
           if row>column then
              A[row,column] := 1/(row*column)
           end;
           if row<column then
              A[row,column] := 0
           end
        end;
     end;
     return(A)
  end:
```

Note that the only new thing that I used here is the command,

```
A := matrix(n,n)
```

This command generates a new matrix with n rows and n columns. Initially, this matrix is filled by zeros. The next few lines of code will fill this matrix with the right elements. Finally, in the last line, we force the procedure diagMatrix to return the produced matrix as one piece. You can try to use the command diagMatrix with a large number as the parameter n. However, note that in this procedure, there is no protection against incorrect input values. Perhaps you could try to improve the procedure. Now, let us see what this procedure can do.

- diagMatrix(10)

$$
\begin{pmatrix}
1 & 0 & 0 & 0 & 0 & 0 & 0 & 0 & 0 & 0 \\
\frac{1}{2} & 1 & 0 & 0 & 0 & 0 & 0 & 0 & 0 & 0 \\
\frac{1}{3} & \frac{1}{6} & 1 & 0 & 0 & 0 & 0 & 0 & 0 & 0 \\
\frac{1}{4} & \frac{1}{8} & \frac{1}{12} & 1 & 0 & 0 & 0 & 0 & 0 & 0 \\
\frac{1}{5} & \frac{1}{10} & \frac{1}{15} & \frac{1}{20} & 1 & 0 & 0 & 0 & 0 & 0 \\
\frac{1}{6} & \frac{1}{12} & \frac{1}{18} & \frac{1}{24} & \frac{1}{30} & 1 & 0 & 0 & 0 & 0 \\
\frac{1}{7} & \frac{1}{14} & \frac{1}{21} & \frac{1}{28} & \frac{1}{35} & \frac{1}{42} & 1 & 0 & 0 & 0 \\
\frac{1}{8} & \frac{1}{16} & \frac{1}{24} & \frac{1}{32} & \frac{1}{40} & \frac{1}{48} & \frac{1}{56} & 1 & 0 & 0 \\
\frac{1}{9} & \frac{1}{18} & \frac{1}{27} & \frac{1}{36} & \frac{1}{45} & \frac{1}{54} & \frac{1}{63} & \frac{1}{72} & 1 & 0 \\
\frac{1}{10} & \frac{1}{20} & \frac{1}{30} & \frac{1}{40} & \frac{1}{50} & \frac{1}{60} & \frac{1}{70} & \frac{1}{80} & \frac{1}{90} & 1
\end{pmatrix}
$$

Of course, you can use this procedure with much larger parameter n. ■

You can find a number of procedures related to matrices in the library linalg. You can access these procedures by importing them or by using the slot operator ":::". For example, this command will produce a matrix 5 × 5 with random elements of the integer type.

- linalg::randomMatrix(5, 5, Dom::Integer)

$$
\begin{pmatrix}
250 & 320 & 939 & 509 & 717 \\
-263 & 164 & 843 & 129 & 555 \\
-524 & -103 & 126 & -773 & 432 \\
287 & -540 & -553 & 329 & -100 \\
315 & 886 & -396 & 498 & 119
\end{pmatrix}
$$

You will find more about matrices and their applications in the later chapters of this book when we talk about linear algebra.

5.9 Procedures with a Variable Number of Input Parameters

Imagine that you have developed a procedure that performs an operation for two objects; say, two numbers. This can be something quite simple, like this one that produces the sum of the squares of the input numbers.

- ```
  squares := proc(n,m)
  begin
      return(n^2 + m^2)
  end:
  ```

This nice procedure works quite well, if we use exactly two numbers. For example,

- ```
  squares(1,2)
  ```
  5

But what about three numbers? Suppose that we wish to produce the sum of the squares of three numbers.

- ```
  squares(1,2,3)
  ```
  5

As you can see, the result is completely wrong. The third number was completely ignored. Well, perhaps we will be able to get the right output for a single number?

- ```
  squares(1)
  ```
  Warning: Uninitialized variable 'm' used;
  during evaluation of 'squares'
  Error: Illegal operand [_power];
  during evaluation of 'squares'

As you can see, with just one input parameter, the result is even worse. We got an error message.

It is completely natural to ask such procedures to produce output for any number of input values. This works in standard arithmetic, where the summation operation produces output for any number of input values:

- 1 + 2 + 3 + 4 + 5

  15

and

- 3 + 16 + 34 + 45 + 9870 + 3456

  13424

It would be good if we could implement this type of behavior in our procedures. This is possible, but before we do it, we need to find out how our procedure checks the number of input values. This can be done using the MuPAD function args().

The function args(), when used inside a procedure, will produce the following values:

$$\text{args}(x)=\begin{cases} \text{number of arguments} & \text{if } x = 0 \\ i\text{-th argument} & \text{if } x = i < \text{args}(0) \\ \text{arguments } i\text{-th, ..., } j\text{-th} & \text{if } x = i..j \\ \text{all arguments} & \text{if there is no input parameter} \end{cases}$$

For example, if we execute the command squares(2,4,3,6,7), the function args used inside squares procedure will produce the following:

args()=2,4,3,6,7

args(0)=5

args(3)=3

args(5)=7

args(2..3)=4,3

Knowing this, we can try to apply the args() function to the squares procedure. First, we will remove both input parameters from the procedure declaration. This is because the procedure should work for any number of input parameters. Then, we will add a new local variable n that will keep track of the number of input parameters. Inside the procedure body, we will add the assignment n:=args(0).

Finally, instead of adding the squares of two numbers, we will extend this operation to all the input parameters. A loop will do this quite well. We need to add two more variables to carry the loop operations. This will be the loop index i, and the variable result that will represent the final output.

Here is the final procedure.

- ```
  squares := proc()
    local n, i, result;
    begin
        n := args(0);
        result := 0;
        for i from 1 to n do
            result := result + args(i)^2
        end;
        return(result)
    end:
  ```

Now you can check how this procedure works for different inputs:

- squares(1,2,3,234,45676)

  2086351746

- squares()

  0

- squares(123)

  15129

You can check a few more inputs. It will work as long as the input values make sense. For example,

- squares(a,b,3,c,4,d)

  $a^2 + b^2 + c^2 + d^2 + 25$

Now it is your turn. Try to find a few more examples of functions that will work for any number of input values. For example, a function genFact(n,m,p,..q) that will produce $n!m!p!..q!$, i.e. a product of the factorials of all the input values. Do not forget, you will need to use type checking to verify if all the input parameters are positive integer numbers. Otherwise, things could go completely wrong. If you do not have time to do this procedure now, I will remind you about it at the end of the chapter.
  ```

5.10 Recursive Procedures

Mathematicians like mathematical induction. This idea was very useful for centuries and gave us a way to prove things that were difficult or impossible to prove in any other way. Recursion is a computing implication of mathematical induction. In many programming situations, recursive procedures are very useful to simplify the code of a program. However, if you decide to use recursion in programming, you must be aware that recursion can be much slower than iterative processes and it may take more computer memory. But why not try it?

Let us start with the definition of a recursive procedure.

A procedure, say, myproc is recursive if in order to calculate myproc(n) it calls itself to calculate myproc(n-1), and so on until it reaches the bottom limit of the calculations, say, myproc(0) or myproc(1).

Here is an example. The well-known function factorial can be written in a recursive form:

$$n! = \begin{cases} n \cdot (n-1)! & \text{if } n > 0 \\ 1 & \text{if } n = 0 \end{cases}$$

This way,

$$
\begin{aligned}
5! &= 5 \cdot 4! \\
&= 5 \cdot (4 \cdot 3!) \\
&= 5 \cdot (4 \cdot (3 \cdot 2!)) \\
&= 5 \cdot (4 \cdot (3 \cdot (2 \cdot (1!)))) \\
&= 5 \cdot (4 \cdot (3 \cdot (2 \cdot 1))) = 120
\end{aligned}
$$

You certainly will agree that the idea looks very interesting and could be very useful. Let us see how recursion can be implemented.

Example 5.6 A recursive procedure

The sequence $\{x_n\}$ is given by the formula

$$x_n = \begin{cases} 1 & \text{if } n = 1 \\ 3 & \text{if } n = 2 \\ (x_{n-1})^3 (x_{n-2})^2 & \text{if } n > 2 \end{cases}$$

We can easily produce the first few terms of this sequence:

$x_1 = 1, \; x_2 = 3$

$x_3 = 3^3 1^2 = 27$

$x_4 = 27^3 3^2 = 177\,147$

$x_5 = 177147^3 27^2 = 4052\,555\,153\,018\,976\,267$

Now the question is, how do we evaluate x_{20}? You will agree that calculating it by hand would be a rather tedious task. However, the last line of our formula suggests that this is a recursive definition; therefore, we must try to develop a recursive procedure that will calculate not only x_{20} but also x_{21}, etc.

Before we start developing our procedure, we need to decide which programming structure would best suit this problem. Observe that we have a few conditions to verify. We need to check if the input number, the index of the term to be calculated, is an integer number, and then we need to check whether this number is positive, equal to 1, equal to 2, or anything else. In each case, we need to produce the correct result. Take a look at how this can be done using the elif construction:

```
• recExamp := proc(n: Type::Integer)
  begin
      if n<=0 then
          error("Input n>0")
      elif n=1 then
          return(1)
      elif n=2 then
          return(3)
      else
          return(recExamp(n-1)^3 * recExamp(n-2)^2)
      end
  end:
```

Now you may check if our hand-made results were correct, or rather — if our procedure gives the correct output.

- `recExamp(3)`

  27

- `recExamp(4)`

  177147

- `recExamp(5)`

  4052555153018976267

Now that we are sure that we got what we expected, we may try something bigger. But be careful, I still haven't told you another important detail. Therefore, for now do not use numbers larger than 10. Anyway, for $n = 7$ we get:

- `recExamp(7)`

  1496308304356787524377994530022358811144\
  1291564856033869973087894344880666632032\
  4917045206101357176493140254308746271765\
  3611228639157078727001769235840187737685\
  6420649022545170512350837153768140461315\
  5744459736220270729547670205355547134881075

For a larger n, you will get a much bigger number, and the calculation time will be very long. For example, for $n = 20$ you could wait as much as a few hours. This will depend on what kind of computer you are using.

Now, let us explain what is going on — why is the calculation time of recExamp so long? Just look:

recExamp(7) will need recExamp(6) and recExamp(5),

recExamp(6) will need recExamp(5) and recExamp(4), etc.

However, the result for recExamp(5) calculated in line 1 will not be reused in line 2. The procedure will simply forget it, so it will be calculated again. The same goes for recExamp(4), it too, will be calculated many times, as will all the others. You can try to write down how many times each result will be calculated because of our procedure's short memory. This is really a waste of time. It would be nice if the procedure could remember all the previously calculated terms and reuse them later without wasting our time. This can easily be done by adding the option `remember` to our procedure. Just like this:

```
• recExamp := proc(n: Type::Integer)
  option remember;
  begin
      if n<=0 then
          return("Input n>0")
      elif n=1 then
          return(1)
      elif n=2 then
          return(3)
      else
          return(recExamp(n-1)^3 * recExamp(n-2)^2)
      end
  end:
```

Now try `recExamp(7)`, or even `recExamp(10)`. I will not copy these results here, because certainly you would not want to pay for a book where a number of pages is filled with lines and lines of digits. However, for smaller values of n, you can now get results in a matter of seconds or minutes, and not hours like before. Note that `recExamp(10)` produces a number that is large enough to fill two pages. So, calculations for $n = 12$ or more may still be difficult to carry out, due to computer memory restrictions. ∎

5.11 Using Libraries

Many functions and procedures implemented in MuPAD are a part of its kernel. These procedures do not require any special attention. They are ready for use every time we start up MuPAD. A number of other procedures, however, are placed in MuPAD's libraries instead. The standard library of MuPAD procedures, `stdlib`, is also available for use immediately after starting up MuPAD. However, the procedures from the other libraries are slightly less accessible. In order to use them in your calculations, you first need to export them to the current notebook, or access them directly from their library. For example, in the library `combinat` (combinatorics), there is a procedure called `permute(n)`, which produces all the possible permutations of the numbers 1..n. You cannot use this procedure directly, like here:

- permute(3)

  permute(3)

As you can see, MuPAD does not recognize the procedure. In order to use it, you will need to export it from the library or call it directly from the library. Let me show you how this can be done.

In order to call the function permute(n) directly from the library combinat, you need to use the statement combinat::permute(n), which orders MuPAD to go to the library combinat and execute the procedure permute with the given parameter *n*. For example:

- combinat::permute(3)

  [[2, 3, 1], [3, 2, 1], [1, 3, 2], [3, 1, 2], [1, 2, 3], [2, 1, 3]]

This time, MuPAD was able to produce all the possible permutations of the numbers 1, 2, 3 for us.

Calling functions directly from the library can be a tedious task. Every time you do this, you must type in the library name, then the slot operator ":: ", and finally the function name. However, there is another way — you can export some or all the functions from a library into your notebook. This way you will be able to reduce the amount of typing and your statements will be shorter. To export several functions, you need to use this command:

export(*library_name, function1, function2, ..., function_n*)

For example, in order to export the procedure permute from the library combinat, you would need to use this command:

- export(combinat,permute)

This would then enable you to execute the statement:

- permute(4)

  [[3, 4, 2, 1], [4, 3, 2, 1], [2, 4, 3, 1], [4, 2, 3, 1], [2, 3, 4, 1],
  [3, 2, 4, 1], [3, 4, 1, 2], [4, 3, 1, 2], [1, 4, 3, 2], [4, 1, 3, 2],
  [1, 3, 4, 2], [3, 1, 4, 2], [2, 4, 1, 3], [4, 2, 1, 3], [1, 4, 2, 3],
  [4, 1, 2, 3], [1, 2, 4, 3], [2, 1, 4, 3], [2, 3, 1, 4], [3, 2, 1, 4],
  [1, 3, 2, 4], [3, 1, 2, 4], [1, 2, 3, 4], [2, 1, 3, 4]]

You can export as many procedures as you wish using the export statement. For example,

- export(combinat, cartesian, permute, powerset)

Finally, you can also export an entire library at once, with the command,

$$\text{export}(\textit{library\_name})$$

For example, this command will export the complete plot library into your notebook:

- export(plot)

You might wonder how many libraries there are in MuPAD, and what is stored inside each one. You will probably not find a book or an article where all the libraries and their contents are listed. However, there is a lot of information about MuPAD's libraries in its help files. In order to get a list of all the MuPAD libraries you would need to execute this statement:

- info()

  – Libraries:

Ax,	Cat,	Dom,	Graph,	RGB,
Series,	Type,	adt,	combinat,	detools,
fp,	generate,	groebner,	import,	intlib,
linalg,	linopt,	listlib,	matchlib,	module,
numeric,	numlib,	ode,	orthpoly,	output,
plot,	polylib,	prog,	property,	solvelib,
specfunc,	stats,	stdlib,	stringlib,	student,
transform				

Now, once you know the names of all the MuPAD libraries, you can check what is inside each one of them. For example,

- info(stdlib)

  Library 'stdlib': the basic functionality of MuPAD

  No Interface.

  – Exported:

D,	Im,	O,	Re,
RootOf,	_implies,	_in,	_notsubset,
_subset,	_xor,	anames,	assert,
assign,	assume,	asympt,	bytes,
card,	coeff,	coerce,	collect,
combine,	complexInfinity,	conjugate,	contains,
content,	contfrac,	degreevec,	denom,

```
diff,          discont,       doprint,       expand,
export,        expose,        expr,          expr2text,
factor,        fclose,        fopen,         fprint,
freeze,        gcd,           gcdex,         genident,
getprop,       ground,        hastype,       icontent,
igcd,          igcdex,        ilcm,          infinity,
info,          int,           interpolate,   irreducible,
is,            isprime,       ithprime,      lcm,
lcoeff,        length,        lhs,           limit,
linsolve,      lmonomial,     lterm,         map,
maprat,        max,           min,           normal,
nthcoeff,      nthmonomial,   nthterm,       numer,
package,       pade,          partfrac,      piecewise,
plotfunc2d,    plotfunc3d,    powermod,      print,
product,       protocol,      radsimp,       random,
rationalize,   read,          readbytes,     rectform,
revert,        rewrite,       rhs,           rtime,
select,        series,        setuserinfo,   simplify,
solve,         split,         strmatch,      substring,
sum,           system,        taylor,        tcoeff,
testeq,        text2expr,     time,          unassume,
undefined,     unexport,      unfreeze,      universe,
version,       zip
```

Now you can take a look at this list, and see if you can recognize the meaning of each procedure. If you wish to obtain more information about a given procedure, you will need to use the commands that I described in one of the previous chapters. For example,

- `info(taylor)`

  taylor – a library procedure [try ?taylor for help]

- `info(plot::Line2d)`

  plot::Line2d – graphical primitive for 2D lines

  – Interface:

  ```
  plot::Line2d::Color,
  plot::Line2d::Frames,
  plot::Line2d::From,
  plot::Line2d::FromX,
  plot::Line2d::FromY,
  plot::Line2d::Legend,
  plot::Line2d::LegendEntry,
  plot::Line2d::LegendText,
  plot::Line2d::LineColor,
  ```

```
plot::Line2d::LineStyle,
plot::Line2d::LineWidth,
plot::Line2d::ParameterBegin,
plot::Line2d::ParameterEnd,
plot::Line2d::ParameterName,
plot::Line2d::ParameterRange,
plot::Line2d::TimeBegin,
plot::Line2d::TimeEnd,
plot::Line2d::TimeRange,
plot::Line2d::Title,
plot::Line2d::TitleAlignment,
plot::Line2d::TitleFont,
plot::Line2d::TitlePosition,
plot::Line2d::TitlePositionX,
plot::Line2d::TitlePositionY,
plot::Line2d::To,
plot::Line2d::ToX,
plot::Line2d::ToY,
plot::Line2d::Visible,
plot::Line2d::VisibleAfter,
plot::Line2d::VisibleAfterEnd,
plot::Line2d::VisibleBefore,
plot::Line2d::VisibleBeforeBegin,
plot::Line2d::VisibleFromTo
```

- `info(linalg::randomMatrix)`

  linalg::randomMatrix – a procedure of domain type 'DOM_PROC'

Note that for procedures that had already been exported into the notebook, as well as procedures from `stdlib`, you can use the command without the library name. Otherwise, you will have to use this syntax:

> info(*library_name*::*procedure_name*)

In order to get more detailed help about a particular procedure, you would have to use one of these commands:

> ?*procedure_name*
>
> ?*library_name*::*procedure_name*

5.12 User Defined Libraries of Procedures

If you are like me, you will certainly soon start developing procedures for various mathematical topics. After a while, you will need to think about how to organize your procedures, how to make your own libraries, and finally, how to make these libraries available to your colleagues and other people. There are at least three ways of doing this in MuPAD. Here, I will only describe the simplest method. The two other methods are much more complicated and describing them is beyond the scope of this book. However, do not worry—help is on the way, and very soon you may expect changes in MuPAD that will allow you to produce professional libraries of user-defined procedures. Meanwhile, the method described here is quite simple, but very effective.

Example 5.7 User defined library of MuPAD procedures

Let us start by developing a notebook with a few procedures. In fact, you can put not only procedures but a number of other things, like declarations of some constants or variables, functions, etc., into your library. In order to keep things short, I will only declare two procedures, two new constants and one function here. However, you can develop as many MuPAD declarations as you wish. Here is my notebook:

```
• MyPI := 3.14;
• MyE := 2.72
• MyExp := x -> MyE^x
• average := proc()
  local n, i, result;
  begin
     n := args(0);
     result := 0;
     for i from 1 to n do
        result := result + args(i)
     end;
     result := result/n;
  return(result)
  end:

• squares := proc()
  local n, i, result;
```

```
begin
   n:=args(0);
   result := 0;
   for i from 1 to n do
      result := result + args(i)^2
   end;
   return(result)
end:
```

After testing the constants and procedures declared here, remove all the unnecessary commands and declarations from the notebook. Leave only the things that you want to pack into your library. Now, use the command,

- reset()

This will remove MuPAD data from the computer memory. All the unnecessary procedures, variables and constants are now definitely forgotten. You only have your notebook, and whatever was left in it. You should now save your notebook for future use and modifications.

Now, you must prepare the location where you will collect all your libraries of procedures. In order to do this, use Windows Explorer or the My Computer icon on the desktop of your Windows. Go to **Program Files**, then to the folder **SciFace**, and finally, open the MuPAD folder (MuPAD Pro 2.x or MuPAD Pro 3.x, etc.). Depending on which version you have installed, the last "x" would be replaced by the appropriate digit. Once inside the MuPAD Pro folder, you need to create a new folder. You can call it **userlib** or anything else that you can remember easily. This is where you will save all your libraries.

Now you can return to MuPAD. Open the notebook which will be your new library, remove the reset() statement and enter these two commands at the end of the notebook:

- WRITEPATH := "userlib";
  write(Text,"myfunctions.mu")

You can replace the name of the file, in this case myfunctions.mu, with any other name that will be relevant for your file. The first statement you added tells MuPAD to use the new folder userlib as the default folder for all user-defined libraries.

The second statement, write(Text,"myfunctions.mu"), will dump all currently defined objects and exported procedures into a text file. However, you may wish to only save a few selected procedures. In that case, you would use the command,

- write(Text, "myfunctions.mu", proc1, proc2, ..., procN)

where proc1, proc2, ..., procN would be replaced by the names of the procedures and other things that you wish to save.

Now, run all the commands in your notebook. You can do this by choosing **Notebook>Evaluate>Any Input** from the menu. Having executed the final command, MuPAD will generate for you a text file containing the declarations of all the procedures that you have left in the notebook.

The next time you start up MuPAD, you will need to start with these two commands:

- READPATH := "userlib";
  read("myfunctions.mu")

This makes **userlib** the default folder from where MuPAD will read the user-defined libraries, and then loads **myfunctions.mu**. You can now start using your procedures just as you would any procedure exported from MuPAD library. For example,

- squares(12,3,4,78,9)

  6334

- average(12,3,4,78,9)

  106/5

- MyPI+MyE

  5.86

- MyExp(3.985)

  53.92089381

Note, instead of saving your files into a folder that is located in the MuPAD directory you can write your libraries to another folder, for example to c:\userlib. You will need to create such folder first and then use statements:

READPATH := "c:\\userlib":

WRITEPATH := "c:\\userlib":

If you use MuPAD for teaching, then you certainly will want to avoid, as much as possible, unnecessary commands that might be confusing for your students. Another important issue is the maintenance of a computer lab for your class. In that case, you should think about having a directory on the network server from where students will be able to read the libraries you made for them or even with them. This is quite easy. Here is how I did it on my university network.

STEP 1: Choose the network drive for your libraries. In my case the server hard disk was mapped to the letter h:, and I made a new folder userlib on the server's hard disk. Thus the READPATH and WRITEPATH commands looked as follows:

```
READPATH := "h:\\userlib":

WRITEPATH := "h:\\userlib":
```

STEP 2: Open a new file in MuPAD—this can be done with the menu **File > New Source**. Now, type in these two commands. Note that if you are using MuPAD for Windows, you need to use "\\" instead of "\". Save the file as *.mu. In my case, I used the name userpref.mu and I saved it in MuPAD's root directory.

STEP 3: In the MuPAD menu **View > Options**, click on the **Kernel** tab. Here, in the box **User-defined start-up file**, type in the path to your new preferences file or click the three dots to the right and navigate to your *.mu file. In my case, the **Kernel** dialog box looked like the fig. 5.1

STEP 4: Copy the user preferences file to each computer in the lab and make the appropriate changes in the **Kernel** tab of each MuPAD installed in the lab. After this is done, you will only need to update the files on the network drive. Both you and your students will be able to execute the commands write and read without additional READPATH or WRITEPATH commands.

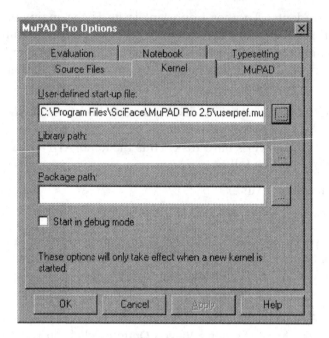

Fig. 5.1 Linking the user start-up file

I hope this method will be enough for you to produce your own libraries that you and your colleagues can use while working in mathematics or physics. Note that all the notebooks saved as *.mu are text files. You can always open them with MuPAD and edit again. ■

5.13 Final Comments on MuPAD Coding Guidelines

Throughout the pages of the first five chapters of this book, I have been giving you many hints about how to format MuPAD programs. All the rules that we went through are very useful for creating readable and well-formatted programs. Before going on to the next chapter, I will introduce a few more rules and summarize them all. This will help you a lot in writing more complicated programs later on.

5.13.1 Magic Numbers

In many statements from this and the previous chapter, I have used certain numbers to specify the range of iterations in the for loop, and in some other programming constructions. Experienced programmers, like my friend Cay Horstmann, call them magic numbers. A magic number is a constant that is embedded in the code without a constant declaration. However, experienced programmers never use magic numbers in any serious program. Instead, they declare global or local constants that will replace the magic numbers. For example, instead of using,

* if x < 456 then print("Hi") else print("hello")

or,

* for i from 23 to 456 do ...

They would prefer to write,

* START := 23;
  END := 456
  for i from START to END do ...

Getting into this habit will help you a lot when you wish to make further modifications in your program. Tracing constants inside the code is extremely difficult, and you can easily miss them. On the other hand, modifying a constant declared at the beginning of a program or a procedure is quite simple. So, let us agree,

Rule 8   No more magic numbers in MuPAD programs.

5.13.2 Comments

While developing MuPAD notebooks, you do not really need to add comments. You can mix code and text without any problem. However, once you start writing libraries of procedures or even just larger procedures, you will need to comment your work. Comments are patches of text left inside of MuPAD code. This text may contain important information about the role of the particular part of the code, about the whole procedure and its behavior, or finally, about you and date when you created the procedure. For example, it might be difficult for you to remember after a few months what this procedure does:

* squares := proc()
  local n, i, result;
  begin
      n:=args(0);
      result := 0;

```
      for i from 1 to n do
         result := result + args(i)^2
      end;
      return(result)
   end:
```

In order to save your time, you can comment it like this:

```
• squares := proc()
   /* Author - Mirek Majewski,
      Zayed University, Abu Dhabi, Emirates
      Date: 22/11/2001
      This procedure calculates the sum of squares of
      a given sequence of numbers.
      It will work for any quantity of input numbers. */

   local n, i, result;
   begin
      n := args(0);
      result := 0;
      for i from 1 to n do
         result := result + args(i)^2
      end;
      return(result)
   end:
```

Here, everything contained between /* ... and ...*/ will be treated by MuPAD as a comment. You can use also // in order to comment just a single line.

If you also think using comments can be useful, then we can agree on another rule:

Rule 9   Comment your programs whenever you think it may be important.

5.13.3 Braces

Braces, sometimes also called brackets, are important components of any program or procedure. However, there are many situations where you will feel lost while using braces. Therefore, we usually follow the rule:

Rule 10   Opening and closing braces must line up, either horizontally or vertically.

In MuPAD programs we generally use round braces, "(" and ")". However, from time to time you will have to use curly braces (for sets) and square braces (for lists). Furthermore, keywords in programming constructions such as if, while, repeat, etc. should also be considered as braces delimiting the lines of the code that you place between them. According to rule 10, we should line them up, either vertically or horizontally. For example, the construction

```
for i from 1 to n do r := r + args(i)^2 end;
```

makes sense, as long as we are able to fit it into a single line. In many places in this book, however, I have had to split such commands into multiple lines, due to limited width of the page. Thus, I had to line up for and end vertically. Just like here:

```
for i from 1 to n do
    result := result + args(i)^2
end;
```

The same rule applies to any type of brackets. For example, you can use something like this,

```
[[1,7,31],[3,2,6],[4,6,7],[0,9,8]]
```

But if the line is getting too long, or you wish to emphasize the structure, you should write this instead:

```
[
    [1, 7, 31, 14],
    [3, 2,  6, 28],
    [4, 6,  7, 42],
    [0, 9,  8, 65]
]
```

If you think that you waste too much of space this way, you can simplify this structure down to this form:

```
[[1, 7, 31, 14],
 [3, 2,  6, 28],
 [4, 6,  7, 42],
 [0, 9,  8, 65]]
```

Formatting all the programming constructions when you need to use a procedure name and a long sequence of input parameters is a particular problem. This is a case where you will have to break rule 10. Otherwise, you would end up with very strange-looking code. For example, does this look good?

```
• tree := plot::Lsys(
                      PI/4, "F",
                      "F"="BR[+HF][-HF]HRF",
                      "R"="BRR",
                      "F"=Line,
                      "R"=Line,
                      "B"=RGB::Brown,
                      "H"=RGB::ForestGreen
                ):
```

In situations like this, it is necessary to break rule 10 and format the code this way instead:

```
• tree := plot::Lsys(
      PI/4, "F",
      "F"="BR[+HF][-HF]HRF",
      "R"="BRR",
      "F"=Line,
      "R"=Line,
      "B"=RGB::Brown,
      "H"=RGB::ForestGreen
  ):
```

Here, the first character of the declared variable tree plays the role of the starting bracket. The code is still very well formatted, and does not force you to type in hundreds of space characters.

In other programming languages, there are different rules for program formatting, and sometimes there are no rules at all. However, in modern programming languages like C++ or Java, having standard coding rules is very important. This is because quite often, programs written in these languages are developed by teams of people, and it is important that everybody uses the same coding rules. So, keep in mind that your students and colleagues should not have problems reading and understanding your programs. Such problems will definitely not happen if we all use the same coding rules.

5.14 Chapter Summary

In this chapter, you have learned a number of things about writing and using MuPAD procedures. You have also learned a bit about MuPAD's libraries and creating your own libraries. You will find out more about particular libraries in later chapters. For example, over the next four chapters you will use the plot library to create various

MuPAD plots. In later chapters, you will also find information about other libraries.

In this chapter, we have introduced a number of MuPAD syntax rules and procedures. Here is a list of them:

5.14.1 MuPAD Procedures

Declaration of a Procedure

proc_name := proc(*input parameters*)
local *local_variables*;
begin
    here is the body of the procedure, i.e. your code
end_proc

Declaration of a Sequence

f(*n*) $ *n=n1 .. n2*
 f(*n*) $ *n* in *set_of_values*

Declaration of a List

[*el1, el2, el3, el4, ... , el_n*]
[f(*n*) $ *n=n1 .. n2*]

Declaration of a Set

{*el1, el2, el3, ..., el_n*}
{f(*n*) $ *n=n1 .. n2*}

Declaration of a String

"*use quotes to declare a string*"

Declaration of a Matrix

matrix([[*a1, a2, ..., an*], [*b1, b2,...,bn*], ... , [*d1,d2,...,dn*]])

Exporting Procedures from Libraries

export(*library_name, function1, function2, ..., function_n*)
export(*library_name*)

Obtaining Information about Procedures

info(*library_name::procedure_name*)

?*procedure_name*

?*library_name::procedure_name*

Domain Related Procedures

Dom::*domain_name*(*object*) - *declares an object from a domain domain_name*

domtype(*object*) - *returns the domain of a given object*

Type Related Procedures

assume(*x*, Type::*type_name*) - *declares type of the variable x*

Type Semantical Checking (by value)

getprop(*x*) - *returns the mathematical properties of x (checks value)*

is(*x*, Type::*type_name*) - *checks if x has a given property (checks value)*

testtype(*x*, Dom::*domain_name*) - *checks if value of x belongs to the given domain (checks value)*

Type Syntactical Checking (by form)

type(*x*) - *returns the syntactical type of x*

testtype(*x*, Type::*type_name*) - *checks if x looks like objects of type Type::type_name*

Various MuPAD procedures

error("*error message here*") - *display the error message*

delete *variable_name*   - *delete variable*

delete *procedure_name*   - *delete procedure*

WRITEPATH := "*userlib*"   - *declare write path for user libraries*

write("*myfunctions.mb*")   - *write user library into default folder*

READPATH := "*userlib*"   - *declare read path for user libraries*

read("*myfunctions.mb*")   - *read library from the default path*

/* *comment is here* */

// *This is a single line comment*

Checking Procedure Input Parameters

$$\text{args}(x)= \begin{cases} \textit{number of arguments} & \text{if} \ \ x = 0 \\ \textit{i-th argument} & \text{if} \ \ x = i \leq \text{args}(0) \\ \textit{arguments i-th, ..., j-th} & \text{if} \ \ x = i..j \\ \textit{all arguments} & \text{if} \ \ \textit{there is no input parameter at all} \end{cases}$$

5.14.2 MuPAD Code Formatting Rules

In this chapter, we have finished defining the rules that will help you in formatting MuPAD code. Here are all the rules introduced in this and in the previous chapters.

Rule 1 There are always spaces after keywords and surrounding binary operations.

Rule 2 Always use a variable to store every MuPAD object that you are going to use later.

Rule 3 Constant names are UPPERCASE, with an occasional UNDER_SCORE, while variables names are always in lowercase.

Rule 4 Do not put two MuPAD commands in the same line.

Rule 5 Use indents to emphasize the nesting level in your programs.

Rule 6 The names of MuPAD functions, procedures and variables should be descriptive and not unreasonably long.

Rule 7 Variable and procedure names are in lowercase, with occasional upperCase characters in the middle.

Rule 8 No more magic numbers in MuPAD programs.

Rule 9 Comment your programs whenever you think it may be important.

Rule 10 Opening and closing braces must line up, either horizontally or vertically.

5.15 Programming Exercises

The number of exercises for this chapter is not very large. Generally, you could take any exercise from chapter 4 and instead of developing a MuPAD program, you could develop it as a procedure. You will also find many occasions to write procedures or even develop libraries of procedures in later chapters. Therefore, here I will only include a few exercises that will give you a chance to practice writing procedures without referring to the other chapters in this book.

Writing Simple Procedures

1. Write a procedure that will produce the sum of the arithmetic series $a_n = a_0 + nd$. The input parameters shall be the first term of the sequence a_0, the difference d and the number of terms n. Protect your procedure against various possible errors.

2. Write a procedure that will produce the sum of the geometric series $b_n = b_0 q^n$. The input parameters shall be the first term b_0, the ratio q, and n. Protect your procedure against some possible errors.

3. Write down a library of procedures to calculate the areas of some 2D shapes — a triangle, a rectangle, a circle, a regular polygon, a trapezoid, the segments and sectors of a circle, and so on. Protect your procedures against all possible errors. Save the whole library using a suitable name, e.g. areas.mu.

4. Create a library of procedures to calculate the volumes and surface areas of a cube, cylinder, dodecahedron, icosahedron, octahedron, regular cone, hexahedron, regular pyramid, tetrahedron, sphere, etc.

5. Develop a procedure to calculate the binomial coefficients $\binom{n}{r} = \frac{n!}{r!(n-r)!}$. You can develop it in such a way that for the two given numbers n and k, it will produce the term $\binom{n}{r}$. You may also ask your procedure to develop complete rows of a Pascal triangle for the given number of rows n. Finally, your procedure can develop a square matrix $(n+1) \times (n+1)$, where the binary coefficients will be located under the diagonal, and the zeros in the upper right section of the matrix. This is how it might look

for $n = 3$.

$$\begin{pmatrix} 1 & 0 & 0 & 0 \\ 1 & 1 & 0 & 0 \\ 1 & 2 & 1 & 0 \\ 1 & 3 & 3 & 1 \end{pmatrix}$$

6. The sequence v_n is defined by the formula

$$v_n = \begin{cases} 1 & \text{if } n = 1 \\ p & \text{if } n = 2 \\ pv_{n-1} + qv_{n-2} & \text{if } n > 2. \end{cases}$$

Develop a recursive procedure that will calculate the terms of this sequence for the two given numbers p, q. The terms should be calculated for the given number n.

Write a program that will produce all the terms from $n = 1$ up to the given number N. The terms of the sequence can be produced in the form of a sequence, a list or a matrix.

7. Write a procedure genFact(n,m,p,...,q) to produce $n!m!p!...q!$. You should develop it in such a way that you will be able to use any number of input parameters. You can develop it as an iterative procedure; however, a recursive approach is also possible.

Chapter 6

Introduction to MuPAD Graphics

I have already demonstrated, within this book, two types of MuPAD plots. However, these are only a small tip of the huge iceberg that is the MuPAD plots family. The collection of MuPAD graphical objects will not only make mathematicians happy, but mathematically-oriented artists too. You can use MuPAD to make plots of equations, geometric objects, and even fractals. You can develop plots of single objects, or scenes with multiple objects and impressive animation. Finally, if you are interested in curve programming, you can develop some very interesting patterns using turtle graphics and L-systems. Let us start exploring MuPAD graphics by summarizing the graphs we have obtained previously.

6.1 Obtaining Quick Plots

The descriptions of MuPAD's graphical objects are located in several places. A few fundamental plotting procedures are loaded when MuPAD starts up. These procedures are useful if you wish to make a quick plot without loading an additional library or bothering with special parameters. However, the descriptions of a majority of MuPAD's graphical objects and graphical procedures are contained elsewhere, in the plot library. We will discuss this library in detail later on in this chapter. For now, let us concentrate on quick plots.

When starting MuPAD, the following three plotting procedures are exported into your notebook: plot, plotfunc2d, plotfunc3d. The procedure plot is the key engine for all plots and we will use it frequently later on.

Statements that use these procedures can have a number of options. We will cover them in details in the later sections of this chapter.

6.1.1 Procedure plotfunc2d

Let us start with 2D plots. The procedure plotfunc2d plots the graph for functions of one variable in a given interval. This is the quickest way to make a graph of a function while teaching or learning mathematics with MuPAD or just experimenting with various functions.

The minimal syntax of statement calling this procedure is as follows:

$$\text{plotfunc2d}(f_1(x), f_2(x), \ldots, f_n(x), x = x1..x2)$$

The formulae $f_1(x), f_2(x), \ldots, f_n(x)$ denote the functions that you wish to plot. You can put in as many functions as you like. The interval $x = x1..x2$ determines the plot area. Here is an example showing the usage of this procedure.

Example 6.1

Let's make a graph of the parabola $x = y^2$. In order to plot its graph, we have to represent it in the form of the equation $y = f(x)$. By solving the equation of the parabola in respect to the variable y, we obtain the equations of the two functions $y = \sqrt{x}$ and $y = -\sqrt{x}$, which represent the top and bottom branches of the parabola curve. Thus, we will plot the equations of both functions in one plot:

- `plotfunc2d(sqrt(x), -sqrt(x), x = -1..2):`

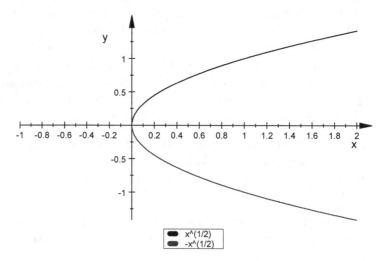

You will easily realize that the part of the graph above the axis x-axis is the plot of the function $y = \sqrt{x}$, while the bottom part of the curve is the plot of $y = -\sqrt{x}$.

By default, MuPAD uses different colors for each function when plotting them. The graph of the first plotted function is blue, the next one is red, then the next is green, and so on. In this book, for the purpose of printing, all the colors were converted to black and white. ∎

Even this simple plotting procedure can be very useful when exploring properties of functions or solving equations. Let us examine one such example.

Example 6.2

Suppose that we need to graphically solve the equation $x^3 = \cos x$. In order to do this, we will isolate both sides of the equation and represent them as separate functions. Thus, the two functions we're looking at are $y = x^3$ and $y = \cos x$. Now we can plot both of them in one picture. To start our investigations, we will use the interval $-2 < x < 2$. Hence, we can begin with the command:

- `plotfunc2d(x^3, cos(x), x = -2..2)`

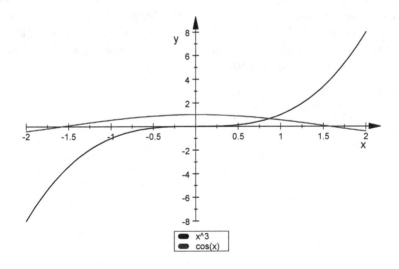

Looking at the obtained graph, we can clearly see that we need to narrow down our search to the interval $0.8 < x < 1$. Thus, in the next step we will use the command:

- `plotfunc2d(x^3, cos(x), x = 0.8..1)`

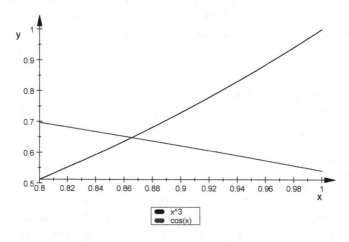

We can stop here if this level of accuracy is enough for us. Otherwise we may continue, choosing narrower intervals, e.g. $0.86 < x < 0.88$ or even smaller, for example $0.864 < x < 0.866$. In some situations it would be useful to display a vertical grid in our plots in order to help us locate the intersection point. For example, when we plot our graph for the interval $0.864 < x < 0.866$ it is difficult to read the position of the point of intersection from the picture. Therefore, the additional parameter XGridVisible=TRUE may help a lot.

- `plotfunc2d(x^3, cos(x),`
  `    x = 0.864..0.866,`
  `    XGridVisible = TRUE`
  `)`

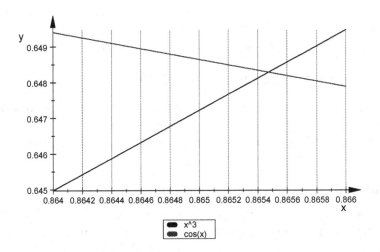

As you can see from the last picture, when we zoom further into the graph, the labels on the x-axis axis become a bit crowded. We can avoid this by printing them vertically, like here:

- ```
plotfunc2d(x^3, cos(x), x = 0.8654..0.8656,
    XGridVisible = TRUE,
    XTicksLabelStyle = Vertical
)
```

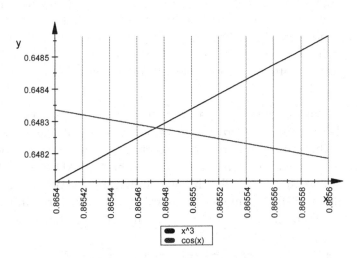
```

Now, when things are back in order, we can read the graph and calculate the point of intersection. We may conclude that the solution for our equation is $x = (0.86546 + 0.86548)/2 = 0.86547$ ∎

6.1.2 Your First Animation with MuPAD

In the introduction to this chapter, I mentioned that we can develop graphs with animation in MuPAD. This option is available since MuPAD version 3.0. In fact, we can animate almost everything in our graphs. Let us explore a very basic example.

Example 6.3

We know quite well that the graph of the function $y = \sin ax$ changes depending on the value of parameter a. Drawing multiple graphs for different values of a is a good way of demonstrating this fact. However, in such a case, animation could be even more useful. Let us start off with a basic statement to plot the function $y = \sin ax$, where $x = -\pi..\pi$. To this statement, we will add information about the range for the parameter a, which will be something like $a = 0..10$. This gives us this statement:

- `plotfunc2d(sin(a*x), x = -PI..PI, a = 0..10)`

The resulting plot doesn't look very impressive; it's just an empty plot. If you examine it carefully, you will notice that the horizontal axis x-axis is not black but blue. This suggests that we have there the graph of a horizontal line that represents the function $\sin ax$, where $a = 0$. So, we do have something. Now, let's check where the rest is—the animation.

Double click on the obtained graph. In this moment, we enter into a new world—the Virtual Camera (VCam) environment (fig. 6.2). We will explore this environment in the next chapter. For now, we have to concentrate on our animation. On the top of the MuPAD screen, there is a new toolbar that looks like the one shown in the enclosed figure 6.1. We will call it the animation toolbar. In fact, it is very similar to animation toolbars in many graphics or multimedia packages.

Fig. 6.1 Animation toolbar

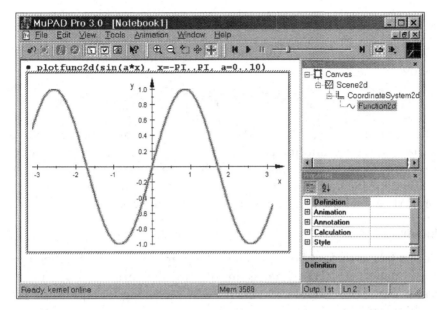

Fig. 6.2 MuPAD Virtual Camera environment

The meaning of the icons in animation toolbar, starting from the left, is as follows: go to the beginning of the animation, start playing the animation, stop playing the animation, a slider to preview the animation by hand, go to the last frame of the animation, play the animation in a loop, and finally, a menu to change the speed of the animation.

Now, click on the start animation button (the single arrow) and see what MuPAD has produced for us. Check all the stages of this animation and think about how you could use such a simple animation to illustrate other examples from mathematics or science.

When you finish experimenting with this animation, click anywhere outside the picture frame, and you will return back to your notebook. ∎

Later, we will spend more time experimenting with animation, but now it is time to become more familiar with quick plots in 3D. For the same reasons as above, I will only give you a very brief description of the plotfunc3d procedure.

6.1.3 Procedure plotfunc3d

The minimal syntax of the statement calling this procedure is as follows:

$$\text{plotfunc3d}(f_1(x,y), f_2(x,y), \ldots, f_n(x,y), x = x1..x2, y = y1..y2)$$

The functions $z = f_1(x,y)$, $z = f_2(x,y)$, ..., $z = f_n(x,y)$ are functions of two variables x and y, and the parameters $x = x1..x2$, $y = y1..y2$ determine the range of variables x and y. Notice however, that we do not provide a range for the variable z. If we were to add here the range $z = z_1..z_2$, MuPAD would treat it as a parameter for animation.

Example 6.4

Let us take a look at how to obtain the graph of a surface defined by the equation $z = \sin xy$, intersecting it with the surface $z = xy$ in the same plot. This simple statement will be enough to produce the required graph:

- ```
  L := 5:
  plotfunc3d(sin(x*y), x*y, x = -L..L, y = -L..L)
  ```

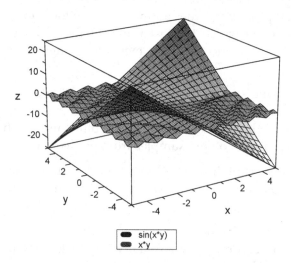

Of course we can still add a few more surfaces to this picture, or change the color and other parameters for each surface, but we will explore these possibilities in the later sections of this chapter. ∎
  ```

6.2 General Concept of MuPAD Plots

As I wrote at the start of this chapter, most types of graphical objects are described in the plot library. Allow me to explain what I mean and to make our language more precise.

MuPAD follows the well-known in computer science concept of object-oriented programming. This is especially visible in the case of the plot library. The plot library contains the descriptions of some types of graphical objects, for example, 2D and 3D circles, curves defined by parametric equations, 3D surfaces, etc. We may call every type described in the plot library a class. For example, Circle2d can be considered as a class of all circles on a plane, or a recipe for building a circle on a plane. Therefore, every time we declare a circle A:=plot::Circle2d(3,[1,1]) we build an object from the Circle2d class. This object is almost a physical object, which can be colored, moved from one place to another or duplicated.

Using the classes described in the plot library is quite easy, and you can do this in a few ways.

1. Suppose you are going to develop a number of different types of plots in the same document. In this case, it would be worthwhile to export the entire plot library into the current MuPAD session. This can be done by executing the statement:

    export(plot):

    Then you will then be able to use plot classes without giving their location.

2. If you are only going to develop a few types of graphical objects, you can export only these classes that you need by listing them in the export command. Here is how you would export the two classes Function2d and Point2d from the plot library:

    export(plot, Function2d, Point2d)

3. Finally, if you need to make a single graph, then you could use a command like this:

    a := plot::Function2d(x^2, x = 1..2)

This way, you use the class Function2d by pointing directly to its location in the plot library.

For the purpose of this book, I will frequently use the second method. In many cases while starting a new example, I will export all the graphical classes that I need for the example, and then I will use them. This will make the code of all the examples much simpler, and at the same time you will get into the good habit of exporting only the classes that you really need. However, in most of the examples with a single declaration of a plot class I will declare the plot object with the plot:: prefix.

As you would have already noticed, we can obtain a few different graphs in a single picture. We do not need to stick to objects of the same type. We can mix, in one picture, the graphs of objects of various different types. This is similar to what happens in a photographic workshop, where the photographer places a few objects on a table and then he makes a photograph. In MuPAD graphics, this analogy to a photographic workshop goes even further. First, we develop each object to be plotted. We decide about its colors, its location and how it will be plotted. Then, we add a background to the scene, the appropriate coordinate axes, and a few other elements. Finally, we place the camera in the right place and make a picture. Such a picture can be placed in a frame, alone or with a few other pictures. We will call this frame the canvas. Therefore, when developing graphics with MuPAD, we have the following hierarchy:

1. Canvas — the container for one or more scenes

2. Scenes — containers for a number of graphical objects

3. Sets of graphical objects or a single object.

This all might look like the example below.

Example 6.5

In this example, I will demonstrate step by step how to develop a scene and obtain a picture of it.

First, we have to export from the plot library the two classes that will be needed in this example: Function2d and Scene2d.

- export(plot, Function2d, Scene2d):

Now, we can declare two objects—Object1 and Object2. Each of them will have some assigned properties.

```
• Object1 := Function2d(sin(3*x), x=-PI..PI,
      Color = [0.2, 0.2, 0.2],
      LineWidth = 0.8
  ):

  Object2 := Function2d(x, x = -PI..PI,
      Color = [0, 0, 0],
      LineWidth = 0.15
  ):
```

Now we develop a scene using these two objects and some environment properties.

```
• MyScene := Scene2d(Object1, Object2,
      BackgroundColor = [0.8, 0.8, 0.8],
      Scaling = Constrained
  ):
```

When all these objects and the scene design are ready, we can make a picture of the scene.

```
• plot(MyScene)
```

Finally, we've got a picture that we can use in further investigations, or just stick it into our electronic photo album.

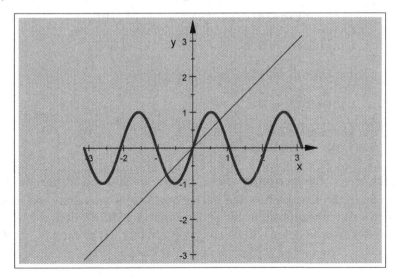

Note that in this example, we did not use the canvas object. In fact, we do not need to use it at all. MuPAD will create it automatically as a container for the produced scene. In this last picture, you can see the canvas as a white frame around the picture.

You may be wondering whether you will have to be formal, and follow all these scene creation steps every time you wish to develop a plot. Of course not—you can also work like a photographer making photos during a Sunday walk with his family. You see something, open your camera, and click. For instance, our example above may be simplified to this form:

```
Object1 := plot::Function2d(sin(3*x), x = -PI..PI,
    Color = [0.2, 0.2, 0.2],
    LineWidth = 0.8
):
Object2 := plot::Function2d(x, x = -PI..PI,
    Color = [0, 0, 0],
    LineWidth = 0.15
):
MyScene := plot::Scene2d(Object1, Object2
    BackgroundColor = [0.8, 0.8, 0.8],
    Scaling = Constrained
):
plot(MyScene)
```

You could simplify it even more:

```
plot(
    plot::Function2d(sin(3*x), x = -PI..PI,
        Color = [0.2, 0.2, 0.2],
        LineWidth = 0.8
    ),
    plot::Function2d(x, x = -PI..PI,
        Color = [0, 0, 0],
        LineWidth = 0.15
    ),
    BackgroundColor = [0.8, 0.8, 0.8],
    Scaling = Constrained
)
```

In each case, the resulting picture may look a bit different. For example, in the first picture the background color was assigned to the scene; on the other hand, in the second case we didn't define a scene, so the background color will be assigned directly to the canvas.   ■

Before we proceed further let us try to capture the difference between a scene and a canvas. This knowledge could be useful for developing more complex mathematical images.

Example 6.6

Let us once again consider the function $y = \sin ax$, where a is a parameter. Animation is a good way to show how the graph changes for different values of a. Unfortunately, we cannot include animations in printed books and articles. Therefore, it could be useful to print multiple pictures showing plots for different values of a. This is a good reason to develop a single canvas with multiple pictures.

Like before, we will start by exporting the required classes. Then we will develop the objects to be plotted, and finally we will build a simple scene for each object. Like this:

```
• export(plot, Function2d, Scene2d):

Fun1:=Function2d(sin(x),    x = 0..2*PI):
Fun2:=Function2d(sin(2*x),  x = 0..2*PI):
Fun3:=Function2d(sin(3*x),  x = 0..2*PI):
Fun4:=Function2d(sin(10*x), x = 0..2*PI):

S1 := Scene2d(Fun1, Header = "a=1"):
S2 := Scene2d(Fun2, Header = "a=2"):
S3 := Scene2d(Fun3, Header = "a=3"):
S4 := Scene2d(Fun4, Header = "a=10"):

plot(S1, S2, S3, S4, BorderWidth = 1)
```

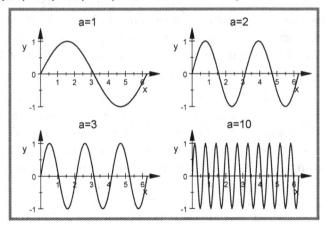

In this way, we can develop pictures with multiple scenes. Later on, I will show you how these multiple scenes can be arranged into a well organized grid with the appropriate number of columns and rows. ■

In all the examples that we have developed so far, I have tried to use various plotting options as little as possible. So, we have only sporadically used different color options or line thickness declarations. This was almost all that we have needed until now. In fact, with this minimum knowledge of options used in plots, we can still do a great job. However, it is good to know more about plot options in order to see what we can really produce with MuPAD. Therefore, let us now explore the options that we can use while plotting graphical objects and scenes. The scope of this book doesn't allow us to be very detailed, so we will only choose the most important options. You do not need to memorize them, but it is important to know what you can do with plots. In the next chapter, I will show you how you can take advantage of any of these options without typing even a single keyword.

6.2.1 Object Attributes & Options

In MuPAD 3.0, the plot library contains the declarations for more than 60 classes and more than 300 different plotting options. It is thus very important not to get lost in this jungle and to find an easy way of dealing with all these classes and options.

In one of the previous examples, I created two objects. These objects were two curves on a plane. I assigned a color and line width to each of them. However, the number of parameters in an object declaration can be much larger. When you export graphical procedures into your notebook, all of these parameters have their default values, which will be used for plotting unless you change them.

First, let's consider this — do we even need to change these parameters? Well, imagine that the objects to be pictured are people working in the same bank or some other company, and all of them wear the same company uniform. The picture of such a group would be very boring, and it would be quite difficult to distinguish who is who. With mathematical objects, it is even more difficult. How can you distinguish a few similar curves? In order to do this, we need to change the colors of curves, their thickness and sometimes also the way we plot them.

Well, you may ask how you can find out what the default uniform of a graphical object is. First, you need to know the type of your object. Suppose, for example, that this is Function2d. The statement info(plot::Function2d) will produce the names of all the options defined in MuPAD for Function2d, as follows:

- info(plot::Function2d)

  plot::Function2d – graphical primitive for 2D function graphs

  – Interface:

  plot::Function2d::AdaptiveMesh,
  plot::Function2d::Color,
  ...

The output is much longer. However, I removed large part of it to save the space.

This list is the first step in gathering information about a given class. For a beginner, most of the names in this list may look strange and sometimes completely obscure. However, after experimenting with a few plots, the meaning of most of these options will become quite obvious.

First, let us analyze the syntax used here. It looks as though every class in the plot library is a container for some other elements and properties. This is still the object-oriented concept that that we mentioned before. Having any class or object of this class, we can access its components using the slot operator "::". Thus, if you wish to access the color of the plotted function 2D, you have to either use its complete path plot::Function2d::Color, if class Function2d has not been imported already, or just the partial path Function2d::Color, if Function2d has been imported.

In the plot library, there are two very useful procedures that can help us to work out the default values of plot options and how to declare these default values. These are procedures plot::getDefault and plot::setDefault. For example,

- plot::getDefault(plot::Function2d::Color)

  [0.0, 0.0, 1.0]

This statement produces the default color used for plotting functions 2D. Similarly,

- `plot::getDefault(plot::Function2d::LineWidth)`

  0.35

This statement produces the default width of the line in millimeters used to draw the plotted functions.

We can use the `plot::setDefault` procedure to set the default parameters for plots within the given notebook. For example,

```
plot::setDefault(
    plot::Function2d::LineWidth = 0.7*unit::mm,
    plot::Function2d::Color = [0.5,0.5,0.5],
    plot::Function2d::TitlePositionX = 0.4,
    plot::Function2d::TitlePositionY = 0.7
):
```

This statement will set the default parameters `LineWidth` to 0.7 mm, the color of the lines to gray 50%, and the title position to $x = 0.4$ and $y = 0.7$ coordinate system units.

Once this is done, every declared object of class `Function2d` will be plotted using these new parameters.

- ```
  MyFunction := plot::Function2d(sin(x), x = -1..1,
      Title = "Function sine"
  ):
  plot(MyFunction)
  ```

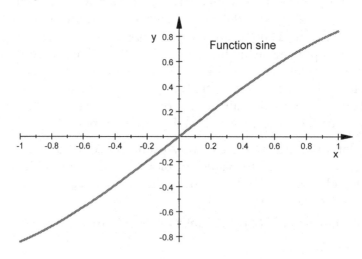
  ```

If you are tired of producing long list of options for every type of graphical object used in your notebook, then there is an easy way to find out and type in the name of each class from the plot library and the names of its options. For example, suppose that you wish to draw a filled circle on a plane and you do not remember the name of the right class and the options that you can use for it. To do this, in any input region, type in plot:: and press [Ctrl]+[Space]. You will get a small popup box (see fig. 6.3), where you can scroll and choose the right name. When you've got the right name, e.g. plot::Circle2d, type in the slot operator again, and press [Ctrl]+[Space] once more. This will produce for you a popup box with the names of all the options available for Circle2d, like in fig. 6.3.

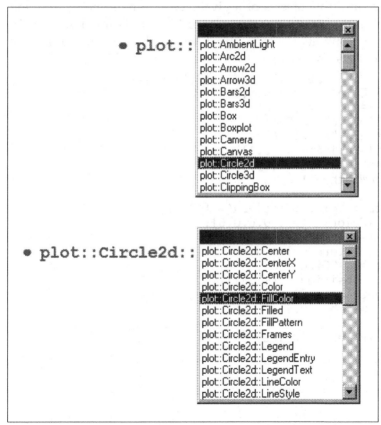

Fig 6.3 Use [Ctrl]+[Space] to select required object and option

When you choose what you need from this list, you get the full path to the required option, for example `plot::Circle2d::FillColor`. This is enough to find out, for example, what the default color for the interior of a filled circle is:

- `plot::getDefault(plot::Circle2d::FillColor)`
  [1.0, 0.0, 0.0]

Now you can easily assign a new color to all the circles in your notebook by using this statement:

```
plot::setDefault(
    plot::Circle2d::FillColor = [0,0,1]
):
```

If you have multiple objects of a given class in the same picture, the global declaration of options doesn't work well. Quite often we have to change the options for each object separately. We can do it when declaring objects, like this:

```
Circle1 := plot::Circle2d(1,[2,1],
    FillColor = [0.8,0.8,0.8],
    Filled = TRUE,
    FillPattern = Solid,
    Title = "Circle A",
    TitlePosition = [2,1],
    LineColor = [0,0,0],
    LineWidth = 1.5
):
Circle2 := plot::Circle2d(1,[-2,-1],
    FillColor = [0.6,0.6,0.6],
    Filled = TRUE,
    FillPattern = Solid,
    Title = "Circle B",
    TitlePosition = [-2,-1],
    LineColor = [0,0,0],
    LineWidth = 1.5
):
plot(Circle1, Circle2)
```

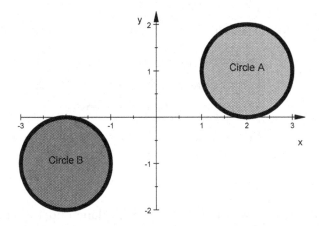

The way of declaring options presented here for circles is universal for all the classes from the plot library and the objects representing them. So, whenever you wish to create any plot object, export its class from the plot library or use its class name with the `plot::` prefix, then declare its required parameters (the radius and center for a circle, the equation for a curve, etc.), and finally declare the values for other parameters.

Note that the names of classes in the plot library always start with a capital letter. The names that start with a lowercase letter are names of methods, i.e. operations on graphical objects. Therefore, `Function2d` and `Dodecahedron` are names of objects, while `copy` and `setDefault` are names of methods. Observe that in the plot library, there are many names that suggest operations rather than objects, for example `Scale2d` or `Translate3d`. However, when we apply `Scale2d` to an object we get back another object. Therefore, it is convenient to think about `Scale2d` as a class of objects that are scaled copies of existing objects.

6.3 Canvas, Scene and Coordinate System Options

`Canvas`, `Scene2d` and `Scene3d` are three different classes. However, they share a number of common features, and so it is convenient to describe them together. In most cases, we do not declare the canvas, we just put the objects or scenes to be plotted inside the plot command. We do need to create the canvas object if we wish to control it directly, for example to change its frame color or size.

The simplest form of declaration for these three objects is as follows:

> plot::Canvas(objects,options)
>
> plot::Scene2d(objects, options)
>
> plot::Scene3d(objects, options)

Let us develop a simple example showing the use of Canvas and Scene2d, and pointing out a few important things.

Example 6.7

In our example, we will use a circle on a plane and the graph of a function. We will also declare two scenes, one with both the circle and the function graph, and the other with only the circle. Therefore,

```
• export(plot, Circle2d, Function2d, Scene2d):
• C1 := Circle2d(1, [2,3]):
  C2 := Function2d(sin(x), x = 0.. PI):
  C3 := Scene2d(C1, C2, BackgroundColor = [0.8, 0.8, 0.8])
  C4 := Scene2d(C1, BackgroundColor = [0.7, 0.7, 0.7]):
```

Now, let us plot both scenes in one plot. We do not declare the canvas here, but it is nonetheless created automatically.

```
• plot(C3,C4)
```

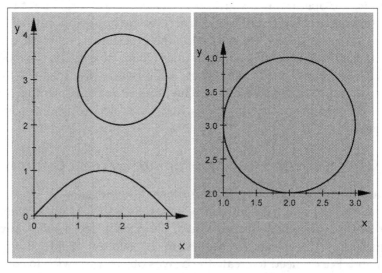

The obtained picture shows the simplest way to use Scene2d and plot two or more scenes on the same canvas. Note that we can declare plot options on various levels. For example, the color of the line can be declared using setDefault for the whole notebook, or we can declare it for each object separately; finally, we can declare the color of the line inside the scene or even inside the canvas declaration. There is a rule — if the color was not declared for a given object, then this object will be plotted according to the color declaration in the larger object. For example, if we did not declare a color for the circle, then the circle will have the line color that was declared in the scene or in the canvas, or the default color. In other words, the circle will be plotted with the first color available for it. The same rule is valid for many other options and objects.

Let us see what will happen if we execute the last plot(C3,C4) statement with additional declarations regarding line color and width. Here, we will declare white lines and a thickness of 0.7 mm. Can you explain why both plotted objects are this time white?

- plot(C3, C4, LineColor = [1,1,1], LineWidth = 0.7)

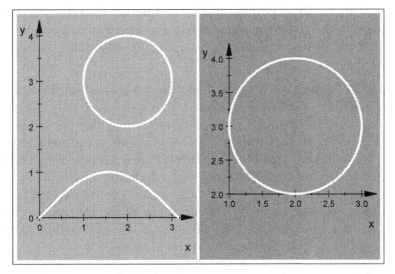

When using the plot command, we can mix scenes and objects. For example, plot(C1, C2, C3) can be used to plot the circle C1, the graph of a function C2, and the scene C3 with both the circle C1 and the function C2. All the scenes will be plotted as independent scenes on

the same canvas. However, objects that are not scenes will be grouped
into one scene, like here:

- `plot(C1, C2, C3)`

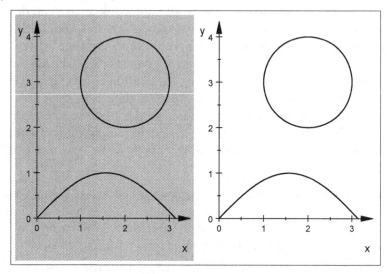

Note that the order of plotting objects is different than the one that is
suggested in the `plot` command. First, the ready scenes are plotted,
and then the scene obtained by grouping the remaining individual
objects.   ∎

In the rest of this chapter, we will analyze a few selected options that
can be used in objects to be plotted. We will cover only the most
important options. We will pay special attention to the options that are
important to an average MuPAD user—a teacher or a student. As we
know, some options can be declared on various levels; for example,
the color of the line can be declared while declaring a function, or later
in the scene declaration, or even later in the canvas declaration. In
such cases, we will cover the given option in the most appropriate
place. So, for example, we will talk about line color while talking
about options for objects using linear components.

6.3.1 Options for Canvas

Here are listed the most important options for Canvas along with their
default values. For most of these options, it is obvious how their
values can be declared.

OutputUnits = mm, *the default unit for all plots*

Height = 80, *the height of canvas*

Width = 120, *the width of canvas*

Layout = Tabular, *how multiple plots on the same canvas will be placed*

Columns = 0, *the number of columns on canvas, 0=automatic*

Rows = 0, *the number of rows on canvas, 0=automatic*

Spacing = 1, *the distance between scenes on the same canvas, in mm*

BackgroundColor = [1.0, 1.0, 1.0], *the color for the background*

BorderColor = [0.5, 0.5, 0.5], *the color for the border*

BorderWidth = 0, *the size of the border, in mm*

Margin = 1, *the width of the canvas margin, in mm*

BottomMargin = 1, *the width of bottom margin, in mm*

LeftMargin = 1, *the width of left margin, in mm*

RightMargin = 1, *the width of right margin, in mm*

TopMargin = 1, *the width of top margin, in mm*

Header = NIL, *the canvas header declaration*

HeaderFont = ["sans-serif", 12], *the font for the header and its font size in points*

HeaderAlignment = Center, *the location of the header*

Footer = NIL, *the canvas footer declaration*

FooterFont = ["sans-serif", 12], *the font for the footer and its font size in points*

FooterAlignment = Center, *the location of the footer*

As you can easily see, the canvas is a rectangle with margins, a border, a header and a footer. The margin is a part of plotting area surrounding the plots, and it has the same color as the rest of the canvas. On the other hand, the borders can have a different color.

Color declarations

In MuPAD, we declare colors using the RGB color space. In short, this means that each color definition will contain three numbers, defining the quantity of red, green and blue ink in the given color. In computer graphics, there are two different ways of declaring RGB colors. One popular way is to describe the quantity of RGB components using integers between 0 and 255. Another system uses numbers from the interval [0,1] to describe RGB components. This system is used, for example, in the very powerful POV-Ray 3D modelling programming language. MuPAD uses the same system. In order to define a color in MuPAD we have to give a list of three numbers, for example [0.23, 0.89, 0.999]. Remember, all the numbers must be between 0 and 1. In this system, 0 means that there is no given component at all, and 1 means that there is 100% of a given component. For instance, [1,0,0] means red, [0,1,0] means green, [0,0,1] means blue, and [1,1,1] means — well, guess what?

There are a few ways to declare the color value of a color option. The easiest way is like this:

$$\text{OptionName} = [\#R, \#G, \#B]$$

just like this

```
BackgroundColor = [0.5256, 0.7865, 0.5432]
BorderColor = [0.2304, 0.8765, 0.0987]
```

You can use also predefined colors from the RGB library. Just type in this command to see what is available:

*   info(RGB)

Here is the output that you will get. I only copied a few lines here.

RGB values for color names

− Interface:

```
RGB::AliceBlue,
RGB::AlizarinCrimson,
RGB::Antique,
RGB::Aquamarine,
RGB::AquamarineMedium,
RGB::AureolineYellow,
RGB::Azure,
.............................
RGB::YellowBrown,
RGB::YellowGreen,
RGB::YellowLight,
RGB::YellowOchre,
RGB::Zinc
```

For other objects, like surfaces or curves, we may use a more sophisticated way to define color. For example, we can use a function to add color to a 3D surface. We will talk about this method later.

Colors in MuPAD can be transparent. In order to declare a transparent color you have to use four numbers where the last one sets transparency. For example declaration,

```
FillColor = [1, 0, 0, 0.5]
```

will declare red color 50% transparent. Unfortunately transparency works only in 3D graphics.

Units

In MuPAD, the default unit for all plots is the millimeter (mm). This means that length and width of objects is by default given in mm. However, we may change it to use other units that are sometimes more suitable for us, like inches (inch) or points (pt). For example,

```
plot::setDefault(
   plot::Canvas::OutputUnits = unit::inch
)
```

This declaration will make inches the default units for our plots. Just remember that 1 inch is equal to 2.54 cm or to 72.27 points, and 1 mm is equal to 2.845275619 points. In MuPAD, units are grouped in the unit library. If you wish to see what units were implemented in your version of MuPAD, type in unit:: and then press [Ctrl]+[Space].

A declaration like the one above will implement changes for the whole canvas. However, we can also change the units for just a single plot element. For example, this declaration

- C1 := plot::Circle2d(1, [2,3], LineWidth = 2*unit::pt):

will draw our circle with a line 2 points in width. On the other hand, this declaration

- plot(C3,C4,LineColor=[1,1,1],LineWidth=0.4*unit::inch)

will use a line width of 0.4 inches for any object that does not have a declared line width.

Canvas Layout

By default, canvas layout is tabular without a declared number of columns and rows. This means that MuPAD will arrange the plots on a canvas in the form of a table, where the number of rows and columns will be adjusted to the existing space. However, we can force MuPAD to organize our table according to our needs, for example to use 3 rows and 2 columns for 6 plotted scenes. We can also declare the layout as Horizontal, where all the scenes will be organized in a table with one row only, or Vertical, where all the scenes will form one vertical column.

Headers and Footers

Headers and footers are two optional elements for any graph. The header will be placed above the plotted graph, and the footer below. We can use any available font in any size for headers and footers. We can align the header and footer to the right, the left, or in the center, which is also where they are placed by default.

Here is a very simple example showing how to declare headers and footers and put them in the desired place.

- ```
C1 := plot::Circle2d(1, [0,2]):
C2 := plot::Circle2d(1, [2,0]):
C3 := plot::Function2d(x, x = -1..3):

S1:=plot::Scene2d(C1,C3,BackgroundColor=[0.8,0.8,0.8]):
S2:=plot::Scene2d(C2,C3,BackgroundColor=[0.7,0.7,0.7]):

plot(S1,S2,
    Header = "Developing symmetry",
    HeaderAlignment = Left,
    Footer = "Circle & Line",
    FooterAlignment = Right,
    BorderWidth = 1*unit::mm
)
```

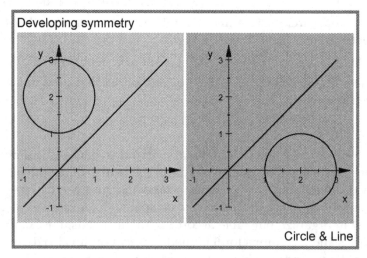
```

6.3.2 Options for Scene2d

Many options used in Scene2d are the same as for a canvas. We will therefore concentrate here mostly on new elements, which are the legend and the background transparency option.

BorderWidth = 0, *the width of the scene border*

BorderColor = [0.5, 0.5, 0.5], *the color of the scene border*

Height = 80, *the height of the scene*

Width = 120, *the width of the scene*

Margin = 1, *the width of the margin, in mm*

BottomMargin = 1, *the width of the bottom margin, in mm*

LeftMargin = 1, *the width of the left margin, in mm*

RightMargin = 1, *the width of the right margin, in mm*

TopMargin = 1, *the width of the top margin, in mm*

Header = NIL, *the header's text declaration*

HeaderFont = ["sans-serif", 12], *the header's font declaration*

HeaderAlignment = Center, *the position of the header*

LegendVisible = FALSE, *determines whether the legend is visible or not*

LegendFont = ["sans-serif", 8], *the font and font size for the legend*

LegendPlacement = Bottom, *the vertical position of the legend*

LegendAlignment = Center, *the horizontal position of the legend*

Footer = NIL, *the footer declaration*

FooterFont = ["sans-serif", 12], *the footer's font*

FooterAlignment = Center, *the alignment of the footer*

BackgroundTransparent = FALSE, *the transparency of the background*

BackgroundColor = [1.0, 1.0, 1.0], *the background color, in this case white.*

Legend is a completely new feature of the 2D scene. This is a small box, by default located at the bottom of the picture, showing which color was used for which graph. The example below shows how to develop a graph of two functions and add an appropriate legend for them. Note that before you can add the legend, you need to declare text for it. Like for headers and footers, the text for a legend needs to be declared as a string using quotation marks, for example, Legend="sin(x)".

```
• C1:=plot::Function2d(sin(x),x = -PI..PI,
     LineColor = [0, 0, 0],
     LineWidth = 0.5,
     Legend = "sin(x)"
  ):
  C2:=plot::Function2d(sin(2*x),x = -PI..PI,
     LineColor = [0.5, 0.5, 0.5],
     LineWidth = 1,
     Legend = "sin(2x)"
  ):
```

```
plot(C1,C2, LegendVisible = TRUE)
```

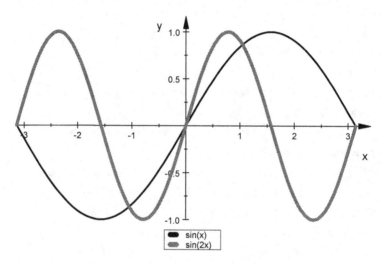

By default, the background for a graph is white. We can change this color, or we can make it transparent, which will show the color of the canvas.

6.3.3 Options for Scene3d

We already know most of the Scene3d options from Scene2d. However, Scene3d is a very sophisticated object that more closely resembles our photographic studio than Scene2d. We can deal with a real camera and work with lights. Another new option for us is the possibility to change the aspect ratio of the plotted scene.

BorderWidth = 0, *the width of the border*
BorderColor = [0.5, 0.5, 0.5], *the color of the border*

Width = 120, *the width of the scene in mm*

Height = 80, *the height of the scene in mm*

Margin = 1, *the width of the margin*

TopMargin = 1, *the width of the top margin*

RightMargin = 1, *the width of the right margin*

BottomMargin = 1, *the width of the bottom margin*

LeftMargin = 1, *the width of the left margin*

CameraDirection = NIL, *the direction vector of the camera*

CameraDirectionX = NIL, *the x coordinate of the direction vector*

CameraDirectionY = NIL, *the y coordinate of the direction vector*

CameraDirectionZ = NIL, *the z coordinate of the direction vector*

YXRatio = 1, *the ratio of the scene in the plane XY*

ZXRatio = 2/3, *the ratio of the scene in the plane XZ*

Lighting = Automatic, *defines how the scene will be illuminated; other options are* Explicit *and* None

Header = NIL, *the header declaration*

HeaderFont = ["sans-serif", 12], *the header's font and font size*

HeaderAlignment = Center, *the header's location*

Footer = NIL, *the footer declaration*

FooterFont = ["sans-serif", 12], *the footer's font and font size*

FooterAlignment = Center, *the footer's location*

LegendVisible = FALSE, *determines whether the legend is visible or not*

LegendFont = ["sans-serif", 8], *the font and font size for the legend*

LegendPlacement = Bottom, *the location of the legend*

LegendAlignment = Center, *the alignment of the legend*

BackgroundTransparent = FALSE, *the transparency of the background*

BackgroundStyle = Flat, *the style of the background; other options are* TopBotom, LeftRight, Pyramid

BackgroundColor = [1.0, 1.0, 1.0], *the color of the background*

BackgroundColor2 = [0.75, 0.75, 0.75], *the second color of the background for non* Flat *styles*

Here is a very simple example showing how some of the above options will change the plot. Let us look at the result first, and then we will discuss the options used here.

```
• surf:=plot::Function3d(sin(x*y),
    x = -PI..PI, y = -PI..PI
  ):
  scene1:=plot::Scene3d(surf,
    YXRatio = 2,
    ZXRatio = 0.5,
    BackgroundStyle = Pyramid,
    CameraDirection = [-2,1,1]
  ):

  plot(scene1)
```

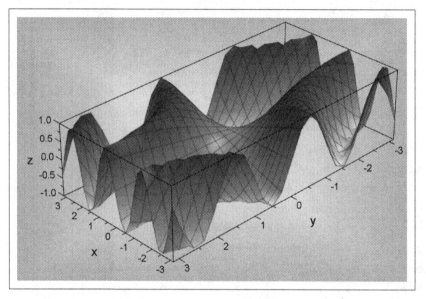

As you can clearly see, by using options YXRatio and ZXRatio, we were able to expand our graph in one direction and squeeze it in the vertical direction. By adding the declaration CameraDirection= [-2,1,1], we can control more precisely how the camera will be set up. The triplet [-2,1,1] here represents the coordinates of the default camera vector, i.e. the vector pointing along the camera axis from the camera focal point towards the camera's position. Note that this vector does not define the location of the camera. The default camera has many other options that were not listed or even mentioned here. Moreover, in MuPAD users can declare their own cameras. For more detailed information about camera declaration, you can check the help documentation, or simply wait until the next chapter.

6.3.4 The Coordinate System in 2D and 3D

There is one more object that needs to be mentioned while discussing 2D and 3D scenes. This is the coordinate system. Every time we create a new plot, the coordinate system will be added automatically. We do not need to worry about it as long as we do not require a more sophisticated coordinate system than the default one. Below are shown most of the options for CoordinateSystem2d and for CoordinateSystem3d. However, I will provide only very short explanations and only for a few of them. Discussing all these options is far beyond the scope of this book.

Options for CoordinateSystem2d

Viewing Box section

ViewingBox = [Automatic..Automatic, Automatic..Automatic], *the viewing box declaration*

ViewingBoxXMin = Automatic, *the viewing box X range declaration*

ViewingBoxYMin = Automatic, *the viewing box Y range declaration*

ViewingBoxXMax = Automatic, *the viewing box X range declaration*

ViewingBoxYMax = Automatic, *the viewing box Y range declaration*

ViewingBoxXRange = Automatic..Automatic, *the viewing box X range declaration*

ViewingBoxYRange = Automatic..Automatic, *the viewing box Y range declaration*

Scaling = Unconstrained, *the scaling used to display pictures, the other value is* Constrained

Axes of the coordinate system

CoordinateType = LinLin, *the type of coordinates used, in this case linear for all axes*

Axes = Automatic, *the type of axes used; other possible values are* Box, Frame, Origin, None

AxesTitleFont = ["sans-serif", 10], *the font and font size for labels on the axes*

AxesLineWidth = 0.18, *the line width for the axes*

AxesTips = TRUE, *use arrows for axis endings*

AxesOrigin = [0, 0], *sets the origin point for the axes*

AxesOriginX = 0, *sets the x-axis origin point*

AxesOriginY = 0, *sets the y-axis origin point*

AxesInFront = FALSE, *determines whether the axes are displayed in front of or behind the plotted objects*

AxesVisible = TRUE, *sets the visibility option for the axes*

XAxisVisible = TRUE,

YAxisVisible = TRUE,

AxesLineColor = [0.0, 0.0, 0.0], *sets the axes line colors*

AxesTitleAlignment = End, *sets the position of the axes titles*

XAxisTitleAlignment = End,

YAxisTitleAlignment = End,

AxesTitles = ["x", "y"], *sets labels for the axes*

XAxisTitle = "x",

YAxisTitle = "y",

YAxisTitleOrientation = Horizontal,

Grid section

GridLineColor = [0.6, 0.6, 0.6], *the grid color, in this case gray 60%*

GridLineWidth = 0.1, *sets the width of the grid lines*

GridLineStyle = Solid, *the style of the grid lines*

GridVisible = FALSE,

XGridVisible = FALSE,

YGridVisible = FALSE,

SubgridLineWidth = 0.1,

SubgridLineColor = [0.8, 0.8, 0.8],

SubgridLineStyle = Solid,

Ticks section

TicksLength = 2, *sets the length of the ticks*

TicksVisible = TRUE, *sets the visibility of the ticks*

TicksLabelsVisible = TRUE, *determines whether the ticks labels are displayed or not*

XTicksVisible = TRUE, *sets the visibility of the ticks on x-axis*

XTicksLabelsVisible = TRUE, *determines whether the ticks labels are displayed or not on the x-axis*

YTicksVisible = TRUE, *sets the visibility of the ticks on x-axis*

YTicksLabelsVisible = TRUE, *determines whether the ticks labels are displayed or not on the y-axis*

XTicksLabelStyle = Horizontal, *sets how the ticks labels will be displayed; other values are* Vertical, Diagonal, *and* Shifted

XTicksNumber = Normal,

YTicksLabelStyle = Horizontal, *sets how the ticks labels will be displayed; other values are* Vertical, Diagonal, *and* Shifted

YTicksNumber = Normal,

TicksLabelStyle = Horizontal,

TicksNumber = Normal,

TicksLabelFont = ["sans-serif", 8], *the font and font size for the ticks labels*

As you have noticed, we are dealing here with four groups of parameters — the viewing box, the axes of the coordinate system, the grid and the ticks on the axes.

The viewing box is one of the most important options for all types of plots. It will cause the plotted object to be displayed only in the declared area. For example, the declaration ViewingBox=[-2..2, -1..1] would force MuPAD to display only the area of the plot that is restricted by $x = -2..2$ and $y = -1..1$. Here is a very important information for the users of older versions of MuPAD. Older versions of MuPAD allowed users to declare range for the variable y when declaring a plot of a function $y = f(x)$. However, starting from version 3.0, we can only declare the range for the variable x, while any declaration for y will be treated as a parameter for animation. Therefore, in order to restrict the y range to $-1..1$, declare the ViewingBox or set ViewingBoxYRange=-1..1.

The viewing box is quite important when plotting objects with high peaks up or down. For example, it can be that a function has some discontinuities and its values are getting very large. In such a case, MuPAD tries to fit as much as possible of the given function on the graph. This leads to distorted pictures, where the most interesting parts of the graph are very small. By defining the range for the viewing box, we can force MuPAD to display only the part that is important to us.

The axes section contains options that allow us to define how the coordinate axes will look.

In the grid section, we can find options for the declaration of a grid on our pictures. This is very useful if we wish to read data from the pictures, especially the points of intersection of curves.

Finally, the ticks section describes how the ticks on the axes and their labels will be displayed.

The options for a 3D coordinate system are similar to the options for 2D. The most important difference is that there are three coordinate axes. Therefore, for any option that in the above table referred to axes 0X and 0Y, we would add an identical option for the axis 0Z. For example, having the 2D options,

XAxisVisible = TRUE,

YAxisVisible = TRUE

in the 3D coordinate system we would add,

ZAxisVisible = TRUE.

Another important change is to use, where appropriate, three coordinates instead of two. For example, the following option AxesTitles=["x","y"] was valid in 2D, but in 3D we should use the option AxesTitles=["x","y","z"].

6.4 Options Related to Primitives

In computer graphics, the word *primitives* means all the simplest building components to be plotted. Primitives are used to build more complex objects, etc. In MuPAD, we will consider as primitives the graphs of functions, surfaces, points, circles, other geometric objects, etc. Most of these primitives use similar options and option parameters. Therefore, it makes a lot of sense to talk about them within the same section.

In this section, I will describe some of the most common options. It would be quite difficult to develop a detailed survey of all the options available for each plot object, and certainly it would require more space than we can afford in this book. Therefore, I will only concentrate on the options that an average MuPAD user may need. There is another reason that allows me to skip a lot of details here. As you remember, in the earlier pages of this book, I promised to show you how we can change the values of all options without typing in a single word. Therefore, you only need to understand what we are talking about and to have a general idea about MuPAD plotting options.

Here is the list of the most common options for MuPAD primitives. Note that not all the options can be applied to any given primitive. For example, it doesn't make any sense to talk about line width when plotting points, or about the filling color when plotting the graph of a function.

6.4.1 Basic Options for Primitives

Color = [0.0, 0.0, 1.0], *the main color for most of 2D primitives*
Color = [1.0, 0.0, 0.0], *the main color for most of 3D primitives*
DiscontinuitySearch = TRUE, *checks function 2D for discontinuity*

Fill colors for patches in 3D plots

FillColor = [1.0, 0.0, 0.0], *the first basic color for plotting surfaces*
FillColor2 = [0.392193, 0.584307, 0.929395], *the second basic color for surfaces*
FillColorFunction = NIL, *here, you can declare a color function*
FillColorType = Dichromatic, *the type of filling used; it can be* Flat, Monochrome, Dichromatic, Rainbow *and* Functional
Filled = TRUE, *the patches are filled with color*

Legend options

Legend = NIL, *the legend for a given primitive*
LegendEntry = TRUE,
LegendText = NIL,

Line options

LineColor = [0.0, 0.0, 0.0, 0.25], *the first basic color for plotting lines*
LineColor2 = [1.0, 0.078402, 0.576495], *the second basic color for plotting lines*
LineColorFunction = NIL, *here, you can declare a color function*
LineColorType = Flat, *the type of coloring used for the line, can be* Flat, Monochrome, Dichromatic, Rainbow *and* Functional
LineStyle = Solid, *the style of the line, other options include* Dashed *and* Dotted
LineWidth = 0.35,

Mesh options

Mesh=121, Mesh = [25, 25], *the mesh values for 2D and 3D plots*
MeshVisible = FALSE,
Submesh = [0, 0],

Animation options

Frames = 50, *the number of frames in an animation*
ParameterName = NIL, *the name of the animation parameter*
ParameterBegin = NIL, *the starting value of the animation parameter*
ParameterEnd = NIL, *the ending value of the animation parameter*
ParameterRange = NIL, *the range for animation parameter (combines the two options above)*
TimeBegin = 0.0, *the starting time for the animation's time counter*
TimeEnd = 10.0, *the end time for the animation*
TimeRange = 0.0..10.0, *the time range for the animation (combines the two options above)*
VisibleAfterEnd = TRUE, *the animated object will be visible after finishing the animation*
VisibleBeforeBegin = TRUE, *the animated object will be visible before starting the animation*

Point options

PointSize = 1.5, *the default size of points, here 1.5 mm*
PointStyle = FilledCircles, *points will look like filled circles; other options are* Diamonds, FilledDiamonds, Squares, Filled Squares, Circles, FilledCircles, Crosses, XCrosses *and* Stars
PointsVisible = FALSE, *hide or display points in 3D objects mesh*

Title options

Title = NIL,
TitleAlignment = Center,
TitleFont = ["sans-serif", 11],
TitlePosition = NIL,
TitlePositionX = NIL,
TitlePositionY = NIL,
TitlePositionZ = NIL,

Asymptotes for graphs of functions

VerticalAsymptotesColor = [0.5, 0.5, 0.5], *the color of asymptotes*
VerticalAsymptotesStyle = Dashed, *the line style for asymptotes*
VerticalAsymptotesVisible = TRUE,
VerticalAsymptotesWidth = 0.2,

Now, when we know what options we may use in MuPAD plots, let us see what we may expect to get from these options.

6.4.2 Working with Color

Color is probably the most spectacular thing in MuPAD plots. This is especially the case with 3D plots. In a printed black and white book, there is no way to show what you can do with colors. Because of this, I will only show you here some very basic examples. You may find much better examples on the book web site, or you can develop them on your own.

Example 6.8

In this example, we will develop a color function and show how to use it to color a 3D surface.

In order to use the color scheme defined by a function, you have to define the function that will produce a list of three or four numbers between 0 and 1. For example:

- g :=
```
(x,y,z)->[abs(sin(x)),abs(sin(y)),abs(sin(z)),abs(sin(z))]
```
In such a color function, the produced values will describe the amount of red, green and blue color components using the RGB color space. The fourth value will be for surface opacity. Using the value 0 for this will make the patches of the surface completely transparent, and value 1 will make them completely opaque. Values between 0 and 1 will give the surface partial transparency.

In our example, we will produce a surface that is completely opaque.

- g := (x,y,z)->[abs(sin(x)), abs(sin(y)), abs(sin(z))]:
- f2 := plot::Function3d(
    ```
    x^2 + y^2, x = 0..2*PI, y = 0..2*PI,
    FillColorFunction = g
    ```
  ):

```
plot(f2)
```

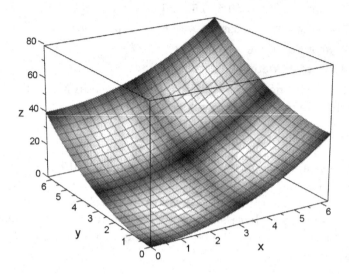

You may easily see that the function $g(x,y,z)$ fills the 3D space with a color pattern. Each point in 3D has its own defined color. The color of our surface is just the intersection of the surface with the color space defined by the function $g(x,y,z)$.   ■

In the above example, I used a simplified concept of color defined by a function. The general form of this definition is slightly more complex. Later, we will have many opportunities to play with color in MuPAD. For now, let us move on to other options related to objects.

6.4.3 Mesh and Other Options

Like many other mathematical packages, when plotting the graph of a curve, MuPAD first determines the number of points on the curve and then connects each pair of two consecutive points. The points used in this process are called a mesh. A very similar process is applied for plotting objects in 3D space. The syntax of the Mesh option for objects in 2D is

Mesh = #

where # is a positive integer determining the number of points used to divide the interval $a < x < b$. The points a and b are also counted as mesh points. Note that using this option for 2D graphs, we only give a single integer.

For 3D objects, the syntax is similar:

$$\text{Mesh} = [\ \#1,\ \#2\ ]$$

Example 6.9

The Mesh option is used to determine the accuracy of a plot. Let's see how it looks for a simple function. In this example, we will use the function $y = \sin(x)$ with three different mesh values: 5, 10, and 100.

- export(plot, Function2d):

```
F1 := Function2d(
    sin(x), x = 0..2*PI, y = -1..1,
    Mesh = 5,
    Color = RGB::Red,
    LineWidth = 0.5
):

F2 := Function2d(
    sin(x), x = 0..2*PI, y = -1..1,
    Mesh = 10,
    Color = RGB::Blue,
    LineWidth = 0.25
):

F3 := Function2d(
    sin(x), x = 0..2*PI, y = -1..1,
    Mesh = 100,
    Color = RGB::Black,
    LineWidth = 0.15
):

plot(F1, F2, F3)
```

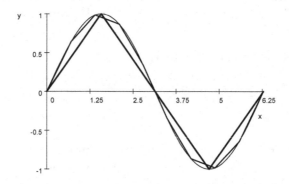

The obtained graph shows clearly that small values for Mesh produce very inaccurate plots. However, the larger the value of Mesh, the longer we have to wait for a plot. ■

Option LineWidth

Apart from color, we can also use different line thicknesses to distinguish between different curves. This can be useful when creating, for example, lecture notes that will be printed in black and white, where it makes more sense to use a different thickness for individual lines than to use different colors. The syntax for LineWidth is

LineWidth = #

where # is a real number in the default unit of measurement. The default value for this option is 0.35 mm.

Option LineStyle

LineStyle is another option that allows us to distinguish between different curves in the same plot. This option can only have three values: Solid, Dashed and Dotted. For most of the objects the default value is Solid.

Option Filled

The option Filled can be applied only to objects that are represented by a closed curve, like an ellipse or a polygon. You can choose between two values: TRUE and FALSE, where TRUE is the default value for majority of objects.

Options Title and TitlePosition

We can attach a title to any graph. The title can be declared as follows:

Title = "*Here goes the title of my graph*"

and its position,

TitlePosition = [#1, #2]

The numbers #1 and #2 are Cartesian coordinates of the title on the plot coordinate system. This way, we can add additional information to our plots.

Here is a very simple example showing the use of the Title option.

```
G := plot::Function2d(sin(x), x = -PI..PI,
    Title = "sin(x)",
    TitlePosition = [2.6, 0.8]
):
H := plot::Function2d(cos(x), x = -PI..PI,
    Title = "cos(x)",
    TitlePosition = [2.8,-0.8]
):
plot(G,H)
```

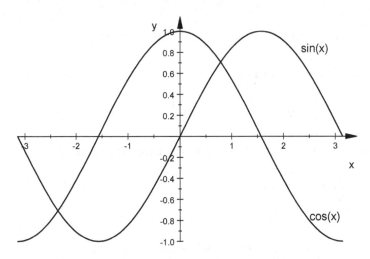

6.5 Operations on Graphical Objects

In this section, we will concentrate on a few operations that can be performed on existing graphical objects. Such operations can be used to translate, rotate and scale objects, or just to produce a copy of an object and modify some of its properties.

Imagine that you have an object, and you wish to make 10 almost identical copies of it. Of course, you could always use the cut-and-paste method to make 10 instances of your object and then modify each of them separately. However, your notebook would grow significantly, and you would have to scroll through many screens to see the whole notebook. Instead of this, you can use a specific command to make a copy of the object and modify only what you really need. The two operations copy and modify are very useful for this. In fact, modify is just a more convenient version of copy, so we will only cover here the modify operation.

6.5.1 Operation modify

The operation modify is a part of the plot library. We can export it just like any other plot command, or use with the plot:: prefix. Its syntax is very simple:

modify(*existing object, modifications*)

For example, suppose that we have declared an object:

```
• f := plot::Function2d(
    sin(x), x = -PI..0,
    LineStyle = Solid,
    LineWidth = 0.15,
    Color = [0, 0, 0]
  ):
```

Now we can use function f to make another function g

```
g := plot::modify(f,
    XRange = 0..PI,
    LineWidth = 1,
    Color = [0.5,0.5,0.5]
  ):
```

and finally, we can plot both functions in one graph:

```
plot(f,g)
```

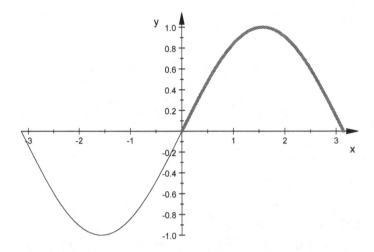

The situation presented here was not especially complex, but you may find the function modify very useful when plotting multiple curves on the same plot.

In later chapters I will show how we can translate, rotate or scale our objects.

6.6 Chapter Summary

In this chapter, I introduced the major principles of MuPAD graphics. We only discussed a few basic commands for plotting mathematical objects. However, we spent a lot of time talking about the plotting properties of the objects — the so-called options, and the possible values of these options. Let's summarize all the things that you learned in this chapter.

6.6.1 MuPAD Command Summary

plotfunc2d($f_1(x)$, $f_2(x)$, ...,$f_n(x)$, $x = x1..x2$, $y = y1..y2$)

plotfunc3d($f_1(x,y)$, $f_2(x,y)$, ..., $f_n(x,y)$, $x = x1..x2$, $y = y1..y2$)

export(plot) - *exports the whole* plot *library*

export(plot, Function2d, Point2d) - *exports* Function2d *and* Point2d *only*

plot::copy(*existing object*) - *makes a copy of an existing object*

plot::modify(*existing object, new options*) - *a copy of an object with new options*

6.6.2 MuPAD Syntax

Declaring Canvas, Scene2D **and** Scene 3D

> plot::Canvas(objects,options)
>
> plot::Scene2d(objects, options)
>
> plot::Scene3d(objects, options)

Options for Canvas

OutputUnits = mm, *the default unit for all plots*

Height = 80, *the height of canvas*

Width = 120, *the width of canvas*

Layout = Tabular, *how multiple plots on the same canvas will be placed*

Columns = 0, *the number of columns on canvas*

Rows = 0, *the number of rows on canvas*

Spacing = 1, *the distance between scenes on the same canvas, in mm*

BackgroundColor = [1.0, 1.0, 1.0], *the color for the background*

BorderColor = [0.5, 0.5, 0.5], *the color for the border*

BorderWidth = 0, *the size of the border, in mm*

Margin = 1, *the width of the canvas margin, in mm*

BottomMargin = 1, *the width of bottom margin, in mm*

LeftMargin = 1, *the width of left margin, in mm*

RightMargin = 1, *the width of right margin, in mm*

TopMargin = 1, *the width of top margin, in mm*

Header = NIL, *the canvas header declaration*

HeaderFont = ["sans-serif", 12], *the font for the header and its font size in points*

HeaderAlignment = Center, *the location of the header*

Footer = NIL, *the canvas footer declaration*

FooterFont = ["sans-serif", 12], *the font for the footer and its font size in points*

FooterAlignment = Center, *the location of the footer*

Options for Scene2d

BorderWidth = 0, *the width of the scene border*

BorderColor = [0.5, 0.5, 0.5], *the color of the scene border*

Height = 80, *the height of the scene*

Width = 120, *the width of the scene*

Margin = 1, *the width of the margin, in mm*

BottomMargin = 1, *the width of the bottom margin, in mm*

LeftMargin = 1, *the width of the left margin, in mm*

RightMargin = 1, *the width of the right margin, in mm*

TopMargin = 1, *the width of the top margin, in mm*

Header = NIL, *the header's text declaration*

HeaderFont = ["sans-serif", 12], *the header's font declaration*

HeaderAlignment = Center, *the position of the header*

LegendVisible = FALSE, *determines whether the legend is visible or not*

LegendFont = ["sans-serif", 8], *the font and font size for the legend*

LegendPlacement = Bottom, *the vertical position of the legend*

LegendAlignment = Center, *the horizontal position of the legend*

Footer = NIL, *the footer declaration*

FooterFont = ["sans-serif", 12], *the footer's font*

FooterAlignment = Center, *the alignment of the footer*

BackgroundTransparent = FALSE, *the transparency of the background*

BackgroundColor = [1.0, 1.0, 1.0], *the background color, in this case white.*

Options for Scene3d

BorderWidth = 0, *the width of the border*

BorderColor = [0.5, 0.5, 0.5], *the color of the border*

Width = 120, *the width of the scene in mm*

Height = 80, *the height of the scene in mm*

Margin = 1, *the width of the margin*

TopMargin = 1, *the width of the top margin*

RightMargin = 1, *the width of the right margin*

BottomMargin = 1, *the width of the bottom margin*

LeftMargin = 1, *the width of the left margin*

CameraDirection = NIL, *the direction vector of the camera*

CameraDirectionX = NIL, *the x coordinate of the direction vector*

CameraDirectionY = NIL, *the y coordinate of the direction vector*

CameraDirectionZ = NIL, *the z coordinate of the direction vector*

YXRatio = 1, *the ratio of the scene in the plane XY*

ZXRatio = 2 / 3, *the ratio of the scene in the plane XZ*

Lighting = Automatic, *defines how the scene will be illuminated; other options are* Explicit *and* None

Header = NIL, *the header declaration*

HeaderFont = ["sans-serif", 12], *the header's font and font size*

HeaderAlignment = Center, *the header's location*

Footer = NIL, *the footer declaration*

FooterFont = ["sans-serif", 12], *the footer's font and font size*

FooterAlignment = Center, *the footer's location*

LegendVisible = FALSE, *determines whether the legend is visible or not*

LegendFont = ["sans-serif", 8], *the font and font size for the legend*

LegendPlacement = Bottom, *the location of the legend*

LegendAlignment = Center, *the alignment of the legend*

BackgroundTransparent = FALSE, *the transparency of the background*

BackgroundStyle = Flat, *the style of the background; other options are* TopBotom, LeftRight, Pyramid

BackgroundColor = [1.0, 1.0, 1.0], *the color of the background*

BackgroundColor2 = [0.75, 0.75, 0.75], *the second color of the background for non* Flat *styles*

Options for CoordinateSystem2d

Viewing Box section

ViewingBox = [Automatic..Automatic, Automatic..Automatic], *the viewing box declaration*

ViewingBoxXMin = Automatic, *the viewing box X range declaration*

ViewingBoxYMin = Automatic, *the viewing box Y range declaration*

ViewingBoxXMax = Automatic, *the viewing box X range declaration*

ViewingBoxYMax = Automatic, *the viewing box Y range declaration*

ViewingBoxXRange = Automatic..Automatic, *the viewing box X range declaration*

ViewingBoxYRange = Automatic..Automatic, *the viewing box Y range declaration*

Scaling = Unconstrained, *the scaling used to display pictures, the other value is* Constrained

Axes of the coordinate system

CoordinateType = LinLin, *the type of coordinates used, in this case linear for all axes*

Axes = Automatic, *the type of axes used; other possible values are* Box, Frame, Origin, None

AxesTitleFont = ["sans-serif", 10], *the font and font size for labels on the axes*

AxesLineWidth = 0.18, *the line width for the axes*

AxesTips = TRUE, *use arrows for axis endings*

AxesOrigin = [0, 0], *sets the origin point for the axes*

AxesOriginX = 0, *sets the x-axis origin point*

AxesOriginY = 0, *sets the y-axis origin point*

AxesInFront = FALSE, *determines whether the axes are displayed in front of or behind the plotted objects*

AxesVisible = TRUE, *sets the visibility option for the axes*

XAxisVisible = TRUE,

YAxisVisible = TRUE,

AxesLineColor = [0.0, 0.0, 0.0], *sets the axes line colors*

AxesTitleAlignment = End, *sets the position of the axes titles*

XAxisTitleAlignment = End,

YAxisTitleAlignment = End,

AxesTitles = ["x", "y"], *sets labels for the axes*

XAxisTitle = "x",

YAxisTitle = "y",

YAxisTitleOrientation = Horizontal,

Grid section

GridLineColor = [0.6, 0.6, 0.6], *the grid color, in this case gray 60%*

GridLineWidth = 0.1, *sets the width of the grid lines*

GridLineStyle = Solid, *the style of the grid lines*

GridVisible = FALSE,

XGridVisible = FALSE,

YGridVisible = FALSE,

SubgridLineWidth = 0.1,

SubgridLineColor = [0.8, 0.8, 0.8],

SubgridLineStyle = Solid,

Ticks section

TicksLength = 2, *sets the length of the ticks*

TicksVisible = TRUE, *sets the visibility of the ticks*

TicksLabelsVisible = TRUE, *determines whether the ticks labels are displayed or not*

XTicksVisible = TRUE, *sets the visibility of the ticks on x-axis*

XTicksLabelsVisible = TRUE, *determines whether the ticks labels are displayed or not on the x-axis*

YTicksVisible = TRUE, *sets the visibility of the ticks on x-axis*

YTicksLabelsVisible = TRUE, *determines whether the ticks labels are displayed or not on the y-axis*

XTicksLabelStyle = Horizontal, *sets how the ticks labels will be displayed; other values are* Vertical, Diagonal, *and* Shifted

XTicksNumber = Normal,

YTicksLabelStyle = Horizontal, *sets how the ticks labels will be displayed; other values are* Vertical, Diagonal, *and* Shifted

YTicksNumber = Normal,

TicksLabelStyle = Horizontal,

TicksNumber = Normal,

TicksLabelFont = ["sans-serif", 8], *the font and font size for the ticks labels*

Object options

Color = [1.0, 0.0, 0.0], *the main color for all 3D primitives*

DiscontinuitySearch = TRUE, *checks function 2D for discontinuity*

Fill colors for patches in 3D plots:

FillColor = [1.0, 0.0, 0.0], *the first basic color for plotting surfaces*

FillColor2 = [0.392193, 0.584307, 0.929395], *the second basic color for surfaces*

FillColorFunction = NIL, *here, you can declare a color function*

FillColorType = Dichromatic, *the type of filling used; it can be* Flat, Monochrome, Dichromatic, Rainbow *and* Functional

Filled = TRUE, *the patches are filled with color*

Legend options

Legend = NIL, *the legend for a given primitive*

LegendEntry = TRUE,

LegendText = NIL,

Line options

LineColor = [0.0, 0.0, 0.0, 0.25], *the first basic color for plotting lines*

LineColor2 = [1.0, 0.078402, 0.576495], *the second basic color for plotting lines*

LineColorFunction = NIL, *here, you can declare a color function*

LineColorType = Flat, *the type of coloring used for the line, can be* Flat, Monochrome, Dichromatic, Rainbow *and* Functional

LineStyle = Solid, *the style of the line, other options include* Dashed *and* Dotted

LineWidth = 0.35,

Mesh options

Mesh=121, Mesh = [25, 25], *the mesh values for 2D and 3D plots*

MeshVisible = FALSE,

Submesh = [0, 0],

Animation options

Frames = 50, *the number of frames in an animation*

ParameterName = NIL, *the name of the animation parameter*

ParameterBegin = NIL, *the starting value of the animation parameter*

ParameterEnd = NIL, *the ending value of the animation parameter*

ParameterRange = NIL, *the range for animation parameter (replaces the two options above)*

TimeBegin = 0.0, *the starting time for the animation's time counter*

TimeEnd = 10.0, *the end time for the animation*

TimeRange = 0.0..10.0, *the time range for the animation (replaces the two options above)*

VisibleAfterEnd = TRUE, *the animated object will be visible after finishing the animation*

VisibleBeforeBegin = TRUE, *the animated object will be visible before starting the animation*

Point options

PointSize = 1.5, *the default size of points, here 1.5 mm*

PointStyle = FilledCircles, *points will look like filled circles; other options are* Diamonds, FilledDiamonds, Squares, Filled Squares, Circles, FilledCircles, Crosses, XCrosses *and* Stars

PointsVisible = FALSE,

Title options

Title = NIL,
TitleAlignment = Center,
TitleFont = ["sans-serif", 11],
TitlePosition = NIL,
TitlePositionX = NIL,
TitlePositionY = NIL,
TitlePositionZ = NIL,

Asymptotes for graphs of functions

VerticalAsymptotesColor = [0.5, 0.5, 0.5], *the color of asymptotes*
VerticalAsymptotesStyle = Dashed, *the line style for asymptotes*
VerticalAsymptotesVisible = TRUE,
VerticalAsymptotesWidth = 0.2,

6.7 Programming Exercises

1. Use the procedure plotfunc2d to obtain the graphs of the functions $y = x^2$ and $y = \sqrt{2x}$ in the same picture. By narrowing down the range of the plot, zoom in towards the point of intersection of both curves. Obtain the approximate coordinates of this point with an error no larger than 0.001.

2. Use plotfunc2d to obtain a graph of the function $f(x) = \dfrac{16 + \sin x}{3 + \sin x}$. Try to guess the equations of the two tangent lines that do not cross the curve. Plot $f(x)$ and both lines in one picture.

3. Use plotfunc2d to obtain a graph of the function $h(x) = x - 2\cos x$. Try to guess the equations of the two tangent lines. Plot $h(x)$ and its tangent lines in one picture.

4. Use plotfunc2d to show that the lines $y = x$ and $y = -x$ are tangent to the curve $y = x\sin(\frac{1}{x})$.

5. Use plotfunc2d to obtain the graphs for the functions $y = x^2 \sin(\frac{1}{x})$, $y = x^2$ and $y = -x^2$ in the same picture.

6. Use plotfunc2d to show that the equation $\cos x = -\frac{1}{5}x$ has exactly three solutions. Find their approximate values.

7. Use `plotfunc3d` to plot:

   a. a parabolic cylinder $z = 4 - x^2$

   b. a hyperbolic paraboloid $z = y^2 - x^2$

   c. a paraboloid $z = x^2 + y^2$

8. Use `plotfunc3d` to obtain the graph of a hyperboloid of two sheets $z^2 - x^2 - y^2 = 1$. Hint: express each sheet as a separate function.

9. In problems a-e, work out how the graph changes when the value of u changes. Use $u = -3, -2, -1, 0, 1, 2, 3$. You can also develop animation of graphs where u is the animation parameter.

   a. $f(x) = x^3 - 3x + u$

   b. $f(x) = x^3 + ux$

   c. $f(x) = x^3 + ux^2$

   d. $f(x) = x^4 + ax^2$

   e. $f(x) = x^5 + ux^3$

10. Plot a graph of the function $f(x,y) = x^2 + y^2 - x^2y^2$. Use planes $y = a$, where $a = 0, 5, 10$, to see the level curves of the function $f(x,y)$.

11. Plot a graph for $f(x,y) = \dfrac{xy}{x^2 + y^2 + 0.1}$. Use different surface shading options in order to see the changes of the graph near point $(x,y) = (0,0)$.

Chapter 7 _____

Interactive Graphics & Animation with VCam

In the previous chapter, I mentioned many times that you do not need to learn all these plotting options, and that I will show you how to deal with them without using any of the keywords and parameters that we had discussed. Now it is time for me to fulfill my promise. So, let us start exploring interactive graphics development with MuPAD. Certainly, you will still need to type in at least some basic statements where you will declare the objects to be plotted, but this is really all. The rest you will do in interactive mode, by experimenting in the Virtual Camera environment.

7.1 Learning the Virtual Camera Environment

Let us start with a very basic plot. In fact, it's not really important what we're going to plot. We will use the plotted object as an opportunity to explore the Virtual Camera (VCam) environment. So, let us begin with a simple 3D object:

- ```
MySurface := plot::Function3d(
    sin(x*y)/(1+x^2+y^2), x=-2*PI..2*PI, y=-2*PI..2*PI):
plot(MySurface)
```

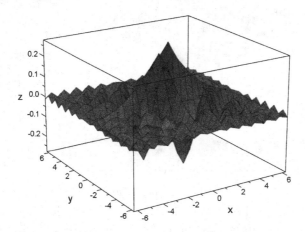
```

If you typed in exactly the above code, the result was a picture like the one above. Now, right click on the picture, i.e. use the right button of the mouse. Doing so will give you a small pop-up menu with most of the options related to the current MuPAD region. Do not worry about them. Choose the very last option, **Graphics Object**. Now you have a choice of two sub-options — **Edit** and **Open**. Both options are similar, but there are some important differences which make the **Edit** option useful for quickly fixing up plots and the **Open** option useful for more serious experimenting. For now you need to know that **Edit** is also accessible by double clicking the picture and it opens the graph inside the current notebook, while **Open** is accessible only through the menu and it opens the graph as a separate file outside the notebook.

Let us choose the option **Open** and see what we got. For most graphs, the VCam window looks like the one shown in Fig. 7.1. In many cases, however, you will observe slight and sometimes even significant differences.

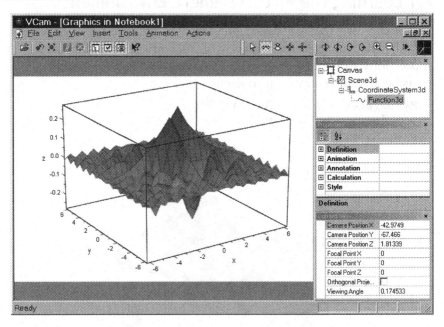

Fig. 7.1 Virtual Camera Environment

In our example, we have a window with four panels — the main one to the left shows the graph, while the remaining three on the right side show the objects used in our scene, object properties and the current view parameters. We will call them—the graph window, objects browser, properties inspector and current view panel respectively.

You can easily see that the objects browser shows the scene in the form of a tree, where you can easily work out the relations between objects in the plot. In the case of our picture, we have the canvas, the scene, the coordinate system and finally the plotted object. The names of all the objects are exactly the same as the names of the plot classes that we use while writing the plot object declarations. If you develop a picture with multiple scenes and multiple objects in each scene, the tree will be more complex, but you will still be able to find out where a given object belongs. A more complex tree can be displayed in a collapsed form, where parts of the tree are hidden. You will be able to see which parts are shown or hidden by looking at the $\boxed{+}$ and $\boxed{-}$ signs to the left of the given object name. The sign $\boxed{-}$ means that there is nothing hidden, while the sign $\boxed{+}$ means that this object is not displaying all of its components. Click on the $\boxed{+}$ sign to expand this part of the tree, and click on $\boxed{-}$ to collapse the tree structure for the given object.

Now, let us take a look at the right-middle panel, the so-called properties inspector. It shows the current properties of the selected object. Click on a few different objects in your scene to see what properties they have. You will notice that properties of most objects are combined in a few groups. In most examples, we're likely to have the following groups—**Definition**, **Animation**, **Annotation**, **Calculation** and **Style**. However, the canvas and scene objects, for example, have only **Annotation**, **Layout** and **Style**, while the coordinate system object has four different groups—**Definition**, **Axes**, **Tick marks** and **Grid lines**. In general, the meaning of these names is as follows:

Property groups for canvas, scene and coordinate system

Definition - *a declaration of the coordinate system, viewing box, scaling*

Annotation - *the explanatory features added to the picture - legend, headers, etc.*

Layout - *the layout of the canvas and scene, size, number of rows, etc.*

Style - *the style of the display, colors, backgrounds,*

Axes - *declarations for the axes of the coordinate system*

Tick marks - *declarations of the tick marks on axes*

Grid lines - *declarations of the grid lines on the picture*

Property groups for primitives

Definition - *the definition of the primitive, for example its equation, range of variables*

Animation - *the declaration of the parameters for an animation*

Annotation - *the explanatory features added to the picture like the title or legend*

Calculation - *the parameters used to calculate the object,*

Style - *colors, the thickness of the lines, points, etc.*

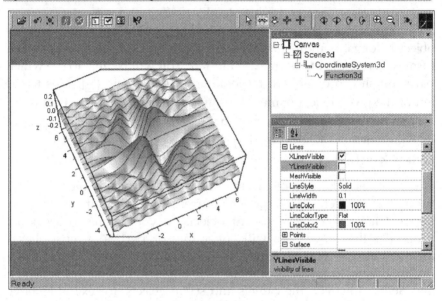

Fig. 7.2 Changing plot object properties

Like in the objects browser, you can click on the ⊞ sign to expand the given group and then change whatever you need or wish to alter. Try a few of these options and see what you can do with them. For example, for Function3d, expand the **Style** group and then the **Lines** subgroup. Now change LineWidth to 0.1 and LineColor to black 100% (no transparency at all), choose only XLinesVisible and then rotate the picture in the graph window. You will get something like figure 7.2. You could go completely wild and make your picture very fancy. It is very important to know what you can really produce using VCam. Notice that all these keywords that you see in the objects browser and properties inspector are the ones that you would normally enter in the declarations of the plot objects.

Before we finish our experiments with VCam, let us see what else we have here. On top of the screen there is a long toolbar, shown in Fig. 7.3.

Fig. 7.3 Virtual Camera toolbar

You can clearly see that we have three groups of icons here. The left one contains various functions—open a new VCam file, reset the viewpoint, fit the contents of the graph to the area of the picture, recalculate the plot (exclamation mark), stop the calculations, and show or hide the objects browser, properties inspector and current view panel.

The second group, in the middle, you will use frequently to switch the mode of manual operations. There are five icons here for five different modes—select an object, rotate, zoom in and zoom out, move the object, and finally get the coordinates of a point on the plotted object.

The last group of icons gives you access to a few basic animations without declaring the animation parameters. We have here rotations in four different directions, an animated zoom in/zoom out, and an icon to select the speed of the animation.

Note that in our current example, we do not have the animation toolbar that we had already used in chapter 6. This is is because in this picture, we did not declare any animation. However, we can still do it. For example, click on Function3D in the objects browser, then go to its properties inspector and expand the **Definition** group (see Fig. 7.4).

Here change the equation to sin(a*x*y)/(x^2+y^2+1), by adding the parameter "a" inside the sine function. In the same properties inspector, type in the letter "a" for ParameterName, then set ParameterBegin to 0 and ParameterEnd to 1 (see Fig. 7.4). That's all. You have declared your first animation of a 3D surface. Now you have to ask MuPAD to produce it for you. Look on the left side of the VCam

toolbar. You will see a button with the exclamation character .

Right now the button is green instead of gray, which means that the current scene should be recalculated. Click on the green button and wait a few minutes or seconds, depending on how fast your computer is. During this time, MuPAD will produce a short movie containing 50 frames of your animation. When the calculations are finished, the VCam screen will be refreshed and the animation toolbar will appear on top of it. This is all. Now you can press the play button and enjoy the animation.

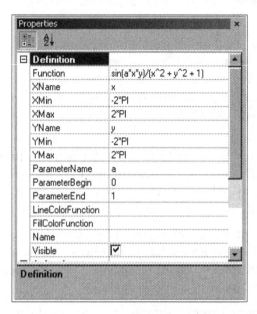

Fig. 7.4 Setting animation parameters

Before we finish this experiment, it would be useful if you would spend some time to explore the VCam menu. There are a few options here that are not available through the buttons on the toolbar. At the moment, you may wish to save your animation as a separate file. In order to do this, go to the **File** menu, choose **Export** and then save it as firstanim.xvz. This will produce a separate file with just this animation. Later on, you can use it without even starting MuPAD. Just click on the name of the file and VCam will open it for you. This is how I store some of my most interesting graphical examples. Note however, while opening an *.xvz file in VCam you cannot recalculate the scene, so you cannot change many things in your graph.

Finally, if you wish to go back to your notebook, close the VCam window and the graph in your notebook will be updated.

Before proceeding to the next section, you can try comparing the two VCam modes, the one that you have learned through the **Open** menu option, and the other one accessible through the **Edit** option. Double-click on the picture in the notebook. This will take you to the **Edit** mode. Here you will also find most of the options that you have learned just a while ago. There will be some differences in the menu, however, and of course the picture will be displayed with the background of the notebook, so you will still see the declarations for the plot object.

7.2 Using VCam to Improve 2D Graphs

In many applications, we have to plot multiple graphs of functions in one plot. Quite often, we deal with functions that have both very small and very large values. This makes it especially difficult to develop a good picture that shows everything we need. Let us take a look at one such example.

Example 7.1

A product of two functions can be investigated by examining the properties of the individual functions. For example, let us consider the function $f(x) = e^x \sin 2x$ or even the family of functions $f_n(x) = e^x \sin nx$, where $n = 1, 2, 3, 4, \ldots$. We know that for any real number x, $e^x > 0$ and $|\sin nx| \leq 1$. Therefore, we can easily conclude that $|e^x \sin nx| \leq e^x$ and

$$-e^x \leq e^x \sin nx \leq e^x$$

We also know that if $nx = k\pi + \frac{\pi}{2}$, then $\sin nx = 1$ and $e^x \sin nx = e^x$. This means that the graph of e^x touches each of the curves $f_n(x)$. We can also conclude that the graph of $-e^x$ touches the bottom part of each of the curves $f_n(x)$. Let us see how we can display this fact on a graph.

First, we need to create two constants that will serve as a range for the plotted functions.

- `L1 := -2*PI:`
  `L2 := +2*PI:`

In the next step, we will create three objects representing the functions $f_n(x)$, where $n = 1, 2$ and 3. We can declare each of these functions separately or declare all of them using the sequence operator.

- ```
  Functions := plot::Function2d(
      exp(x)*sin(n*x), x=L1..L2
  ) $ n=1..3:
  ```

Finally, we need to declare plot objects for the functions e^x and $-e^x$.

```
TopFunction := plot::Function2d( exp(x), x=L1..L2):
BotFunction := plot::Function2d(-exp(x), x=L1..L2):
```

Now, when all the components of our plot are ready, we can produce it.

- ```
  plot(Functions, TopFunction, BotFunction)
  ```

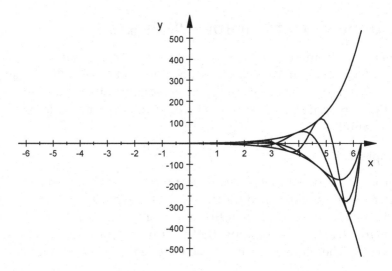

From this picture, it is very difficult to tell which line represents which function. The left part of the picture is completely useless. The right part is very crowded. In fact, there is no way to keep both the left and right parts of this plot in the same picture. The right part with its large values will always make the left part small and completely unreadable. Perhaps it wouldn't be a bad idea to start the whole plot from $x = 0$, or even to split the plot into two separate plots. For this exercise, let us concentrate on the right part of the picture. This means we have to set L1:=0 and perhaps L2:=PI. So, the complete program for our experiments is as follows:

- ```
  L1 := 0: L2 := PI:
  Functions := plot::Function2d(
      exp(x)*sin(n*x), x=L1..L2
  ) $ n=1..3:
  ```
  ```

```
TopFunction := plot::Function2d( exp(x), x=L1..L2):
BotFunction := plot::Function2d(-exp(x), x=L1..L2):

plot(Functions, TopFunction, BotFunction)
```

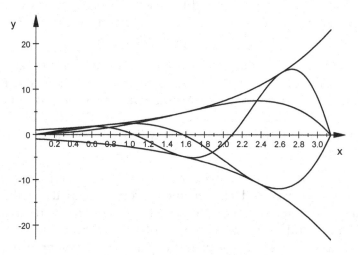

This plot is much better, although we had to sacrifice the left part of it. You can always explore the left part of the graph later. Now, open the plot using the **Open** option and check the structure of the objects in the objects browser. You should have the same structure as the one presented in Fig. 7.5.

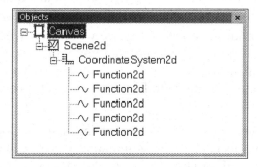

Fig. 7.5 Objects used in the exercise

Note that we have here five Function2d objects, where the last two are for e^x and e^{-x}. Now, choose the last function in the objects browser. You will notice that one of the curves in the main window has a gray shadow. This is the line that you have selected. In the properties inspector, expand the **Style** group and change the LineWidth to 0.5 and LineColor to Gray50. Collapse the **Style** group and expand the **Annotation** group. Here, in the title and legend texts, type in the formula -exp(x) and then set TitlePositionX=3.05 and TitlePositionY=-18.5. This will place the text "-exp(x)" somewhere near the curve in the bottom right part of the graph.

Perform similar operations with the second-last Function2D object. Select it in the objects browser, set its color to gray 50% and its line width to 0.5. Then expand the **Annotation** group for this curve and declare the title and legend texts as exp(x). Finally, set the title position to $x = 3.1$ and $y = 18.5$. This will be all you need to do with both functions.

Using the same way as above, choose each of the remaining three Function2d objects and change their properties to those that you think are best. For example, you can set their colors to red, green and blue. You should add a title to each of these curves. Do you know which of them is which? Select and expand the **Definition** group, and in the first row you will find the formula of the selected object. So, it is easy to find out which function you have selected in the picture.

When declaring the position for the title of each function, you may use the query tool, that is, the button with a cross-like figure at the right end of the VCam toolbar . Now, your mouse pointer has changed to the cross. Click on the picture while pressing the right mouse button, and move it to the place where you wish to place the text of the title. Near the mouse pointer, you will see a small rectangle with the coordinates of this point. Write them down and use them to set the position of the title text. Don't forget to also check the legend entry and add a legend text.

Finally, it is time to make some global changes. In the objects browser, select Scene2d and expand the **Annotation** group. Check the option LegendVisible. While here, you can also decide where the legend will be placed. Now, select CoordinateSystem2d in the objects browser and expand the **GridLines** group. Here check the options XGridVisible and YGridVisible. You can also check the options XSubgridVisible and YSubgridVisible. You may also make some other changes that you think will improve your picture. However, I believe that the picture we've got is good enough to illustrate our example. Fig. 7.6 shows what we have finally produced.

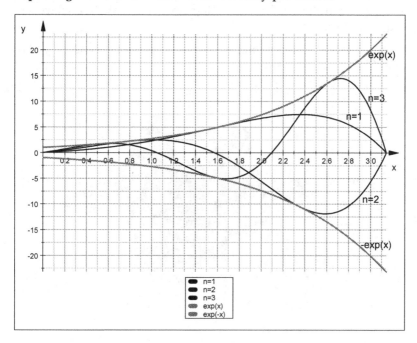

Fig. 7.6 The final plot after changes in VCam

There is no doubt that we still have a lot of opportunities to improve the above picture. For example, we can add a title to this scene or change the type of the coordinate system used. Feel free to experiment with all the possibilities and to implement all the possible ideas you find relevant.   ∎

7.3 Using VCam to Improve 3D Graphs

3D plots, like 2D plots, quite often require some careful treatment before we will be able to use them as illustrations for mathematical problems. Let us explore a very simple example and see what we may face while developing 3D plots.

Example 7.2

One of the nicest examples of 3D surfaces that we use in calculus is the surface of the function $f(x,y) = (x^2 - y^2) \cdot e^{-x^2-y^2}$. We prove, for example, that the function attains two local minimum values and two local maximum values. Let us produce a graph of this function and show how to emphasize its interesting properties. We will start with a very simple declaration:

- ```
  FExp := plot::Function3d((x^2 - y^2)*exp(-x^2 - y^2)):
  plot(FExp)
  ```

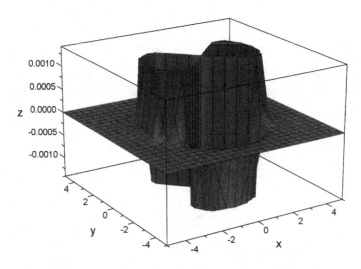

The first thing we notice is that the graph doesn't show the shape that we know from calculus textbooks. This is because we did not declare ranges for the variables x, y and MuPAD has choosen bad ranges. However, just looking at the picture, we can easily see that taking x and y from the interval $[-3, 3]$ will be enough for us. Therefore, let us add the range for both variables and plot the graph again. We may also add a horizontal plane to our picture, which we will later use to make horizontal cross-sections of the graph.

- ```
  FExp := plot::Function3d(
      (x^2 - y^2)*exp(-x^2 - y^2), x=-3..3, y=-3..3
  ):

  PL := plot::Function3d(0, x=-3..3, y=-3..3):

  plot(FExp, PL)
  ```

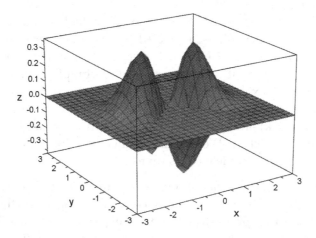

This time we have produced something that more closely resembles pictures from our calculus textbooks. There are still a few things that we may want to improve, however. For example, the spiky parts at the top and bottom of the graph do not look good. I would also like to use the plane to show the level curves of the surface, so that I can find out the maximum and minimum values of the function from the picture. So, let us open up VCam (right mouse click, choose the option **Graphics Objects** from the menu, and then **Open**) and try to improve the graph as well as to add some missing features.

First, let us for a short while remove the plane from the picture. This will allow us to experiment with the graph of $f(x,y)$. In the objects browser, choose the second Function3d and then expand the **Definition** group in the properties inspector. Here uncheck the option Visible. This will hide the plane from the picture.

Now lets us improve the spiky parts of the graph. In the objects browser, select the first Function3d and then expand its **Calculation** group in the properties inspector. Here, change XSybmesh and YSubmesh values from 0 to 1. This will add one point between each two consecutive points of the current mesh. For this example, it will be just enough. For other, more complex examples, you may need to use larger values. Note that in our plot we have 25 mesh lines in each direction. Such a change is therefore very significant, because by adding 1 more point in each mesh segment, we add 24 new lines in each direction. These lines will not be displayed but they will still exist. This means that MuPAD will have to calculate the value of the function for $(25 + 24) \cdot (25 + 24) = 2401$ points, where originally it only had to calculate it for $25 \cdot 25 = 625$ points. This makes a significant difference in the time it takes to produce the graph. If you add more points, you will extend the time of calculations quite significantly. Now, press the recalculate button in VCam (the green button with an exclamation mark) and wait for the new graph. Now we've got a nice looking, smooth picture. We can also change its colors, and a few other minor parameters, but I leave this up to you.

In the next stage, let us change the coordinate system, so that it will be more readable. Choose the CoordinateSystem3d in the objects browser, and in its properties inspector expand the group **GridLines**. Here, check all the options for grid and subgrid visibility. This will display a gray grid behind and under the graph. You can change the color of the grid and the width of the line, for example GridLineWidth=0.2. This will make the grid a bit stronger. Next, you can expand the **Axes** group and change the width of the axes to AxesLineWidth=0.3 and change their color, for example to red. This will distinguish them from the rest of the picture. This should give you a picture similar to this one.

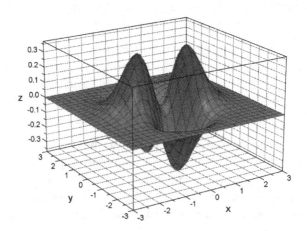

In the next step, we will try to make use of the plane that at the moment is hidden. Select the second Function3d and check the Visible option in its **Definition**. We've got our function back on the plot. Now, in the properties inspector, expand the **Style** group and next the **Surface** subgroup. Here change the FillColorType to monochrome and FillColor to any color that you like but with only 50% opacity. It may take a while to find the right color, but you will eventually get it.

In the final stage, we are going to add animation to show the level curves. Let us expand again the **Definition** group for the plane, change the equation here to a single letter "a", and add the animation parameter declarations ParameterName=a, ParameterBegin=-0.4 and ParameterEnd=0.4. This will produce a nice animation where the plane moves from the bottom of the z-axis to the top, displaying the level curves of the graph. In a static black and white picture it will look like this:

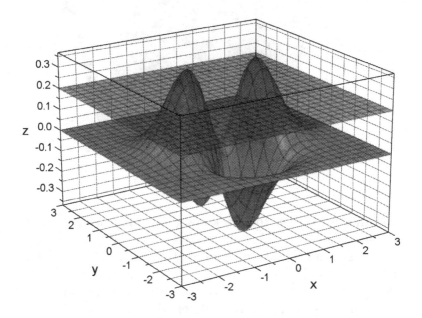

You may consider your job to be done. The animation shows exactly what we wanted to see. While playing the animation, we can rotate the graph and observe the level curves. Finally we can stop the animation at any time and zoom from the left or right side to it so that we can almost exactly read the minimum and maximum values of the function.   ■

We will finish our experiments with 3D graphics by exploring the possibility of adding our own lights and positioning them according to our needs.

Example 7.3

Let us start with a simple plot.

```
• Surf2:=plot::Function3d(
    cos(sqrt(x^2 + y^2)), x=-10..10, y=-10..10
  ):
  plot(Surf2)
```

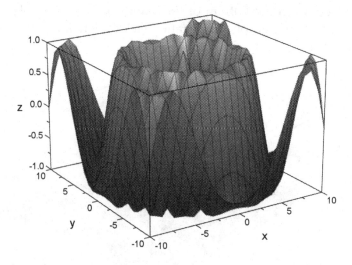

In calculus textbooks, this is known as a hat surface. In fact, it resembles a fancy hat. Let us suppose that we wish to color this surface in such a way that the bottom of the hat will be blue, right side will be red and left side will be yellow. All this can be done by changing the color of the surface to pure white and then applying three different lights from three different directions.

Let us open the picture in VCam. First, we will need to change the color of the Function3d object to white. Therefore, select Function3d in the objects browser, expand its **Style** group and then change the FillColor to pure white and FillColorType to flat. If you wish to have a very smooth surface, you can also change the submesh values to 1 or 2.

Now, we have a boring picture with a white surface with some default light. We need to enable user lights. Select the Scene3D object in the objects browser, expand the **Style** group there and change the option Lighting to Explicit. For the moment we will get rid of the default light and there will be darkness in our photographic studio. Do not worry, this is just for a short while. Now, in the objects browser select the Coordinate3d object or anything below it, and then find the option **Insert** in the VCam menu. Here you have the opportunity to insert a new camera and four types of lights—ambient, distant, point and spot lights. Ambient light is the one that comes from multiple reflections. You can see it in a dark room when no source of light is active. Distant light is the one that is somewhere far away and its rays are almost parallel. A point light is similar to a bulb light. Its rays go in all directions from a single point. Finally, a spot light is similar to the lights you see in a theater. Its rays form a cone and the actor stands exactly at the center of the cone base. In our example, we will use three distant lights. However, you can also experiment with other types of lights.

So, choose the **Distant Light** option from the **Insert** menu. You will find it in the objects browser. Select the new object DistantLight and expand its **Definition** group. Here, set its LightIntensity to 1, LightColor to red, and position to 0, -10, 5. In the same way, add two more distant lights, set their color to yellow and blue, intensity to 1 and positions to -10, 0, 5 (yellow light) and -10, -10, -5 (blue light). The results will certainly look better than what I can show here in a black and white picture.

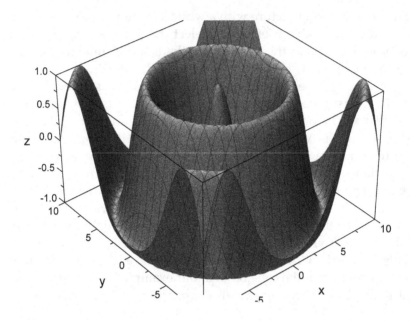

7.4. Interactive Development of Animations

In this book, we have already developed a few simple animations. In the later chapters, there will be a lot of opportunities to develop interesting animations of mathematical objects. The goal of this section is to explore the strategy of building an educational or scientific animation step by step from scratch to the very end.

Developing a good educational animation requires some preparations. It isn't enough to show an object moving from one place to another. The animation should show something that has real educational or scientific value. For instance, an animation showing how the rays of a point light are reflected from a parabolic mirror may have great value, as it explains one of the laws of science. Of course, there are many animations that do not illustrate anything and we consider them as a form of animated art rather than as educational animations. But the border between mathematics and art can be quite fuzzy.

When developing an educational animation, we have to prepare a few things before we start coding: the objects that will be used, some of which will be static and some animated; a rough storyboard; a scenario for the animation, and finally we must decide how the objects will be placed and combined in one scene. We can then start coding, testing and improving our work. With VCam, we do not need to worry about object colors and a few other parameters. These can be done later in VCam. So, we can code rough declarations of the objects in MuPAD, and then adjust the details later in VCam. There is one important rule to keep in mind – any object that you need for your animation must be declared using MuPAD code. In VCam, we can only add lights and a camera. We cannot add any new plot objects in VCam. I guess you are now ready to start doing something more concrete.

Example 7.4

When learning trigonometric functions, we have to find out how the graph of a trigonometric function, for example $y = \sin x$, is build. From the school definition of $\sin x$ we know that when we have a right-angled triangle with one angle equal to x, the sine function can be defined by the formula $\sin x = \dfrac{oposite\ side}{hypothenuse}$. Then, we develop a more complex definition of the sine function for all angles, not necessary smaller than 90°. In this definition, we use a circle with a radius equal to 1, along with one point, say P, that slides along the circle, a corresponding point Q drawing the sine curve, and two segments – one segment connecting the center of the circle with the point P, and another one connecting the points P and Q.

Let us build a model illustrating the sine function concept. We've already mentioned all the necessary elements that we need to this model. These are:

1. a circle with a radius equal to 1,

2. the point P sliding along the circle,

3. the point Q sliding along the sine curve,

4. the line segment L1 connecting center of the circle and point P,

5. the line segment L2 connecting points P and Q,

6. the sine curve.

We will also need the axes of the coordinate system, but these will be created by MuPAD automatically. So, we do not need to worry about them. Now we can start drawing the storyboard. You can do it by hand or in any graphics program.

Here is a very basic storyboard for our animation.

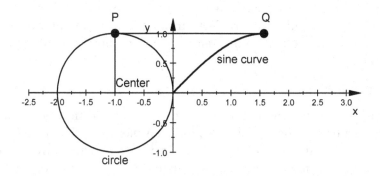

Fig. 7.7 Animation storyboard

The scenario of this animation is very simple:

1. The point *P* starts from the center of the coordinate system and then moves along the circle.

2. The point *Q* moves to the right on the same level like the point *P*.

3. The sine curve is drawn by the point *Q*.

4. Segments *L*1 and *L*2 move according to the corresponding points coordinates.

Now, when we have decided what should happen in our animation, we can start writing object declarations. For the time being, let us set the animation parameter to $a = \pi/2$, and later we will change it to a real animation parameter.

For this animation, we will use three new MuPAD plot classes. These are Point2d, Line2d and Curve2d. We declare a Point2D by giving its coordinates in the form of a list [x,y], and Line2d by giving the coordinates of two points in two lists. Curve2d is a nice class for developing parametric plots. It is convenient to describe a circle using its parametric equation $x = r\cos t$ and $y = r\sin t$. In our example, we will use $x = -1 + \cos t$ and $y = \sin t$. In later chapters, we will talk more about parametric plots. Now, using this very basic information we can complete our examples.

Here are the declarations of our objects:

- `a:=PI/2:`

```
//static object
Circle := plot::Curve2d([-1+cos(t),sin(t)], t=0..2*PI):

//objects to animate
P := plot::Point2d([-1+cos(a),sin(a)]):
Q := plot::Point2d([a, sin(a)]):

L1 := plot::Line2d([-1,0],[-1+cos(a),sin(a)]):
L2 := plot::Line2d([-1+cos(a),sin(a)], [a, sin(a)]):

SineCurve := plot::Function2d(sin(x), x=0..a):
```

Now we can plot all of our objects.

- ```
  plot(Circle, L1, L2,SineCurve, P, Q,
      Scaling=Constrained
  )
  ```

The result is exactly what was presented in Fig. 7.7. Note that the circle is not located in the center of the coordinate system. It is moved one unit to the left, so that the sine curve starts nicely from the center of the coordinate system.

Once you are sure that you've got exactly the same picture, we can start changing the code into an animation. Remove the declaration for "a" from the code, and replace it using the command `delete a`. Then, in every line where we had used the parameter "a", add `a=0..2*PI` at the end. This will change "a" into an animation parameter for every occurrence of "a". You will end up with code similar to what is shown below.

- `delete a:`

```
//static object
Circle := plot::Curve2d([-1+cos(t),sin(t)], t=0..2*PI):

//objects to animate
P := plot::Point2d([-1+cos(a),sin(a)], a=0..2*PI):
Q := plot::Point2d([a, sin(a)], a=0..2*PI):
```
```

```
L1 := plot::Line2d(
    [-1,0],[-1+cos(a),sin(a)],a=0..2*PI
):

L2 := plot::Line2d(
    [-1+cos(a),sin(a)], [a, sin(a)], a=0..2*PI
):

SineCurve := plot::Function2d(
    sin(x),x=0..a,a=0..2*PI
):
```

- ```
  plot(Circle, L1, L2,SineCurve, P, Q,
      Scaling=Constrained
  )
  ```

This is all that we absolutely had to declare as MuPAD code. All the other work can be done with VCam. Let us open the obtained picture in VCam and make the animation more appealing. We will need to change the size of the points, the width of some lines, and perhaps some of the colors.

In the objects browser in VCam, we have the following objects — Curve2d, two Line2d objects, Function2d and two Point2d objects. Now we must decide how these objects will be displayed. Before you do this, play your animation in VCam in order to see how it works. What you would like to change to make it more appealing? In my opinion, though you may have quite a different opinion, it would be good to make the line of the sine curve a bit thicker, for example 0.5 mm, and perhaps color it red; make both points larger, say 2 mm, and also in red. The circle could be black. It would be worth to fill it with a single color, but this could require more changes in the code. So, we have to stay with what we've got now. Finally, we can change the colors and line styles for the two Line2d objects.

You already know how to do all this. So, together we will only make changes for one of the points. Select the second Point2d in the objects browser. This is the point that draws the sine line. In its properties, expand the **Style** and then the **Points** group. Here, you need to change PointSize to 3, which is 3 mm; you also need to set PointColor to red or another color that you prefer. All the other changes you can do by yourself. Fig. 7.8 shows what I've produced, again in black and white.
  ```

While adjusting the parameters of each object, it is very important to consider how the animation will be played. The default parameters won't always be enough. If we are going to display an animation directly in VCam, then we always have the opportunity to set the speed of the animation using the animation toolbar. However, later we may not have such control. So, let us discuss how we can declare parameters for an animation.

For each object to animate, we have to declare the time of animation as well as the number of frames. If we wish to have a smooth animation without any blinking, we should use 12 or more frames per second. About 15 frames per second produces quite a smooth animation. In MuPAD, by default the animation time starts at 0 and finishes at 10 (seconds). This means we should declare the number of frames $10 \cdot 15 = 150$.

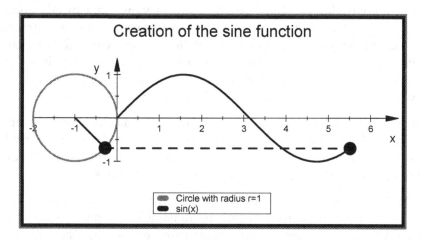

Fig. 7.8 Final animation of the sine curve

You can still add some additional features or elements to this animation. Using similar techniques, you can develop a number of other animations illustrating important concepts in mathematics, science or chemistry. You may now be wondering how to display your animation without MuPAD? We will talk about this issue in the next section of this chapter. ■

7.5 Saving MuPAD Graphs

In this and the previous chapters, we have developed a few interesting graphs and even some animations. Thus, this is a good time to think about how we can preserve our work and show it to others. You can always save your work from inside VCam as an *.XVZ file, and then display it just with VCam. However, there is much more that can be done.

In general we have three types of plots in MuPAD:

1. Static 2D plots

2. Static 3D plots

3. Animations in 2D and 3D

From a computer graphics formats point of view, we can talk about two major formats of images—bitmaps and vector formats. Bitmaps are raster images where each point of the picture is stored as a point with a given color. Thus, bitmaps are just arrays of colored points. The larger the array, the larger the size of the file; the larger the array for a given picture, the more accurate the image.

Pure vector formats store pictures using mathematical representation for lines, points and textures. Therefore, pictures in vector format don't change their quality when we resize them. Formulae don't care about scaling—this is just a change of some parameters.

In MuPAD, static 2D plots are developed in a good vector format. We can copy a picture directly from the notebook and paste it into any document in MS Word or any other word processing package. The quality of this picture will remain always the same. Flash fans can copy such a picture from the notebook and paste it directly on the Flash stage. They will get a perfect Flash file that can be displayed on a web page or as a standalone Flash projector file (*.EXE for Windows or *.HQX for Macintosh).

On the other hand, static 3D plots embedded in the notebook are unfortunately bitmaps in MuPAD 3.0. This is because the new graphics in MuPAD use a very sophisticated and realistic concept of coloring and finishing for the surfaces. This cannot be properly represented in a vector format, and the file would be huge. Therefore,

MuPAD renders the final version of its pictures into a bitmap with a low resolution for the screen display. This means that such a picture copied from a notebook and pasted into a word processing document will not look great in print. In order to save such a picture for printing purposes, we must open it in VCam, and then choose the option **Export** from the menu. We then see a large list of formats that we can use. When saving for printing, you can use bitmap formats like BMP, TIF, PNG, PCX and TGA; furthermore, you should use a resolution of at least 300 DPI. All the pictures representing 3D graphs in this book were saved with a resolution of 600DPI or even 1200DPI. Note that when exporting 3D graphics to a file, we can also use EPS and WMF formats. However, the files produced will be very large and any sophisticated coloring will be simplified, so it will not reflect your work properly. Files saved as JVX can be used with JavaView applets. Finally, files saved as GIF, PNG and JPG can be used for web pages.

Animations, both 2D and 3D, can be exported from VCam into AVI format. This is the most popular format of animations supported by the Windows Media Player. The very first available export option to AVI, i.e. Microsoft Video 1 codec, 15 frames per second and quality 100% produces a very good AVI animation in most cases.

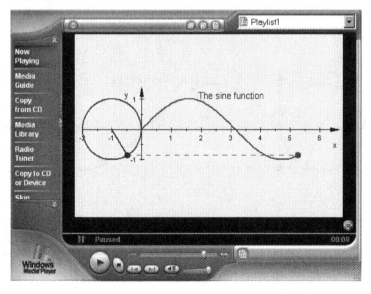

Fig. 7.9 Animation exported from MuPAD as AVI

Using these settings, I was able to save the animation from the last example as an AVI file 2.78 MB in size. This file can be played using the standard Windows Media Player or any other animation player (see Fig. 7.9).

Such an animation can be further edited in Adobe Premiere or another video editing program. We can add a title page with the author's name, other text, background sounds, or narration. We can combine a few such animations into one large movie and display it directly from a computer, CD, or even write it to a video tape.

Another way of displaying MuPAD animations is Flash (versions MX and higher). The AVI file exported from MuPAD can be imported to Flash, provided with a movie control bar, and published in the form of an SWF file for www or as a Flash projector file. Publishing the last example in SWF format, I got a file 644KB in size, and a projector file 1.4 MB in size. A Flash projector file is a regular application and it can be executed in the Windows or Macintosh environment (see Fig. 7.10).

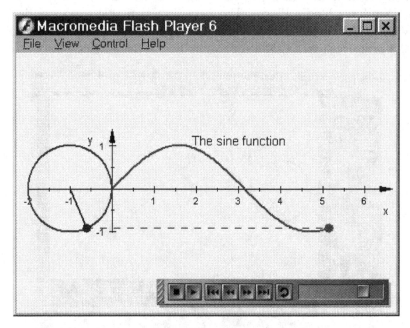

Fig. 7.10 MuPAD animation as a Flash projector

Note that in VCam, you can also export 2D graphics as SVG (Scalable Vector Graphics), which is a great format for vector graphics for the web. Unfortunately, due to some limitations of the SVG format, 2D animations in MuPAD version 3.0 are saved as static pictures. Until this problem is resolved, we have to forget about producing animated SVG files in MuPAD.

7.6 Chapter Summary

In this chapter, we have introduced the major principles of using VCam. We have examined how the parameters of the graphics produced in MuPAD can be adjusted so we can get an effective image or animation. We have also learned how to save MuPAD's static images and animations so we can use them later in printing, displaying on the web or even as standalone Flash applications. There are still many other things that are worth learning about presenting MuPAD graphics. Displaying such graphics on the internet in particular is an exciting adventure. However, we have to leave this topic for another opportunity — a MuPAD graphics book or an Internet article. Right now, it is time to proceed to the next chapter where we will explore some mathematical curves and surfaces.

7.7 Exercises

1. In MuPAD, execute the statement:
   ```
   plot(plot::Function2d(x^2+y^2))
   ```

   Open the obtained picture in VCam using the two methods **Edit** and **Open**. Try to find out the major differences between both environments. Which mode is more convenient for you? Identify the meaning of each icon in the toolbar. What operations are assigned to each icon? Display and hide each panel — objects, properties and current view. What do you see in each of these panels? Select the `CoordinateSystem2d` object in the objects browser. What will be highlighted in the picture? Now, select the `Scene2d` object and check what is highlighted in the picture. Do you see a difference between `Scene2D` and `CoordinateSystem2d`? Explore the properties of all the objects that are displayed in the objects browser. What options are available there?

2. Execute the statement `plot(plot::Point2d([1,2]))`

   Open the picture in VCam in whichever mode you prefer. What are properties of the `Point2d` object? Are they different from the properties of `Function2d` in the previous exercise?

3. Execute the statement `plot(plot::Function3d(x^2+y^2))`

   Open the obtained picture in VCam in whichever mode you prefer. Check what objects you have in your picture this time and what are their properties. Compare the `Function3d` properties with the `Function2d` properties from the previous examples. What are the major differences? Compare the coordinate system from this exercise with the coordinate systems from the previous exercise. Identify the major differences between them.

4. Open in VCam the 3D picture obtained in the previous exercise. How could you change the surface to a pure red color without the default blue color in the bottom part? Which object and which properties would you have to change to produce a pure red color? Explore the color options available in VCam. While changing colors for MuPAD objects, you will find the color opacity option. Change the opacity for one of the colors used in your picture and see how it will affect the surface.

5. Execute the statement

   `plot(plot::Function2d(a*x^2+1, x=-2..2, a=-1..1))`

   What curve is shown on the produced picture? Open the picture in VCam and see what you've got this time — do you see the animation control toolbar there? Play your animation. What concept does this animation explain? Change the above statement to

   `plot(plot::Function2d(x^2+a, x=-2..2,a=0..1))`

   What a conclusion can you draw from this animation? Finally, execute the statement

   `plot(plot::Function2d(x^2-a*x, x=-2..2,a=0..2))`

   What mathematical conclusion can you draw from this animation? Do you think that such animations can be useful in explaining mathematical concepts?

6. Use the equations $sin(a \cdot x)$, $a \cdot sinx$, $sin(x + a)$, and $sin(x) + a$ to develop animations showing the role of the parameter a in each case. For a better effect, use the interval $x = -2\pi..2\pi$.

7. Repeat the previous exercise for the $cosx$, $tanx$ and $cotx$ functions.

8. Repeat the previous exercise for the hyperbolic sine and cosine functions. Use the names `sinh(x)`, `cosh(x)`, `tanh(x)`. What is wrong when when you produce the animation `plot(plot::Function2d(sinh(a*x), x=-2..2,a=0..4))`? Find a way to improve this animation.

9. Execute the statement:

```
plot(plot::Function3d(
    sin(a*x)*cos(a*y), x=-PI..PI, y=-PI..PI,a=-1..1)
)
```

This will produce an animation of a 3D surface. Open it in VCam and play it to see how the surface changes. Do you see how the color of the surface changes while object animates? Is the color assigned to a place on the surface or to the space in 3D?

10. Start with the statement,

```
Pt:=plot::Circle2d(0.5,[1,a*(3-a)], a=0..3):
```

```
plot(Pt)
```

This will produce a circle jumping up and down from the x-axis. You will have to convert it into an animation of a jumping red ball with a 5 mm radius. In the beginning of the animation, the ball should touch the floor, not intersect it. Therefore, you will have to adjust the formula for the y coordinate of the ball. You will have to make this modification at the code level. Also, add a horizontal line that will serve as the floor. Then, open your animation in VCam and adjust all the necessary parameters so that the ball has a radius of 5 mm and is colored red. Play your animation a few times in order to see if everything is correct. Did you notice that this animation is very slow? How could you make it play faster? There are at least two ways to speed it up.

11. Develop an animation that will illustrate the definition of the

tan(*x*) function. You can use a similar concept to the one that we used earlier in the example with the function sin*x*.

12. The statements

```
L1:=plot::Function2d(x^3-4*x, x=-3..3)):
```

```
plot(L1)
```

produce a picture of a polynomial of order 3. Develop an animation that will contain the mentioned curve and a horizontal line L2 intersecting it. The animation should show how many points of intersection a horizontal line may have with L1, and when. Change the animation parameter for the horizontal line L2 so that it will oscillate between the local maximum of the function and the local minimum of the function.

13. Export the animations produced in the previous exercises as AVI files. Play them using Windows Media Player. Check the various compression options in order to compare the quality of the final output. Choose the compression level that produces an animation with, in your opinion, acceptable quality and the smallest possible file. If you have Macromedia Flash MX, you can import one of the AVI animations to Flash and then publish it as Flash SWF file. An evaluation version of Flash MX can be downloaded from the Macromedia web site at www.macromedia.com. Play the animations produced in Flash. What is the difference in file size between AVI and SWF files for the same MuPAD animation?

Chapter 8 _____

Exploring Mathematical Curves and Surfaces

In the previous chapters, we had developed some interesting graphs. However, most of the plots we developed were quite simple. Still, the goal of the last two chapters was only to familiarize ourselves with the concept of MuPAD plots and learn how to apply various object and scene plotting options. Meanwhile, in this chapter we will explore several interesting examples of mathematical graphs.

8.1 Plots in 2D Space & Plot Types

By now, I guess you know most of the secrets of creating graphs and scenes with MuPAD. Now it is time to move on to mathematics and enjoy the creation of mathematical objects. We will start with the most important objects for mathematicians — functions of one variable.

8.1.1 Class Function2d

In the previous chapters of this book, you have already seen examples of plots of functions with one variable. However, there is still a lot to show and experiment with.

Example 8.1

Let's start with a very interesting example from high school mathematics. Let us consider the function $y = \frac{\sin x}{x}$. It is obvious that the function is not defined for $x = 0$. Thus, we often ask our students to calculate the limit of this function as $x \to 0$. This can be quickly worked out if we try to plot this function in the neighborhood of 0. Such results are easy to achieve in the classroom.

- ```
F1 := plot::Function2d(
    sin(x)/x, x = -1..1
):

plot(F1,Scaling = Unconstrained)
```
```

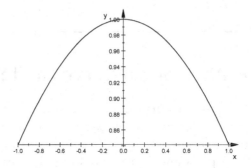

I used Scaling=Unconstrained here, as it allows us to conveniently see what is going on near the point where $x = 0$. It shows exactly how the graph changes as x approaches 0. However, looking at this picture, the student may get impression that our curve intersects x-axis for $x = \pm1$. This can be easily fixed by adding the ViewingBox option to the final plot command.

```
• F1 := plot::Function2d(
      sin(x)/x, x = -1..1
  ):
  plot(F1,
      Scaling = Unconstrained,
      ViewingBox = [-1..1,0..1]
  )
```

Here is the result:

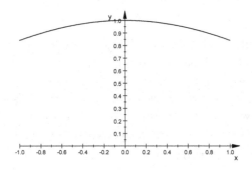

Now, you may clearly see that $f(x) \to 1$ when $x \to 0$. Thus, MuPAD has already fixed the tiny gap that would occur for $x = 0$.

Have you ever tried experimenting with the function $f(x) = \frac{\sin x}{x}$ when $x \to \pm\infty$? Of course, this is difficult, as we cannot draw graphs close to

±∞. But we can use a little trick to solve this problem. Let's substitute $x = \frac{1}{u}$ in our function. This gives us the new function $g(x) = u\sin\frac{1}{u}$, which behaves near 0 in a manner similar to what the previous function would do near ±∞. Experiments with this graph can be quite interesting.

```
• F1 := plot::Function2d(
     u*sin(1/u), u = -1..1,
     LineWidth = 0.25
   ):
   plot(F1,
     Scaling = Constrained,
     ViewingBox = [-1..1,-1..1],
     XAxisTitle = "u",
     YAxisTitle = "y"
   )
```

And here is the resulting graph:

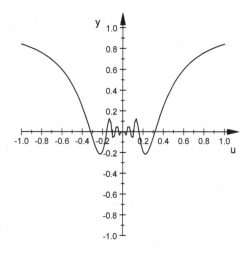

You can certainly see that our curve tends towards 0 when $u \to 0$. However, the graph near 0 is ragged and not very accurate. If you wish to make it more accurate, you will need to change three things: the range of the variable u to $u = -0.2..0.2$, set the Mesh size to Mesh=1000 or even larger, and set the ViewingBox size to ViewingBox=[-0.2..0.2,-0.2..0.2]. This will make a nice graph like this one:

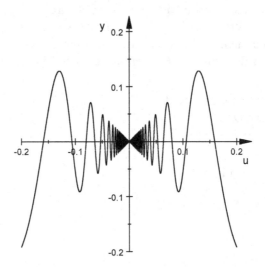

You can easily see that near $u = 0$, the graph is enclosed between the two lines $y = u$ and $y = -u$. Thus, you may conclude that at $\pm\infty$, the graph of $y = \frac{\sin x}{x}$ will be enclosed between the two hyperboles $y = \frac{1}{x}$ and $y = -\frac{1}{x}$.   ∎

8.1.2 Class Curve2d

As a mathematician, you should be familiar with curves given by parametric equations. Such a representation requires two equations, $x = g(u)$ and $y = f(u)$, where u is a parameter that takes real values. Parametric equations are convenient to describe more complicated curves. In many cases, it may be that the parametric representation is the only one available and the curve cannot be described by a single function. On the other hand, you may easily see that any function with one variable, like $f : R \rightarrow R$, can easily be represented by a parametric equation. You simply have to transform the equation $y = f(x)$ into the two equations $x = u$ and $y = f(u)$.

In MuPAD, we have a class called Curve2d, which is used to declare a graphical object given by parametric equations with one variable, and then to obtain its graph. The syntax of its declaration is similar to the syntax of other graphical objects:

```
plot::Curve2d([ g(t), f(t) ], t = t1..t2, options)
```

Notice that the parametric equation of the curve must be enclosed in a square bracket. There are many interesting examples that could be used to demonstrate `Curve2d` plots. Let us take a look at one of them.

Example 8.2

Some of the most popular curves in calculus are the various curves formed as the locus of a point on a circle rotating without slipping outside or inside of a fixed circle. Curves belonging to the first class are known as *epicycloids*, and the other ones as *hypocycloids*. Let us plot a *hypocycloid*. From the above description, we can deduce its parametric formula as,

$$x = (R - r)\cos t + r\cos(R - r)\tfrac{t}{r}$$
$$y = (R - r)\sin(t) - r\sin(R - r)\tfrac{t}{r}$$

where $t = 0..2\pi$, and the letters R, r denote the radius of the large and small (internal) circles respectively. In order to make a picture that will illustrate the above definition, we also need to plot the external circle. For this example, we will use $R = 3$ and $r = 0.5$.

We will start by declaring some constants:

- ```
  R := 3:
  r := 0.5:
  ```

In the next step, we will declare the hypocycloid.

- ```
  Hp := plot::Curve2d(
     [(R-r)*cos(t) + r*cos((R-r)*t/r),
      (R-r)*sin(t) - r*sin((R-r)*t/r)],
     t = 0..2*PI,
     Color = [0, 0, 0],
     LineWidth = 0.41
  ):
  ```

Next, we add the external circle with the radius of 3 units, and we use the red color to display it.

- ```
  Circle := plot::Curve2d(
     [R*cos(t),R*sin(t)],
     t = 0..2*PI,
     Color = [0.2, 0.2, 0.2],
     LineWidth = 0.25
  ):
  ```

Finally, we plot both objects, adding a bit of color, a title and a blue
  ```

grid. You may also consider a slightly more formal approach, and build a scene with both objects.

```
• plot(
    Hp, Circle,
    Scaling = Constrained,
    BackgroundColor = [0.8, 0.8, 0.8],
    Footer = "Hypocycloid",
    GridVisible = TRUE,
    GridLineStyle = Solid,
    GridLineColor = [0.0, 0.0, 1.0]
)
```

Here is our final creation:

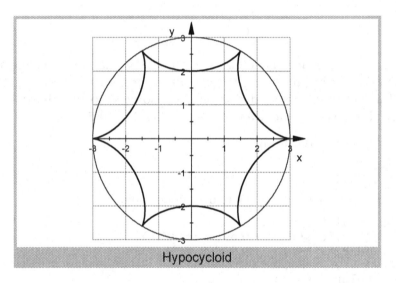

Hypocycloid

Now, you can experiment with the code by changing the sizes of both circles and the values of all the options. By the way, do you know how to change the equation of this *hypocycloid* in order to obtain an *epicycloid*? Try it—it is a very easy and interesting example. Finally, you can develop animation showing how the *hypocycloid* and *epicycloid* are drawn by a point on the rotating circle.   ∎

8.1.3 Developing Polar Plots

Polar plots are obtained when we plot a function in the so-called polar coordinates. In polar coordinates, a point A is represented by two values—the distance r from the origin O and the angle θ (*theta*) between x-axis and the vector $\overrightarrow{OA}$.

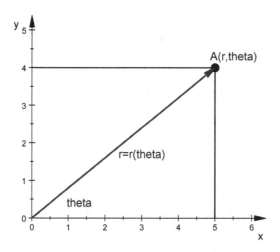

Fig. 8.1 Polar coordinate system

A function in polar coordinates would be represented by the equation $r = f(\theta)$; for example, $r = a(1 + 2\sin(\theta/2))$.

In MuPAD, we use the class Polar to declare graphs of curves in polar coordinates. In order to do this, we use the syntax:

plot::Polar([$r(\theta)$, θ], θ = $t1..t2$, *options*)

You will immediately notice that in the above declaration, we declared a vector [$r(\theta)$, θ] rather than a single function r. This is very convenient, as we may also use parametric equations where both r and θ will be the functions of another parameter, allowing us to deal with parametric functions in polar coordinates. For such equations, we would use the syntax:

plot::Polar([$r(t)$,$\theta(t)$], t = $t1..t2$, *options*)

Curves obtained in polar coordinates are considered to be among the most beautiful of curves. Some of them are already considered classical, like various types of *nephroids*, *cissoids*, and so on, while others were invented quite recently when people started experimenting with computers.

Example 8.3 Freeths nephroid

A curve called the *Freeths nephroid* has, in polar coordinates, the equation $r = a(1 + 2\sin(\theta/2))$, where $\theta = 0..4\pi$ and a is a positive parameter. This equation leads us to a beautiful curve.

```
• Nephroid := plot::Polar(
      [1+2*sin(theta/2), theta], theta = 0..4*PI,
      Mesh = 100,
      Color = [0,0,1],
      LineWidth = 0.40
  ):
   plot(
      Nephroid,
      Scaling = Constrained,
      BackgroundColor = [1, 0.9, 0.9],
      Footer = "Nephroid"
   )
```

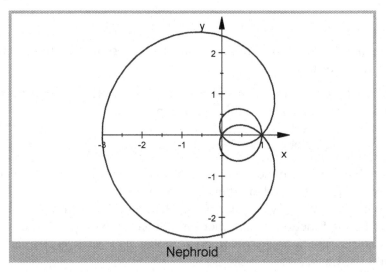

Nephroid

You will find more examples of interesting curves given by polar coordinates in the programming exercises at the end of this chapter.∎

8.1.4 Developing Implicit Plots

Frequently, we deal with mathematical curves that are given in the form of an equation $F(x,y) = 0$. We call the equations of this type implicit equations. Some of these equations can easily be transformed into functions or parametric equations in either Cartesian or polar coordinates. However, sometimes it is impossible to transform such equations into a function or a parametric equation. Obtaining graphs of implicit equations can be a very difficult task. In MuPAD, there is a class that can be used to represent plot objects by implicit equations. Here is the simplest form of its syntax. Later on, we will learn more about the syntax of this class.

plot::Implicit2d($F(x,y)$, $x = x1..x2$, $y = y1..y2$, *options*)

Example 8.4 An Arabesque

In this example I will show you one of the most interesting examples of a curve represented by an implicit equation. Let

$$x \sin x + y \sin y = 0$$

For reasons that will become obvious later when you see the shape of the curve, I will call it an *Arabesque*. In order to obtain this curve, you can start out with very simple MuPAD code. Here it is:

```
• Arabesque := plot::Implicit2d(
      x*sin(x)+y*sin(y), x = -5..5, y = -5..5,
      LineColor = RGB::Black
  ):
  plot(Arabesque)
```

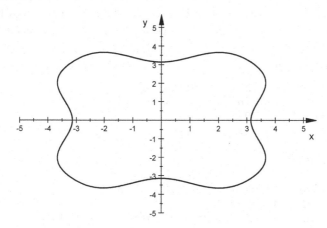

The resulting graph looks rather simple. However, looking at the equation, we may expect that larger values of x and y will also fulfill this equation. For example, every time when $x = y = n\pi$, the left side of the equation is equal to 0. This fact suggests that the line $y = x$ intersects our curve in infinitely many points. Well, let's change the area of the graph to $x = -20..20$, $y = -20..20$ and plot the graph again. This will give us a different picture:

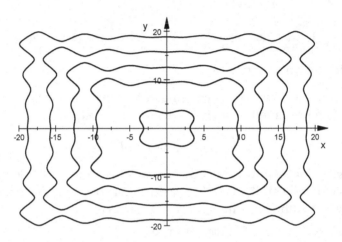

We have obtained four more branches, but the irregular distances between them suggest that something is still wrong. Notice that the distance between the two consecutive points $x = n\pi$ should always be equal to π. Therefore, we shouldn't have different distances between each two consecutive branches. In order to improve the accuracy, we will raise the Mesh value and use Scaled=Constrained to get the proper proportions of the graph. Here is the improved code:

```
• Arabesque := plot::Implicit2d(
      x*sin(x)+y*sin(y), x = -20..20, y = -20..20,
      Mesh = [20,20]
  ):
  plot(
      Arabesque,
      Scaling = Constrained,
      BackgroundColor = [0.8,0.8,0.8],
      LineWidth = 0.4,
      Footer = "Arabesque, x*sin(x)+y*sin(y)"
  )
```

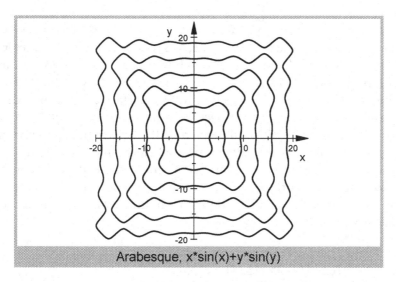

Arabesque, x*sin(x)+y*sin(y)

Finally, we've got a very beautiful curve that somewhat resembles some of the patterns that you see in many Arab decorations. There are many more curves like this one. Try experimenting with the above equation in order to see what sort of interesting curves you may obtain.   ∎

You may perhaps have noticed that implicit plots use a slightly different syntax than other graphical objects. Before we proceed to new objects, let us learn a bit more about the Implicit2d class.

First, let us note that due to the high complexity of the algorithms used to obtain an implicit plot, such plots can be very time-consuming, especially if the shape of the curve is very complex and we need a high accuracy graph. In some situations, the default value of Mesh, which is Mesh=[11,11], might be not enough. We give here two values for the subdivisions on the x-axis and y-axis. Raising the Mesh value to [15,15] or even to [20,20], like in our last example, is usually enough to get a very accurate plot, but it also slows down the calculations significantly. This is especially easy to see when using older computers with a slower processor and low memory.

Another interesting feature of the Implicit2d plot is the Contours option. By default, an Implicit2d plot produces a graph of the equation $F(x,y) = 0$. However, for many mathematical examples, we may wish to produce graphs for $F(x,y) = a$, where a takes some specific values. For example, for $a = 0, 1, 2, 3, 4, 5$. In such cases we will have to use Contours=[0,1,2,3,4,5], where the parameter 0 produces the default implicit plot while other values produce other contours. If you consider this for a moment, you will realize that the contours are in fact the intersections of the surface $z = F(x,y)$ with the planes $z = 0, z = 1, .., z = 5$, projected on the plane XY. Note that each additional contour significantly increases the time required to produce the plot. For example, in order to produce ten contours, we will need at least tenfold more time than for a single implicit plot. You can observe how it works with a simple example. We will use the procedure time to calculate how much time was needed for our implicit plot.

```
time(
    plot(
        plot::Implicit2d(
            sin(x)+cos(y),
            x = -1.5..11, y = -2.5..9.5,
            Contours = [0, 0.25, 0.5, 0.75, 1.0]
        )
    )
)
```

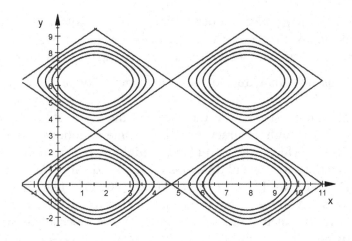

39737

The number below the plot shows the amount of time used to produce the graph, in milliseconds. Therefore, my computer needed 39.737

seconds to produce this graph. On the other hand, it only took 2.293 seconds to develop the single contour $F(x,y) = 0$. This is a very significant difference indeed.

Example 8.5

Sometimes it is quite useful to obtain the graph of a family of curves and their asymptotes. For example, a family of *hyperbolae* and their common asymptotes.

Let's consider the equation of the family of *hyperbolae* $\frac{x^2}{a^2} - \frac{y^2}{a^2} = 1$, where a is an integer number. By moving a^2 to the right side of the equation, we get $x^2 - y^2 = a^2$. Thus, looking at the equation, we may conclude that all these curves will have the same asymptotes $y = \pm x$. Let's see how it looks on a graph. We will start by declaring the variable R, which will be used to keep the size of the plotting window. Then, we will declare the three objects H1 (the family of hyperbolas declared as a contour), As1 and As2 (the two asymptotes). Finally, we will plot all of these objects.

Additionally we may try to produce each contour in a different color. For example, we could use red for $a = 1$, blue for $a = 2$ and green for $a = 3$. Therefore, we will define a color function ColFun that will apply a different color to every curve. Our code will thus look like this:

```
• ColFun := proc(x,y)
    begin
       if x^2-y^2-1 <2 then
           return(RGB::Red)
       elif x^2-y^2-1 <5 then
           return(RGB::Blue)
       else
           return(RGB::Green)
       end_if
    end:

• R := 10:

• H1 := plot::Implicit2d(
       x^2-y^2-1, x = -R..R, y = -R..R,
       Mesh = [5,5],
       Contours = [1,4,9],
       LineColorFunction = ColFun,
```

```
    LineWidth = 0.30
):

As1 := plot::Function2d(
    x, x = -R..R,
    LineWidth = 0.50
):

As2 := plot::Function2d(
    -x, x = -R..R,
    LineWidth = 0.50
):

plot(
    H1, As1, As2,
    Scaling = Constrained
):
```

The resulting picture contains all that we wanted to see—the three *hyperbolae* with $a = 1$, $a = 2$ and $a = 3$, as well as the two asymptotes $y = x$ and $y = -x$.

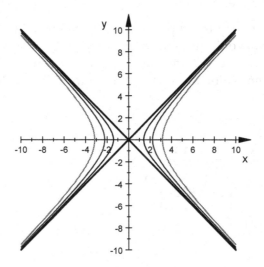

Note that this time, we used a very low value for the Mesh option, and thus we were able to obtain the plot in a reasonable amount of time. ■

8.2 Plots in 3D Space

In this section, we will concentrate on developing 3D graphs of various surfaces. We will start from the most typical case, a function of two variables; later, we will explore parametric surfaces, and then move on to surfaces in cylindrical and spherical coordinate systems. Finally, we will spend some time with curves in 3D. Graphs in 3D are quite interesting, so, if you are anxious to experiment on your own, you may wish to try modifying each equation and piece of MuPAD code. This way, you will learn more about some types of 3D objects.

8.2.1 Class Function3d

We often have to deal with functions with two variables in mathematics. The equations for such functions are usually written in the form $z = F(x,y)$, where x and y are the two real variables. The domain of such a function is a subset of the plane $\mathbb{R}^2$.

You may also consider functions with two variables defined by something more complicated than a single formula. Later on, I will show an example of such a function. Now, let us revise the syntax of the MuPAD statement used to develop graphs for such functions:

plot::Function3d($F(x,y)$, $x = x1..x2$, $y = y1..y2$, *options*)

In this case, the word *options* refers to the many 3D plot options that we analyzed in the previous chapters. We will now finally have an opportunity to use some of these options and experiment with them a bit.

Example 8.6

Let us consider the function $z = \sin(x^2 + y^2)$. This function is a perfect opportunity to experiment with plotting options. We will start with a very basic graph of this function, like this:

- ```
funSin := plot::Function3d(
    sin(y^2 + x^2), x = -5..5, y = -5..5
);

plot(funSin)
```
```

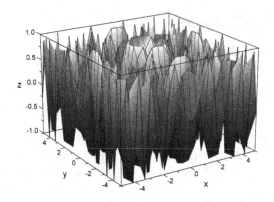

This graph shows that we are dealing with a very interesting surface. However, the spiky mountains in the graph do not tell us very much about the surface. We need to adjust the accuracy of the plot, as well as to change its default colors, add a background, and perhaps adjust the camera settings.

We will start by declaring a color function for our surface. We need to develop a function of three variables *x*, *y* and *z* that will produce RGB color values, i.e. list of three numbers from the interval [0,1]. Such a function can easily be obtained if we use one of the trigonometric functions sin*x* or cos*x* and an absolute value in order to remove all negative values. Below is an example of such a function. You can clearly see that this function fills the 3D space with a color pattern. Thus, what we get on the plot will simply be the intersection of our surface with the color space defined by the function g(x,y,z).

- g := (x,y,z)->[
    abs(sin(PI-x)),
    abs(sin(PI-y)),
    abs(sin(z))
  ]:

The next step is to change the default value of the Submesh option to [5,5]. This will make our surface very smooth. We wish to produce a very realistic graph, therefore it is better to use color patches without lines. This can be achieved by adding the two options XLinesVisible=FALSE and YLinesVisible=FALSE.

We could also use smaller ranges for *x* and *y*, for example x=-3..3, y=-3..3. This will make the graph much clearer.

This example is a good opportunity to see how the camera concept works in MuPAD. Let us declare a very simple camera pointing onto our surface from a bird's perspective, for example from the point [0,5,10]. So, our camera is located 10 units above the plane XY. Let us point the camera at the coordinates [0,0,0], and finally, let us use a lens with an angle of $\pi/9 = 20$ degrees. Here is the complete code of our example. Analyze it carefully to make sure that you know what each option is responsible for.

- ```
g:=(x,y,z)->[abs(sin(PI-x)),abs(sin(PI-y)),abs(sin(z))]:
```
- ```
funSin := plot::Function3d(
    sin(y^2 + x^2), x = -3..3, y = -3..3,
    FillColorFunction = g,
    Submesh = [5,5],
    XLinesVisible = FALSE,
    YLinesVisible = FALSE
):

MyCamera := plot::Camera(
  [0,5,10], [0,0,0], PI/9
):

plot(
    funSin, MyCamera
)
```

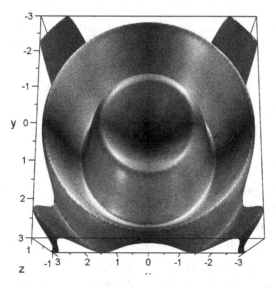

Unfortunately, the black and white printout above does not show the real beauty of this graph. In order to see how it really looks, you should execute the code in MuPAD or check the book web site.

Finally, you can try experimenting with the location of the camera and its angle. You can zoom into your picture, or develop it using a narrower camera angle. Doing so, you will get different views of the same graph.

However, this is not end of the story. Now, you can open the graph in VCam and zoom in on the picture to fill up the plot space. The above picture shows one of the possible final results. It's nice, but maybe you can find other color functions that will make this graph even more interesting.   ■

As I mentioned earlier, using the MuPAD programming language we can define functions that are more complex than those we use everyday in calculus and other mathematical disciplines. Here is one such example.

Example 8.7 Mandelbrot hill

Almost everyone has seen colorful pictures of various fractals. Undoubtedly, the most famous of them is the *Mandelbrot fractal*, sometimes called the *Mandelbrot set*.

The *Mandelbrot set* is a set of points on a complex plane. Just pick a point c on a complex plane and calculate:

$$c_1 = c^2 + c$$

$$c_2 = c_1^2 + c$$

$$c_3 = c_2^2 + c$$

. . .

If the sequence $c, c_1, c_2, c_3, ...,$ remains forever within a distance of 2 from the origin, then the point c is said to be in the Mandelbrot set. If the sequence diverges from the origin, then the point is not in the set. Usually, we prefer to consider a high but finite number of iterations — 100, perhaps even more.

Until now, we didn't use complex numbers in MuPAD. However, if you know a bit about complex numbers, you can easily obtain the formulae describing the above iteration process in real numbers. You have to use the formulae $z = x + iy$, $c = a + ib$, to obtain the formulae representing the real and imaginary parts of z_{n+1}. Take a look here:

$$x_{n+1} = x_n^2 - y_n^2 + a$$
$$y_{n+1} = 2x_n y_n + b$$

Now, for any given point $c = (a, b)$, starting with the initial conditions $x = y = 0$ (or equivalently $x = a$ and $y = b$) and carrying on iterations until $x^2 + y^2 > 4$ or until you reach a high enough number of iterations (say 25), you will determine if c belongs to the *Mandelbrot set* or not. Finally, you can color each point according to the number of iterations obtained for it. For example, you can use color$_1$ for points that run away after one iteration, color$_2$ for points that run away after two iterations, etc., and black for points that did not run away after 25 iterations. This will make a picture similar to this one.

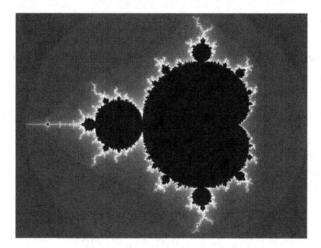

Fig. 8.2 Mandelbrot Fractal

You had perhaps started worrying—this is a 2D picture, so where is the 3D graphics? Well, we can implement it easily. Just imagine that we will produce a function that, for a given point (x, y), will produce its number of iterations, or something dependent on it, while developing the *Mandelbrot set*. We will call this function the *Mandelbrot hill*. Here is the MuPAD procedure defining such a function.

```
• Iterations := 150:
  mandel := proc(x,y)
  local m,a,b,t;
  begin
     if not (testtype(x,Type::Real)and
                testtype(y,Type::Real))
     then
        procname(x,y)
     else
        m := 0;
        a := x;
        b := y;
        while sqrt(x^2+y^2)<2 and m < Iterations do
           t := x;
           x := a + x^2 - y^2;
           y := b + 2*t*y;
           m := m+1
        end;
        return(float(m/Iterations))
     end
  end:
```

Now we will define a nice color function, and finally, we will draw a graph of our *Mandelbrot hill*.

```
• g := (x,y,z)->
     [abs(sin(3*x)), abs(sin(3*y)), abs(sin(3*z))]:

  MandHill := plot::Function3d(
     mandel(x,y), x = -2.2..1.2, y = -1.5..1.5,
     FillColorFunction = g,
     Mesh = [100,100]
  ):

  MyCam := plot::Camera(
     [-3.7, -4, 3], [-0.5,0,0.4], PI/10
  ):

  plot(MandHill, MyCam,
     Scaling = Constrained,
     Axes = None
  )
```

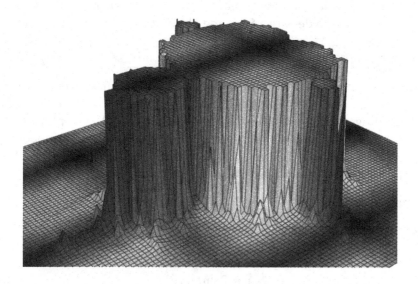

It is your turn now. Can you develop a different color function for the *Mandelbrot hill*? Or perhaps you could look up some descriptions of other fractals and develop a function similar to the *Mandelbrot hill* example?   ■

8.2.2 Class Surface

Like in the case of 2D, not all 3D objects can be described by the single equation $z = f(x,y)$. Quite often, we have to use another way of representing and plotting such surfaces. For example, your surface could be described by a parametric equation where each of its points (x,y,z) is be defined by some other parameters. For instance, a sphere can be represented in 3D by the set of equations

$$x = a \sin u \cos v$$
$$y = a \sin u \sin v$$
$$z = a \cos v$$

where a is the radius of the sphere, $0 < u < 2\pi$ and $0 < v < \pi$. You can find a few other representations of a sphere in calculus or geometry textbooks. Each of them is good, as long as $x^2 + y^2 + z^2 = a^2$.

Many other mathematical surfaces can be represented by parametric equations, i.e. by a list of three equations with two parameters.

MuPAD provides a very easy way to plot such surfaces. The command used to declare a Surface object is as follows:

plot::Surface($[x(u,v), y(u,v), z(u,v)]$, $u = u1..u2$, $v = v1..v2$, *options*)

The above command is more universal than we may expect at first glance. You can easily see that if $x = u$, $y = v$, and $z = z(u,v)$ then we are dealing with something that can be considered as a plot of the function $z(x,y)$. On the other hand, the other two types of functions, $x = x(y,z)$ and $y = y(x,z)$, can also fall into the Surface category. Let us see how Surface can be used to plot a sphere.

Example 8.8

Let's consider a surface described by the following equations

$$x = 2 \sin t \cos u$$
$$y = 2 \sin t \sin u$$
$$z = 2 \cos u$$

where $0 < t < \pi$ and $0 < u < 2\pi$. We can plot this surface in MuPAD as follows:

```
• g := (x,y,z)->[
      abs(sin(5*x)),
      abs(sin(5*y)),
      abs(sin(5*z))
   ]:

   Sphere := plot::Surface(
      [2*sin(t)*cos(u), 2*sin(t)*sin(u), 2*cos(t)],
      t = 0..PI, u = 0..2*PI,
      Mesh = [50,50],
      FillColorFunction = g
   ):

   plot(Sphere, Scaling = Constrained)
```

And here is the resulting plot, after a few small adjustments that were made using the VCam tool.

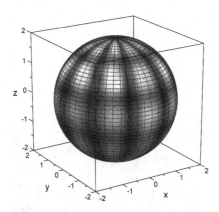

A sphere, even with the nicest colors, is not a very interesting object. You already know spheres quite well. However, you could try experimenting with the range of the two variables *t* and *u*. You will thus be able to see how the changes of either range affect the obtained plot. For instance, what would we get if we used $t = 0..\pi/2$ or $u = 0..\pi$? In the next example, I will show you a more interesting 3D shape. ■

Example 8.9

Let's take the following equations

$$x = \cos 2u \cos(u + v)$$
$$y = \cos 2u \sin(u + v)$$
$$z = \sin v$$

where $-\pi < u < \pi$ and $-\pi < v < \pi$. Can you predict the shape of the surface described by these equations? Probably not. So, let's write the MuPAD code, and plot it.

```
• h := (x,y,z) -> [abs(sin(x)),abs(sin(y)),abs(sin(z))]:
  knot := plot::Surface(
      [cos(2*u)*cos(u+v),
       cos(2*u)*sin(u+v),
       sin(2*v)],
      u = -PI..PI, v = -PI..PI,
      Mesh = [80,80],
      FillColorFunction = h
  ):

  MyCam := plot::Camera([100,70,50], [0,0,0], PI/150):
```

```
plot(knot,MyCam,
    Scaling = Constrained,
    Axes = None
)
```

Look at what we've got. Isn't it beautiful? If you wish to work more on examples like this, just look at the end of this chapter. You will have a great time plotting all those surfaces.   ■

8.2.3 Class Spherical

First, let us revise our knowledge about the spherical coordinates system (see Fig. 8.1). In spherical coordinates, the position of any point P, is determined by the three numbers r, ϕ (*phi*) and θ (*theta*). The number r is the length of the vector $\overrightarrow{OP}$, and θ is the angle between the z-axis and the vector $\overrightarrow{OP}$. Finally, ϕ is the angle between the x-axis, and the vector $\overrightarrow{OP'}$ is the projection of the vector $\overrightarrow{OP}$ on the plane XY.

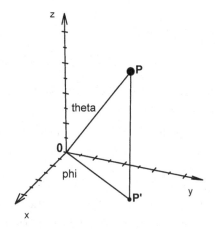

Fig. 8.3 Spherical coordinates

With the help of a few simple transformations, we can easily work out that the relationship between spherical coordinates and Cartesian coordinates can be expressed by these formulae:

$$x = r\cos\phi\sin\theta$$
$$y = r\sin\phi\sin\theta$$
$$z = r\cos\theta$$

After this short introduction to the spherical coordinate system, we are ready to experiment with a few spherical plots in MuPAD. Here is the syntax of the Spherical plot declaration:

```
plot::Spherical([r(u,v), φ(u,v), θ(u,v)], u = u1..u2, v = v1..v2, options)
```

where $r = r(u,v)$, $\phi = \phi(u,v)$ and $\theta = \theta(u,v)$ are the functions of the two variables u and v. This looks quite complicated, so if you wish to simplify things and go back to the well-known case $r = r(\phi,\theta)$, you may suppose that $\phi = u$ and $\theta = v$. This way, you do not need to worry about the additional two parameters u and v. However, it is still worth remembering that this is only a simplified syntax:

```
plot::Spherical([r(φ,θ), φ, θ], φ = φ1..φ2, θ = θ1..θ2, options)
```

Now let us look at an example of a spherical plot.

Example 8.10 A threefold shell

Spherical plots can be quite different from the plots obtained in Cartesian coordinates. This is because their domain is always a spherical angle. Thus, sometimes even very simple equations can lead to a very unusual shape. Here is such an example.

Let's consider the family of plots generated by the equations $r = a^\phi \sin(b \cdot \theta)$. If $a > 1$, then the value of the function will grow as ϕ grows. At the same time, $\sin(b \cdot \theta)$ will cycle between -1 and 1, forming a looped curve. Don't forget, we are in spherical coordinates and not in Cartesian coordinates, so depending on the value of b, we can obtain more or fewer loops. Later you can experiment with different values for a and b. Here, I will show how the plot looks for $a = 1.3$ and $b = 3$.

-
```
h := (x,y,z,phi, theta)->[
    abs(1-sin(10*phi)),abs(sin(10*phi)),0
]:

a := 1.3:
b := 3:

Shell := plot::Spherical(
    [(a^phi)*sin(b*theta),phi,theta],
    phi = -1..2*PI, theta = 0..PI,
    Mesh = [50,50],
    FillColorFunction = h
):

MyCam := plot::Camera(
    [100,70,50],[0,0,-0.5], PI/45
):

plot(
    Shell, MyCam,
    Scaling = Constrained,
    Axes = None
)
```

We ended up with a nice shell with three conchs. This shape is rather unusual, but it looks very interesting. Now, it is your turn. Plot a few more shells using different values for *a* and *b*. You could try, for example, *b* = 1, 2, 3, 4, 5, or even larger values.   ■

8.2.4 Class Cylindrical

In the same way that we use spherical coordinates, we can also use cylindrical coordinates. When using cylindrical coordinates, you may feel like you are using something half-way between spherical and Cartesian coordinates. Here, every point's position, *P*, is described by the three numbers r, ϕ (*phi*) and z. In this case, z is simply the Cartesian coordinate z of *P*, while r and *phi* are the polar coordinates of the point P' (the projection of *P* onto the plane *XY*). In MuPAD, the syntax of the cylindrical plot statement uses a more general concept,

```
plot::Cylindrical([r(u,v), φ(u,v), z(u,v)], u = u1..u2, v = v1..v2, options)
```

where, $r(u,v)$, $\phi(u,v)$ and $z(u,v)$ are the functions of the two variables u and v. Note the interesting opportunities that follow from this syntax. For example, by specifying z as a function of u and v we can break its linear nature. The values of z no longer need to change from A to B. They can oscillate and go back to the same point many times. This will produce rather strange graphs. However, if you do not feel comfortable with this syntax, you can always go back to the well-known school concept of cylindrical coordinates and use the following simplified syntax:

plot::Cylindrical([$r(\phi, z)$, ϕ, z], $\phi = \phi1..\phi2$, $z = z1..z2$, *options*)

As in the case of Spherical plots, we can use the transformations $x = r\cos\phi$, $y = r\sin\phi$ and $z = z$ to convert cylindrical to Cartesian coordinates.

With cylindrical coordinates, as with spherical coordinates, we can produce a number of very interesting surfaces. Here is one of them.

Example 8.11 A coiled tower

Let us plot a very simple function $r = 2\sin(\frac{u^2}{v})$, where $u = -\pi..\pi$ and $v = 0.1..1.5$. For this purpose, we will use the same color function as we did in one of the earlier examples. Here is the MuPAD code:

```
• h := (x,y,z)->[
    abs(sin(x*y)),abs(sin(y*z)),abs(sin(z*x))
  ]:

  Tower := plot::Cylindrical(
    [2*sin(u^2*v), u, v], u = -PI..PI, v = 0..5,
    Submesh = [2,2], Mesh = [100,100],
    FillColorFunction = h
  ):

  MyCam := plot::Camera([5,-30,10], [0,0,2.6], PI/18):
  plot(
    Tower, MyCam,
    Axes = None
  )
```

Here is the result, after a few cosmetic changes in VCam and conversion to black-and-white.

Can you explain why this *coiled tower* becomes so complex in its top section? This is where the value of *v* is close to 5. What would happen if we were to change the range for *v* to 0..1? ∎

8.2.5 Class Implicit3d

We are finally approaching the end of our 3D surfaces story. The class Implicit3d is a new addition to MuPAD, first appearing in version 3.0. It completes the family of mathematical surfaces that can be defined by a single equation or a set of equations. You certainly have noticed that some mathematical equations cannot be easily represented in explicit form $z = F(x,y)$ or in the form of a parametric equation. For example, many *quartics*, i.e. surfaces defined by polynomial equations with three variables of the order 4 are difficult to represent as functions of two variables. I mentioned *quartics* here, as this family contains a number of very interesting surfaces and as I know there are many more still waiting to be discovered. Let us consider one of the many *quartics* from my collection of interesting surfaces and use the Implicit3d plot to get its picture. However, before jumping into this exercise let me show you the syntax of the Implicit3d plot. Here it is:

```
plot::Implicit3d(F(x,y,z), x = x1..x2, y = y1..y2, z = z1..z2, options)
```

Here, $F(x,y,z) = 0$ is an equation of three variables. If the right side of the equation is 0, we can omit it. The equalities $x = x1..x2$, $y = y1..y2$, $z = z1..z2$ are used to declare the range of all three variables, and at the same time the size of the surface's bounding box. Now we are ready to develop our example.

Example 8.12 Kummer's surface

The *Kummer's surface* is a well-known object in algebraic geometry. Its implicit equation has the very nice regular form $x^2y^2 + y^2z^2 + z^2x^2 = A$, where A is a positive number. The equation of the *Kummer's surface* can be solved in respect to any of its variables. We will obtain two equations that represent two halves of the *Kummer's surface*. However, this solutions doesn't look quite as nice as the original equation. Here they are,

$$z_1 = \frac{1}{x^2+y^2}\sqrt{Ax^2 + Ay^2 - x^2y^4 - x^4y^2},$$
$$z_2 = -\frac{1}{x^2+y^2}\sqrt{Ax^2 + Ay^2 - x^2y^4 - x^4y^2} \quad \text{if } x^2 + y^2 \neq 0$$

You certainly agree that neither of the equations looks good, and plotting them even as a Function3d plot is not easy. Therefore, let us see what we can do using the Implicit3d class.

First, let us define the Kummer's equation for $A = 1$. Then we will develop a color function that will change its colors along each axis. Finally, we will declare the *Kummer's surface* with mesh lines. Later you can turn this feature off by changing MeshVisible value to false FALSE. Here is the complete code:

```
• Kummer := x^2*y^2 + y^2*z^2 + z^2*x^2 - 1:

  KummerColor := (x,y,z) -> [
      max(0,sin(2*x)),
      max(0,cos(2*y)),
      max(0,sin(2*z))
  ]:

  KummerSurf := plot::Implicit3d(
      Kummer, x = -3..3, y = -3..3, z = -3..3,
      Mesh = [20,20,20],
      FillColorFunction = KummerColor,
      MeshVisible = TRUE
  ):
```

```
MyCam := plot::Camera([10,10,10],[0,0,0.3],PI/10):

plot(KummerSurf, MyCam,
    Scaling = Constrained,
    Axes = None
)
```

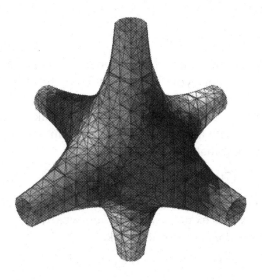

By changing the range for one of the variables to 0..3, we can produce a single half of the *Kummer's surface* and take a look inside the shape. We can also try to develop it with a single color or make its patches transparent. ∎

Implicit plots in 3D space are very slow, due to the complexity of their algorithms. However, quite often this is the only way to produce a surface given by an implicit equation.

Now, when we know how to plot most surfaces, let us move on to curves in 3D. We will start with the Curve3D class.

8.2.6 Classes Curve3d **and** Tube

In this book, you have seen a number of curves drawn in 2D. However, you could also come up with some interesting curves drawn in 3D space. For instance, the famous *Lissajou curves* can be plotted in 3D. You can also find a number of other 2D curves that can be easily converted to 3D.

Here is the syntax of the statement for the Curve3d declaration:

$$plot::Curve3d([x(t),\, y(t),\, z(t)],\, t = t1..t2, options)$$

where $x(t)$, $y(t)$ and $z(t)$ are the functions of the variable t. The curve equation should be given in parametric form.

Example 8.13 A Lissajou curve in 3D

3D curves are not especially exciting when plotted in MuPAD. If you use a single flat color, you may even have trouble distinguishing which part of the curve is closer to you and which is further away. However, you can use a color function to make the graph more appealing. Let's create a quick plot of a curve in 3D. For this purpose, we will take the parametric equation:

$[x,y,z] = [\sin(t)\cos(7t),\, \sin(t)\sin(7t),\, \cos(t)]$, where $t = 0..2\pi$.

In this example, we use the almost the same color function as we did in one of the previous examples. However, we divide each component by 2, thus making the colors much darker.

- `h := (x,y,z)->[abs(sin(x))/2,abs(sin(y))/2,0]:`

```
c1 := plot::Curve3d(
    [sin(t)*cos(7*t), sin(t)*sin(7*t), cos(t)],
    t = 0..2*PI,
    Mesh = 200,
    LineWidth = 0.8,
    LineColorFunction = h
):

plot(c1,
    Scaling = Constrained
)
```

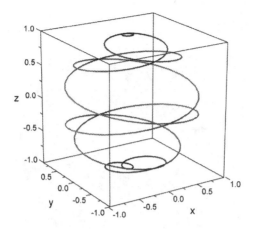

The curve that we have produced is a 3D version of a *Lissajou curve*. By changing its parameters, we may obtain a number of interesting shapes. However, if you wish to produce something more fancy, you should try a Tube plot. Its declaration is almost the same as for a Curve3d plot, but with one additional parameter – the radius of the tube. This means that we produce a tube that wraps around our curve with a given radius. Now, think about the interesting opportunity that this offers. How about declaring the radius in the form of a function of the parameter $r(t)$? Thus, the tube will have a shape with a radius that varies depending on the value of the parameter and the value of the function used for the radius. The syntax of the tube plot is as follows:

$$\text{plot::Tube}([x(t),\, y(t),\, z(t)],\, r(t),\, t = t1..t2, options)$$

Here is our last example again, with an added radius parameter $r = (1.1 + \sin 10t)/10$. I have also added a parameter for animation. The animation will show how the shape of the tube changes depending on the parameter t. Here is the complete code and the final plot. Can you still recognize the shape of our 3D curve?

```
C2 := plot::Tube(
    [sin(t)*cos(7*t),sin(t)*sin(7*t),cos(t)],
    (1.1+sin(10*t))/10,
    t = 0..a,
    Mesh = [200,10],
    a = 1..2*PI
):
```

```
plot(C2, Axes = None)
```

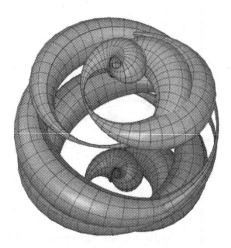

This example concludes the large section related to mathematical curves and surfaces in 2D and 3D space. However, you will still be able to find many other interesting topics and ideas in the later chapters of this book. ■

8.3 Chapter Summary

In this chapter, we have spent a lot of time investigating various ways of plotting curves and surfaces given as functions or parametric equations, in Cartesian, polar, spherical and cylindrical coordinates. Let's make a summary of all the major points covered in this chapter.

Statements for plotting plain curves

plot::Function2d($f(x)$, $x = x1..x2$, *options*)
plot::Curve2d([$g(t)$, $f(t)$], $t = t1..t2$, *options*)
plot::Polar([$r(\theta)$, θ], $\theta = \theta1..\theta2$, *options*)
plot::Polar([$r(t),\theta(t)$], $t = t1..t2$, *options*)
plot::Implicit2d($F(x,y)$, $x = x1..x2$, $y = y1..y2$, *options*)

Statements for plotting curves and surfaces in 3D

plot::Function3d($F(x,y)$, $x = x1..x2$, $y = y1..y2$, *options*)
plot::Surface([$x(u,v)$, $y(u,v)$, $z(u,v)$], $u = u1..u2$, $v = v1..v2$, *options*)
plot::Spherical([$r(u,v)$, $\phi(u,v)$, $\theta(u,v)$], $u = u1..u2$, $v = v1..v2$, *options*)

plot::Spherical($[r(\phi,\theta), \phi, \theta]$, $\phi = \phi1..\phi2$, $\theta = \theta1..\theta2$, *options*)
plot::Cylindrical($[r(u,v), \phi(u,v), z(u,v)]$, $u = u1..u2$, $v = v1..v2$, *options*)
plot::Cylindrical($[r(\phi,z), \phi, z]$, $\phi = \phi1..\phi2$, $z = z1..z2$, *options*)
plot::Implicit3d($F(x,y,z)$, $x = x1..x2$, $y = y1..y2$, $z = z1..z2$, *options*)
plot::Curve3d($[x(t), y(t), z(t)]$, $t = t1..t2$, *options*)
plot::Tube($[x(t), y(t), z(t)]$, $r(t)$, $t = t1..t2$, *options*)

8.4 Programming Exercises

8.4.1 Class Function2d

1. Plot the function $y = 3 \times 2^{-t} \cos 4\pi t$ for $t > 0$. Try to find its asymptotes.

2. Use the Function2d class to produce a graphical solution for the equation $x^2 = \frac{2}{x} - \frac{1}{2}$

3. Plot the function $y = \sin \frac{\pi}{x}$ for $x = -1..1$. Can you explain the strange behavior of the function near $x = 0$?

4. Plot the functions

   a. $y = \dfrac{1 + \cos 6x}{1 + x^2}$

   b. $y = \sin 3x \cos 5x$

   c. $y = \dfrac{\sin x}{1 - x^2}$

   d. $y = \dfrac{1 + x - x^2}{\sin^2 x + 1}$

8.4.2 Class Curve2d

1. The parametric equations $x = \cos at$ and $y = \sin bt$, where a and b are constants, define the family of *Lissajou curves*. Curves like this appear on oscilloscopes in physics laboratories. Try to plot a few *Lissajou curves*. Some of these shapes can be quite interesting.

2. Try to guess the shape of the curve defined by the parametric equations $x = \dfrac{1 - t^2}{1 + t^2}$ and $y = \dfrac{2t}{1 + t^2}$, where $-\infty < t < +\infty$. Plot a graph for it. Were your predictions correct?

3. Plot a graph of the *cycloid* $x = a(t - \sin t)$, $y = a(1 - \cos t)$, where $0 < t < 4\pi$.

4. A curve called an *asteroid* is defined by the equation $x^{2/3} + y^{2/3} = a^{2/3}$. It can be easily proven that its equation can be transformed into the parametric form $x = a\cos^3 t$, $y = a\sin^3 t$. Plot a graph of this curve using its parametric equations. Use $0 < t < 2\pi$.

5. Consider the parametric equation $x = a\cos t - b\cos qt$ and $y = c\sin t - d\sin pt$. Obtain its graph for $a = 16$, $b = 5$, $c = 12$, $d = 3$, $q = \frac{47}{3}$, $p = \frac{44}{3}$ and $t = 0..10\pi$. Try to obtain the most accurate graph. This is a rather difficult case, so you will need to use Mesh=1000.

6. Experiment with the coefficients a, b, c, d, p and q in the previous exercise, in order to obtain different curve shapes.

8.4.3 Polar Plot

1. Use polar plot to explore the functions $r = 2\sin\theta$ and $r = 2\cos\theta$ for $0 \le \theta \le 2\pi$. Use constrained scaling to get correct proportions of the graphs. Can you prove that both these equations describe circles on the XY plane? Can you find the general form of the equation of a circle in polar coordinates?

2. Try to guess the shape of the curve defined by the equation $r = 2 + \sin\theta$ for $0 \le \theta \le 2\pi$, and then plot it in MuPAD. Did you guess its shape correctly? Why does this equation not represent a circle, like both equations from the previous exercise did?

3. Try to guess the shape of the curve given by the equation $r = 2\cos 2\theta$ for $0 \le \theta \le 2\pi$. Now plot it in MuPAD and explain the path of a point on the curve while θ changes from 0 to 2π.

4. The equation of a *lemniscate* curve is given by the equation $r^2 = -4\sin 2\theta$. Examine the table of values of $-4\sin 2\theta$ to see how a graph of this curve might look. Now, find a way to plot its graph in MuPAD. Use the interval $0 \le \theta \le 2\pi$.

5. Find all the points of intersection on the graphs of the equations $r = 1 + \sin\theta$ and $r^2 = 4\sin\theta$.

6. Consider the family of curves defined by the polar equation

$r = \sin n\theta$, where $n = 3..10$ and $0 \leq \theta \leq 2\pi$. Their shape is sometimes called an *n-leaved rose*. Is this name appropriate for the shape of these curves? When it is not appropriate?

7. Plot the *spiral of Archimedes* defined by the equation $r = 3\theta$ for $0 \leq \theta \leq 6\pi$. How do you need to alter the equation in order to make the graph tighter?

8. Obtain a graph of $r = \cos(\frac{7\theta}{4})$ for $0 \leq \theta \leq 10\pi$. Try to guess which of the two numbers 7 and 4 is responsible for the number of leaves. If you are sure of your conclusion, plot a graph for $r = \cos(\frac{7\theta}{3})$. Can you explain the role of the numbers 3 and 4?

9. The equation $r = (A + B\cos n\theta)(C + D\sin p\theta)$, where A, B, C and D are positive integer numbers, describes a family of curves. Plot graphs for a few curves from this family.

10. Try to guess the shape of the graph for the equation $r = (3 + 7\sin 3\theta)\cos 5\theta$. This is indeed very difficult. Obtain its graph for $0 \leq \theta \leq 2\pi$ using MuPAD.

11. The equation $r = e^{\cos\theta} - 2\cos 4\theta$ describes a nice curve sometimes called the *butterfly curve*. Create its graph with MuPAD. Use $0 \leq \theta \leq 2\pi$. It's very interesting to see how the graph will change if you add one more term to the above equation. For example, try to plot $r = e^{\cos\theta} - 2\cos 4\theta + \sin^5(\frac{1}{12}\theta)$ for $0 \leq \theta \leq 24\pi$.

8.4.4 Implicit Plot in 2D

1. Use the implicit plot class to obtain the graph of the *asteroid* $x^{2/3} + y^{2/3} = a^{2/3}$.

2. Use the implicit plot class to obtain the graphs of

   a. $x\cos x + y\cos y = 0$

   b. $x\sin x + y\cos y = 0$

   c. $x\cos x + y\sin y = 0$

   d. $x\cos y + y\cos x = 0$

   e. $x\cos y + y\sin x = 0$

8.4.5 Class Function3d

1. Plot graphs for the following two-variable functions:

    a. $f(x,y) = \sin x \cos y$

    b. $f(x,y) = \sin(x^3 - y^3)$

    c. $f(x,y) = \cos 7y^2 \sin 3x^2$

    d. $f(x,y) = \dfrac{xy(x^2 - y^2)}{x^2 + y^2}$

    e. $z = xy + 5$

    f. $z = 4 - x^2 - y^2$

    g. $z = x^2 + y^2 - 3x + 4y + 2$

    h. $z = 2x^2 + 10x + 3y^3 - 6y^2 + 1$

    i. $z = \dfrac{1}{1 + \cos x + \sin y}$

    j. $z = (x^2 - y^2)e^{-x^2 - y^2}$

    k. $z = e^{(2x - 4y - x^2 - y^2)}$

    l. $z = xye^{-x^2 + y^2}$

    m. $z = xy(x^2 - y^2)$

    n. $z = \dfrac{xy(x^2 - y^2)}{x^2 + y^2}$

8.4.6 Class Surface3d

1. Plot the graphs of the functions represented by these parametric equations:

    a. $(x,y,z) = (u \sin u \cos v,\ u \cos u \cos v,\ -u \sin v)$, where $u = 0..2\pi$ and $v = 0..\pi$.

    b. $(x,y,z) = (\sin u,\ \sin v,\ \sin(u^2 + 2v^2))$, where $u = -3..3$, and $v = -3...3$.

    c. *Steiner's surface* $(x,y,z) = (2 \sin 2u \cos^2 v,\ 2 \sin u \sin 2v,\ 2 \cos u \sin 2v)$, where $u = -\pi/2..\pi/2$ and $v = -\pi/2..\pi/2$.

d. *Enneper's surface* $(x,y,z) = (u - \frac{u^3}{3} + uv^2, v - \frac{v^3}{3} + vu^2, u^2 - v^2)$, where $u = -2..2$ and $v = -2..2$.

e. *Cross Cap surface* $(x,y,z) = (2\sin u \sin 2v, 2\sin 2u \cos^2 v, 2\cos 2u \cos^2 v)$, where $u = -\pi/2..\pi/2$ and $v = -\pi/2..\pi/2$.

f. $(x,y,z) = \left(u\cos v, u\sin v, \cos(uv)\right)$, where $0 \le u \le 2\pi$ and $0 \le v \le 2\pi$.

g. *Dini's surface*
$(x,y,z) = (\cos u \sin v, t \sin u \sin v, \cos v + \log(\tan \frac{v}{2}) + 0.2u)$,
where $u = 0..4\pi$ and $v = 0.001..2$.

h. *Modified Enneper's surface*
$(x,y,z) = \left(u - \frac{u^3}{3} + uv^2, v - \frac{v^3}{3} + u^2v, u^2 + v^2\right)$, where
$u = -3..3$ and $v = -3..3$.

i. *Moebius strip*
$(x,y,z) = \left(\cos u + v\cos \frac{u}{2} \cos u, \sin u + v\cos \frac{u}{2} \sin u, v\sin \frac{u}{2}\right)$,
where $u = 0..2\pi$ and $v = -0.5..0.5$.

j. *Bohemian Dome*
$(x,y,z) = (0.5\cos u, 1.5\cos v + 0.5\sin u, \sin v)$, where
$u = 0..2\pi$ and $v = 0..2\pi$.

k. $(x,y,z) = (\cos 2u \cos(u + v), \cos 2u \sin(u + v), \sin 3v)$, where
$u = -\pi..\pi$, $v = -\pi..\pi$.

l. $(x,y,z) = (\cos u \cos(u + v), \cos u \sin(u + v), \sin 3v))$, where
$u = -\pi..\pi$, $v = -\pi..\pi$.

m. $(x,y,z) = (u\cos v, u\sin v, u + v)$, where $u = -\pi..\pi$ and
$v = -2\pi..2\pi$.

8.4.7 Spherical Plots

1. Use the Spherical class to plot the following functions

   a. $r = 1.2^\varphi \sin A\theta$, where $A = 1,2,3,...7$, $\varphi = -1..2\pi$, and
   $\theta = 0..\pi$.

   b. $r = \theta\varphi$, where $\varphi = -0..2\pi$ and $\theta = 0..\pi$.

   c. $r = \varphi \sin 2\theta$, where $\varphi = 0..\pi$ and $\theta = 0..2\pi$.

d. $r = \pm\sqrt{\cos^2 2\varphi + \sin^2 2\theta}$, where $\varphi = 0..2\pi$ and $\theta = 0..\pi$

e. $r = \pm\sqrt{\cos^2 2\theta + \sin^2 2\varphi}$, where $\varphi = 0..2\pi$ and $\theta = 0..\pi$

f. $r = 1.3^\varphi \tan\theta$, where $\varphi = -1..5\pi$ and $\theta = -\frac{\pi}{4}..\frac{\pi}{4}$

g. $r = 1.2^\varphi \sin\theta^2$, where $\varphi = -1..2\pi$, and $\theta = 0..\pi$

h. $r = 1.2^\varphi \sin\theta^3$, where $\varphi = -1..2\pi$, and $\theta = 0..\pi$

i. $r = \sin(2\theta + 3\varphi)$, where $\varphi = 0..2\pi$, $\theta = 0..\pi$

8.4.8 Classes Curve3d and Tube

1. Plot the following curves in 3D

   a. $(x,y,z) = (3\sin t,\ t,\ 3\cos t)$, where $t = -3\pi..4\pi$

   b. $(x,y,z) = (4\cos t,\ 4\sin t,\ \sin t \cos t)$, where $t = 0..2\pi$

   c. Trefoil knot:
   $(x,y,z) = \left((2 + \cos\tfrac{3}{2}t)\cos t,\ (2 + \cos\tfrac{3}{2}t)\sin t,\ \sin\tfrac{3}{2}t\right)$

   d. $(x,y,z) = (t,\ \sin 5t,\ \cos 5t)$

   e. $(x,y,z) = \left(\sin t,\ \cos t,\ \cos 8t\right)$

   f. $(x,y,z) = \left(t\sin 6t,\ t\cos 6t,\ t\right)$

   g. $(x,y,z) = \left(\cos t\sin 4t,\ \sin t\sin 4t,\ \cos 4t\right)$

2. Plot all the curves from the previous example again, but this time as Tube plots with the radius $r = 1 + a(\sin t + \cos t)$. Choose an appropriate value for the parameter a, so the tubes will not be too thick and the shape of the curves will be still easy to recognize.

A Few Applications of MuPAD Graphics

9.1 Calculus Graphics

In mathematics and science, we often have to represent 3D surfaces in 2D form. In order to do this, we often use vector fields or density plots. In MuPAD we can generate three forms of such representations: the contour plot, the density plot and the vector field. All three of them will be discussed in this chapter.

9.1.1 Plotting Contours

In order to plot the contours of the function $z = f(x,y)$ we must plot a number of horizontal cross-sections, also called contour curves, of the surface. We can plot contour curves using a 3D plot or plot a projection of all the contour curves on the xy-plane. In the latter case, we will get so-called level curves, which form a two-dimensional representation of the three-dimensional surface $z = f(x,y)$, much in the same way a two-dimensional map represents a three-dimensional mountain range. By using different colors for the lower and higher parts of the surface, we can easily determine the height of any particular point on the surface.

Three-dimensional representation is not always as useful as the 2D representation. This representation often fails with more complex surfaces, where some parts of the surface may be hidden behind other parts that are closer to the camera. However, you can also find a number of examples of 3D surfaces for which the three-dimensional plot of contour curves fulfils its role quite well.

First, we must note that in MuPAD 3.0 there is no class that would directly represent contours and level curves. Therefore, in order to produce contours and level curves for a given function $z = f(x,y)$, we will have to work a while; still, we will get them eventually.

Let me remind you that we already mentioned contours while talking about the Implicit2d plot. There, we had an option that allowed us to plot not only the implicit plot of the equation $f(x,y) = 0$ but also the lines representing $f(x,y) = k$, where k is a given number. Therefore, the so-called contours in a 2D implicit plot are, according to mathematical terminology, the level curves of the function $z = f(x,y)$. We also had another example where we used a horizontal transparent plane to intersect it with a surface, allowing us to show the contours of the function. It looks as though we have everything we need to plot level curves and contour curves in 3D.

Example 9.1

Let us start with the function $f(x,y) = \sin xy$. In this example, we will try to investigate how we can obtain contour and level curves for $f(x,y)$. We know that the function sine is periodical and is equal to 0 when $xy = n\pi$. This means that an implicit plot of it will show a number of hyperboles $y = n\pi/x$. In order to see how it might look let us produce the graph of $z = f(x,y)$, and then we will decide which area of the function we wish to use. Here is the graph produced as Function3d for $x = -4\pi..4\pi$ and $y = -4\pi..4\pi$.

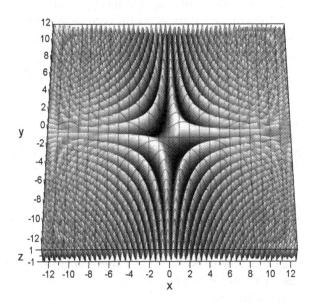

The most interesting part of the graph is close to the point $(0,0)$ where we have a saddle and a number of ranges of hyperbolic hills and hyperbolic valleys starting from it. Therefore, let us concentrate on the area close to the point $(0,0)$, say, for $x, y = -\pi .. \pi$.

Next, we need to decide how many level curves we wish to obtain. We know that the largest value of our function can be 1 and the smallest can be -1. Therefore, five levels for $z = -1, -0.5, 0, 0.5$, and 1 would be just enough.

In order to distinguish between the different levels we must declare a function that will assign different colors to different level curves. This can be done in the same way as in one of the examples that we developed for implicit 2D plots. Here is the complete code for this example, as well as the final result. The color function that I used here makes the lower levels darker and the higher levels lighter. For the purpose of this book, I use black-and-white here. However, you can change the colors to anything you prefer.

- `sf := (x,y)->sin(x*y):`

```
ColFun := proc(x,y)
begin
    if sf(x,y) < -0.75 then
        return(RGB::Black)
    elif sf(x,y) < -0.25 then
        return(RGB::Black80)
    elif sf(x,y) < 0.25 then
        return(RGB::Black60)
    elif sf(x,y) < 0.75 then
        return(RGB::Black40)
    else
        return(RGB::Black20)
    end
end:

LevelCurves := plot::Implicit2d(sf(x,y),
    x = -PI..PI, y = -PI..PI,
    Contours = [-1, -0.5, 0, 0.5, 1],
    Mesh = [25,25],
    LineColorFunction = ColFun
):

plot(LevelCurves)
```

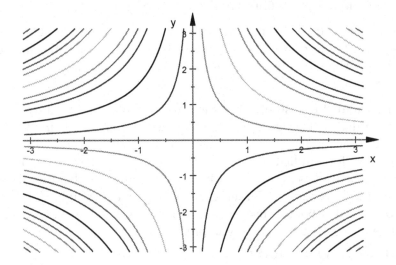

On the black-and-white picture, it is difficult to recognize which level is higher and which is lower. In particular, it is quite difficult to tell how the slope of the surface is formed. A color picture would be much better. For example, you could try to develop this graph using blue colors for the level curves that represent parts of the surface that are below the xy-plane and red colors for parts that are above the it.

Animation is a good way to show the contour curves. In fact, this is the most efficient way to visualize the contour curves. Let us develop an animation that will show us the contour curves for our function. We do not need to change our code too much. We have to replace the Implicit2d plot by Function3d which is much faster to plot. We will then apply a new tool that we hadn't used until now, the ClippingBox. The clipping box forces MuPAD to plot only those parts of the primitives that are inside a given box. For our purpose, we will use the box where $x, y = -\pi..\pi$ and $z = a..a + 0.1$, and $a = -1..0.9$. As you can easily guess, a is our animation parameter, and by applying it to the z-coordinates of the ClippingBox we will cut away both top and bottom parts of the surface showing only the contour curve. Our contour curve will be in fact a thin horizontal cutting of the surface.

Here is the complete code for this example,

```
• sf := (x,y)->sin(x*y):
  h := 0.1:

  LCF := plot::Function3d(sf(x,y), x=-PI..PI, y=-PI..PI,
      Mesh = [50,50]
  ):

  CBox := plot::ClippingBox(
      -PI..PI, -PI..PI, a..a+h, a = -1..1-h
  ):

  plot(LCF, CBox)
```

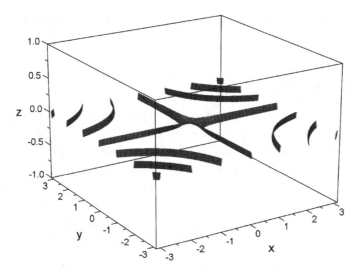

Like in many other examples, a static black-and-white picture does not reflect well the concept that can be shown with animation. Therefore, I suggest that you type in the code of this example in MuPAD, execute it and play the animation to see how it works.   ■

Generally, contour plots in MuPAD are not very accurate. If you wish to get a much more accurate, but of course much more time-intensive plot, you could try using larger values for the Mesh parameter and develop denser contour levels.

A procedure like this will produce a list of *n* numbers from a given minimum value to a given maximum. Such a list can be used as a value for the Contours option.

```
getContours := proc(min, max, n)
local i;
begin
   return(
      [(min + i*(max-min)/(n-1)) $ i=0..(n-1)]
   );
end:
```

Thus, we can easily produce any number of contours for our example by using

- getContours(-1,1,11)

$$[-1,-\tfrac{4}{5},-\tfrac{3}{5},-\tfrac{2}{5},-\tfrac{1}{5},0,\tfrac{1}{5},\tfrac{2}{5},\tfrac{3}{5},\tfrac{4}{5},1]$$

In fact the color function can also be modified to reflect the new number of contours. But this modification I will leave to you for further experiments.

9.1.2 Developing Density Plots

Plotting the density of a surface is another way of showing which parts of the graph are higher than the others. Places that are lower will be plotted using a blue color, while places located higher will have a red color. These of course are just the default colors. You can define your own colors and change the plot completely. Here is the syntax for the density plot,

$$\text{plot::Density}(f(x,y),\ x = x1..x2,\ y = y1..y2,\ options)$$

where the *options* are similar to those used for other plot classes. Let's draw a quick example of a density plot.

Example 9.2

In order to compare the Density plot with the contour plot that we obtained in the last example, we will take the same function as we used previously. Thus, we have:

```
DPF := plot::Density(sin(x*y),
   x = -PI..PI, y = -PI..PI, z = -PI..PI,
   Mesh = [100,100]
):

plot(DPF)
```

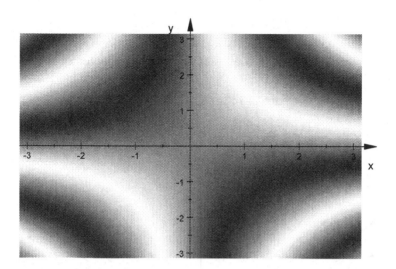

Now, you can compare this plot with the plots from the previous example. Can you identify which parts of the density plot represent the higher parts of the surface and which represent the lower ones?

In this example, we used the Density class in its simplest form. However, we could also use it, for example, to obtain the plot of a matrix or a list of lists of values. ∎

9.1.3 Vector Fields

Vector fields are important in calculus, physics and engineering sciences. A vector field defined in the region U of a plane is actually a vector-valued function f that associates a 2D vector to each point (x, y),

$$f(x, y) = [v(x, y), u(x, y)] = \mathbf{i}v(x, y) + \mathbf{j}u(x, y)$$

We define a vector field in 3D in a similar way. Among the most important applications of vector fields are velocity vector fields, especially those in fluid dynamics.

In MuPAD, we can plot 2D vector fields by producing objects representing the class VectorField2d. The syntax of a 2D vector field object is as follows,

```
plot::VectorField2d([u(x,y), v(x,y)], x = x1..x2, y = y1..y2, options)
```

Example 9.3

Here are the vector field plots for the two functions $f(x,y) = [x,y]$ and $g(x,y) = [\sin(x), \cos(y)]$. In the first case, we plot the vector field for the region $U = [-2,2] \times [-2,2]$.

- ```
  f := plot::VectorField2d(
      [x,y], x = -2..2, y = -2..2,
      Mesh = [15,15],
      Color = RGB::Black
  ):
  ```

  ```
  plot(f)
  ```

Here is the vector field of the function $g(x,y) = [\sin(y), \cos(x)]$ for $x = -5..5$ and $y = -5..5$,

```
f := plot::VectorField2d(
    [sin(y),cos(x)], x = -5..5, y = -5..5,
    Mesh = [25,25],
    Color = RGB::Black
):
```

```
plot(f)
```
```

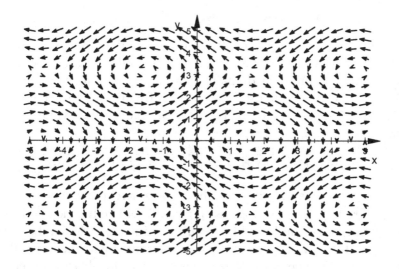

Observe how important the Mesh option is here. By choosing a denser mesh, like we did it in the last case, we were able to obtain a more accurate vector field. This is not especially important while dealing with examples of simple functions, but in most serious applications, the density of the vector field could be quite important.

You can find many interesting examples and applications of vector fields in various university textbooks. Examples from physics, related to fluids or gravitation will be especially valuable. However, even those from a calculus textbook will show you the importance of vector fields. ■

9.1.4 Plotting Areas

In a many situations in calculus, you may need to find the region between two or more curves. For example, you may need to plot the domain of a function with two variables, or solutions of some equations or inequalities. We have a few tools in MuPAD 3.0 that might be useful in such situations. In this section, we will explore the two important classes Inequality and Hatch. Let us start with plots of inequalities.

Imagine that you have to plot the area T, where

$$T = \{(x,y) \,|\, f_1(x,y) > 0 \text{ and } f_2(x,y) > 0 \ldots \text{ and } f_n(x,y)\}$$

For example, you may be looking for a set of the points (x,y) where $x^2 + y^2 < 1$ and $x^2 + y^2 > 0.5$, or something similar. This can be useful when trying to find the domain of a composite function. MuPAD provides us with a very useful tool to deal with such problems. This tool is the `Inequality` class. Let us first analyze the syntax of this function. Here it is:

plot::Inequality($[f1,f2,\ldots,fn], x = x1..x2, y = y1..y2, options$)

The above command will plot the area where the inequalities f_1, $f_2,\ldots,f_n$ are true. The expressions $f_1, f_2, \ldots, f_n$ will be the inequalities of the two variables x and y, while $x = a..b$ and $y = c..d$ define the range for the plotted area. If you do not define your own colors, MuPAD will use its default colors, which in this case are blue, black and red.

Example 9.4 Plotting the domain of a function

The function $y = \sqrt{(1 - x^2 - y^2)(y - x^2)}$ is defined if both the multipliers are positive or both of them are negative. This leads us to two possible cases:

1. $1 - x^2 - y^2 > 0$ (i.e. $x^2 + y^2 < 1$) and $y - x^2 > 0$ (i.e. $y > x^2$)
2. $1 - x^2 - y^2 < 0$ (i.e. $x^2 + y^2 > 1$) and $y - x^2 < 0$ (i.e. $y < x^2$)

Here, I demonstrate how the first area can be represented by the inequality plot.

- p1 := plot::Inequality(
    [x^2 + y^2 < 1 , y > x^2], x = -2..2, y = -2..2,
    Mesh = [200,200]
  ):

  plot(p1, Scaling = Constrained, Axes = Boxed)

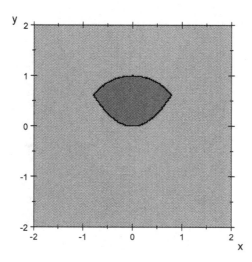

The dark region in the picture is the set of points where $(1-x^2-y^2)(y-x^2) > 0$ and both terms are positive. The light one is where the inequality is false. Finally, the border between both areas is shown as a black line. By default, MuPAD uses blue, red and black colors for the true, false and border areas respectively. It also uses a black color for the areas where it is impossible to determine if the inequality is true or false. Of course, we can define our own colors as options in the Inequality plot, as you will see it shortly.

Now, let us plot the second region of the domain in exactly the same way. All you need to do is change the sign of each plotted inequality. In this plot, we will use a light gray color for the false area and a dark gray color for the true area. In order to do this, we will need to assign appropriate color values to two parameters—FillColorFalse and FillColorTrue.

```
• p2 := plot::Inequality(
      [x^2 + y^2 >1, y<x^2], x = -2..2, y = -2..2,
      Mesh = [200,200],
      FillColorFalse = [0.9,0.9,0.9],//light gray
      FillColorTrue = [0.4,0.4,0.4] //dark gray
   );

   plot(p2,Scaling = Constrained,Axes = Boxed)
```

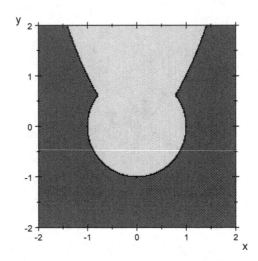

Now you should combine both pictures in order to obtain the domain of our function.   ∎

Another useful tool to plot areas is the Hatch class. We can use it to plot the area between two 2D functions or a 2D function and a line like $y = c$, or the interior of a closed curve given by parametric equations. Let us explore a simple example from a calculus textbook.

Example 9.5

While using integrals to calculate the area enclosed between two or more curves, we often need to plot the area first and then define the limits of the integral. Suppose that we have to find the area of the region bounded by the line $y = 2x$ and the parabola $y^2 = 3 - x$. Let us try to plot this region using the Hatch class. First we need to draw both objects and see how the region is constructed. In fact we have to plot three objects, as the parabola can be represented by two functions, $y = \sqrt{3 - x}$ and $y = -\sqrt{3 - x}$. Let us plot all of our functions using the plotfunc2d command.

- ```
y1 := x -> sqrt(3-x):
y2 := x-> -sqrt(3-x):
y3 :=x -> 2*x:

plotfunc2d(y1, y2, y3, x = -2..3)
```
```

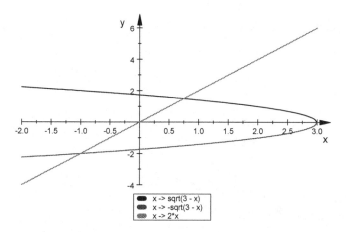

In the next step, we need to calculate the two points of intersection. In this example, we can read their values from the graph, but in other examples it may not be so easy. We need to execute just two short statements:

- solve(2*x = sqrt(3-x),x)

$$\left\{\frac{3}{4}\right\}$$

- solve(2*x = -sqrt(3-x),x)

{-1}

Finally, we know that we should split the region into two subregions, where $-1 \leq x \leq 3/4$ and $3/4 \leq x \leq 3$. So, we can start typing in our code.

```
Y1 := plot::Function2d(sqrt(3-x), x = -1.5..3.5):
Y2 := plot::Function2d(-sqrt(3-x), x = -1.5..3.5):
Y3 := plot::Function2d(2*x, x = -1.5..3.5):

H1 := plot::Hatch(Y3, Y2, -1..3/4,
    FillColor = RGB::Black,
    FillPattern = DiagonalLines
):

H2 := plot::Hatch(Y1, Y2, 3/4..3,
    FillColor = RGB::Black50,
    FillPattern = FDiagonalLines
):
```

```
plot(H1, H2, Y1, Y2, Y3)
```

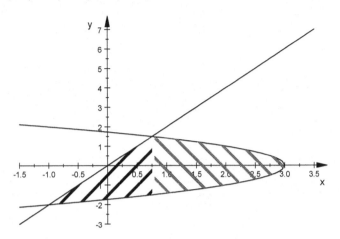

Note that the Hatch object contains only the shaded region between the two objects that were defined as Function2d. If you wish to plot the Hatch area as well as the graph of the functions, you will need to plot all of them like I did it here.

I used two different patterns for the hatch area here. We can also use others, like Solid, HorizontalLines, VerticalLines, CrossedLines and XCrossedLines.

9.1.5 Surfaces of Revolution

Have you ever tried to draw the surface of the revolution of a curve around an axis? If you have tried this, then you certainly know how difficult it is to get a good plot for it. There are two classes in MuPAD that provide an easy way to obtain such plot. These are XRotate and ZRotate. The names of the classes suggest that we are talking here about the rotation of a curve defined by the equation $z = f(x)$ around the x-axis and z-axis respectively.

Example 9.6 Plotting surface of revolution

The object called the *Gabriel's horn* is obtained by revolving the curve $z = 1/x$, for $x \geq 0$ around the x-axis. One of the typical problems in calculus is to prove that the surface area from $x = 1$ to $x = b$ is $S_b \geq 2\pi \ln b$. This in fact means that the total area of *Gabriel's horn* is infinite.

In this example I will show you how you could obtain a quick plot of *Gabriel's horn.*

* ```
horn := plot::XRotate(1/x, x = 1..5,
    Mesh = [20,20],
    AngleRange = -PI..a, a = -PI..PI
):
PL := plot::Function3d(0, x = 1..5, y = -1..1):
plot(horn, PL, Axes = Boxed)
```

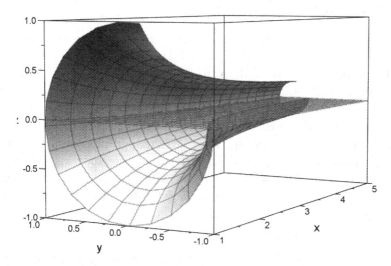

Now, you can rotate the surface using VCam, see it inside and analyze its shape. The function in this example is very simple. However, you could use any other function of one variable from your calculus classes.

In the same way, you could plot the surface obtained by rotating $z = 1/x$ about the z-axis. You would need to replace XRotate with the ZRotate command. Of course, you may try to change the range of the variable $x$ and a few other parameters, but the idea will remain the same. ∎

I did not show you the general syntax for the XRotate and ZRotate commands. This is because the use of these functions is rather simple. However, in order to have all the information, here is the syntax for both commands.
```

> plot::XRotate(f, $x = x1...x2$, *other_options*)
>
> plot::ZRotate(f, $x = y1..y2$, *other_options*)

In both cases, you can use `AngleRange=c..d` to display just the part of the surface of the revolution obtained by rotating the curve from angle c to the angle d.

9.2 Presentation Graphics

In the previous sections of this chapter as well as in the previous chapters of this book, we have plotted data in the form of continuous functions or functions with a limited number of discontinuities. However, in mathematics, science and other disciplines, we also deal with discrete functions. For such functions, both their domain and range are sets of discrete values. Thus, in order to plot them, we need another set of tools. Usually, we plot such graphs using Excel or another spreadsheet package. But what if we produce a sequence of numbers in MuPAD, and then wish to visualize them? Just imagine that you are exploring the sequence $a_n = \left(\frac{n-1}{n+1} \right)^n$ and you wish to see the behavior of a_n as $n \rightarrow \infty$. This can be done quite easily in MuPAD. You will get a nice visual illustration of a_n.

We have to start by declaring the points representing our sequence. This means we have to create a sequence of points with two coordinates (n, a_n). We will then be able to plot our points using the `plot::PointList2d` class or the `plot::Polygon2d` class. In the first case, we will obtain a picture of the terms of our sequence plotted in the 2D Cartesian coordinate system. In the second case, we will get a non-closed polygon. Before we begin our work let us look at the syntax for both classes:

> plot::PointList2d([[$n1, m1$], [$n2, m2$],...,[nk, mk]], *options*)

and

> plot::Polygon2d([[$n1, m1$], [$n2, m2$],...,[nk, mk]], *options*)

Depending on our needs, the `plot::Polygon2d` can be closed and filled. However, for this example we will use a non-closed polygon.

Later we will explore closed and filled polygons.

Let us start by declaring the points for our sequence. Like I said before, we have to create a list of points. In MuPAD a 2D point is in itself a list of two coordinates. Therefore we have,

- an := [[n,((n-1)/(n+1))^n] $ n = 1..25]:

In the next step, we can plot this sequence.

```
• DataList :=plot::PointList2d(an,
      PointSize = 2*unit::mm,
      PointColor = RGB::Red
  ):

  plot(
      DataList,
      GridVisible = TRUE,
      SubgridVisible = TRUE
  )
```

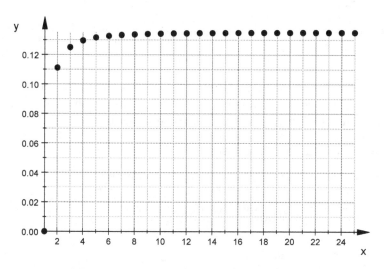

The same sequence, plotted as plot::Polygon2d, will look slightly different.

```
• DataList2 := plot::Polygon2d(an,
      LineColor = RGB::Black,
      PointsVisible = TRUE,
      PointSize = 2*unit::mm
  ):
```

```
plot(
    DataList2,
    GridVisible = TRUE,
    SubgridVisible = TRUE
)
```

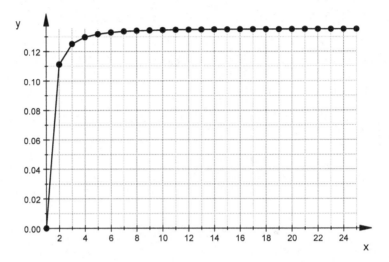

There are a few other ways of plotting discrete data. We can visualize them as bars, piechart, use matrix plots or histograms for statistical data. Let us quickly review some of these possibilities.

The Bars2d and Bars3d classes

The syntax to declare the Bars2d class is following,

$$plot::Bars2d(data, options)$$

and

$$plot::Bars3d(data, options)$$

where *data*, in each case, is a list of lists of numbers. In case of Bars2d, we can use a single list of numbers. In both declarations, *options* may include many of the common plotting options that have already been mentioned, or options specific for bar graphs. We can have separate colors for each bar, bars with shadows, bars plotted vertically or horizontally, bars with or without the ground. Bars can be plotted as boxes, lines, lines and points or just points. Let us see a few small examples. In the first one, we will plot the sequence from the previous example with colors declared as a sequence of colors, shadows for bars and bars as boxes. Here is the complete code.

```
bn := [((n-1)/(n+1))^n $ n = 1..25]:
cl := [[2/3-1/n,2/3-1/n,2/3-1/n] $ n = 1..25]:

MyData := plot::Bars2d(bn,
    Colors = cl,
    Shadows = TRUE,
    BarStyle = Boxes
):

plot(MyData)
```

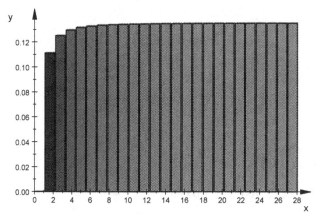

The same plot without shadows, and black bars as lines and points, can be produced as follows.

```
• bn := [((n-1)/(n+1))^n $ n = 1..25]:
  cl := [RGB::Black $ n = 1..25]:
  MyData := plot::Bars2d(bn,
      Colors = cl,
      Shadows = FALSE,
      BarStyle = LinesPoints,
      PointSize = 2
  ):

  plot(MyData)
```

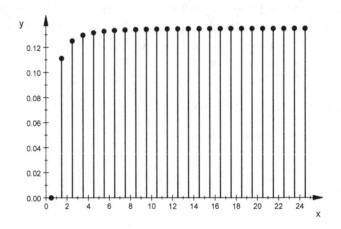

Now, here is a simple example of a Bars3d plot.

```
• MyNumbers := [
    [-2,-5, 5,10, 7,-10],
    [ 3, 2, 6, 5, 2, 18],
    [ 1,-1,-5,15,12,-10]
  ]:
  MyData := plot::Bars3d(
    MyNumbers,
    Colors = [RGB::Black50]
  ):
  plot(MyData, Axes = Boxed)
```

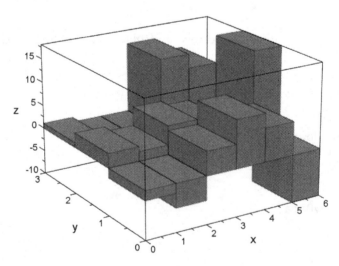

The Piechart2d **and** Piechart3d **classes**

Both these classes can be used to plot a single list of numerical data; you cannot use them with a list of lists of data. The syntax is essentially the same as what you already know:

```
plot::Piechart2d(list_of_data, options)
plot::Piechart3d(list_of_data, options)
```

9.3 Working with Geometry Objects

If you have ever used Cabri or Geometer's Sketchpad before, you could be a little disappointed when learning to work with geometry objects in MuPAD. MuPAD is a command line programming environment, and you won't be able to manipulate objects as much as you could in Cabri. Here, you have to describe objects by giving their coordinates in 2D or 3D space and write programs manipulating these objects. Nonetheless, you can still construct a number of interesting 2D and 3D objects and animate them, thus building some interesting visualizations.

In the first edition of this book, I wrote "the class of geometric objects that you can develop in MuPAD is very basic". Since version 3.0 of MuPAD, this is no longer true. MuPAD now contains a number of new geometry objects and certainly some new ones will be created in the future. So, now we have points, lines, polygons, parallelograms, polyhedra and a few other objects. We also have a few operations that can be used on these objects, such as scaling, rotating, translating and transformations. All this can be done on a 2D plane or in 3D space. We can expect that the number of geometry objects in MuPAD will grow significantly when users start developing their own models and animations.

By issuing the command info(plot), you will be able to find out what geometry classes are included in your version of MuPAD. Here is the list of geometry objects and operations that I have in my copy of MuPAD at the time of writing this text.

Geometry classes

plot::Arc2d - *2D arcs of a circle*

plot::Arrow2d - *arrows in 2D*

plot::Arrow3d - *arrows in 3D*

plot::Circle2d - *circles in 2D*

plot::Circle3d - *circles in 3D*

plot::Cone - *cones*

plot::Cylinder - *cylinders*

plot::Ellipse2d - *ellipses in 2D*

plot::Ellipsoid - *ellipsoids*

plot::Line2d - *segments connecting two points in 2D*

plot::Line3d - *segments connecting two points in 3D*

plot::Lsys - *Lindenmayer systems*

plot::Parallelogram2d - *parallelograms in 2D*

plot::Parallelogram3d - *parallelograms in 3D*

plot::Point2d - *points on a 2D plane*

plot::Point3d - *points in 3D space*

plot::Polygon2d - *open and closed polygons on a 2D plane*

plot::Polygon3d - *open and close polygons in 3D space*

plot::Rectangle - *rectangles on a 2D plane*

plot::Sphere - *spheres in 3D space*

plot::SurfaceSet - *objects defined by mesh of points*

plot::Turtle - *turtle graphics*

Platonic polyhedra

plot::Dodecahedron - *dodecahedrons*

plot::Hexahedron - *hexahedrons*

plot::Icosahedron - *icosahedrons*

plot::Octahedron - *octahedrons*

plot::Tetrahedron - *tetrahedrons*

The commands used to generate geometry objects are very intuitive. For example, in order to declare a point, you only need to give its coordinates and perhaps its color. In order to declare a line, you need to give the coordinates of the two ending points of this line. Let's take a quick tour through some of the classes that represent geometric objects. We'll start with points:

> plot::Point2d(x, y, *options*)
> plot::Point3d(x, y, z, *options*)

We can also declare points using square brackets, like here

> plot::Point2d([x, y], *options*)
> plot::Point3d([x, y, z], *options*)

It is very important to understand the significance of the Point2d declaration at this stage. It produces a geometric object out of an ordered pair of two numbers. You might think that in order to define a point on the plane it is enough to use a declaration like, P:=[2, 3]. You will find that this does not work, however. Just try the following code:

- A := [1,2]:
  plot(A)

This gives you the output

> Error: unexpected arguments: [1, 2] [plot::Canvas::new]

This means that A is a list of two numbers, but it is not a geometry point with all of its properties like color, size and shape.

While declaring points, we describe their color by declaring its RGB coordinates or its name from the RGB library. We can also use the options PointSize and PointStyle to declare the size and shape of points, and we can create labels for each point with the commands Title and TitlePosition. It is easy to declare single points when we only need to declare a few of them. However, in situations where we need a larger set of points, it is very convenient to use MuPAD's programming language. Here is an example of such a situation.

Example 9.7

Imagine that we need to declare a series of points with the coordinates (i, i) and their projections on the x-axis for $i = 1..10$. This can be achieved by declaring the two sequences of points and then plotting all of them with a single plot command. There are a few ways of doing this. Here, I will describe just one of them.

- a := plot::Point2d(
    [i,i],
    PointColor = [i/10, 1-i/10, 1-i/10],
    PointSize = i,
    PointStyle = FilledCircles
) $ i = 1..10:

  b := plot::Point2d(
    [i,0],
    PointColor = [i/10, 1-i/10, 1-i/10],
    PointSize = i,
    PointStyle = FilledCircles
) $ i = 1..10:

  plot(a,b)

And here is the output that we get from MuPAD.

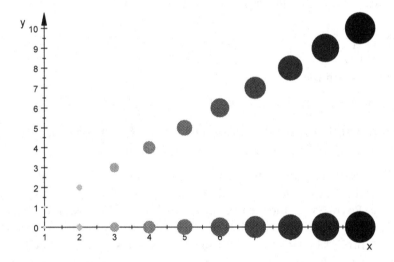

You certainly have noticed that in the above example, we were able to declare not only a different position for each point, but also a different size and color. ∎

Now you can easily find ways to plot sequences of points for geometry constructions in 2D and 3D. Note that when you declare objects that are built out of multiple points, in the final construction each point is a separate object. This means that the objects browser for the scene is crowded by tens or hundreds of points. If you wish to simplify the scene tree and have easier access to other objects, you should use point lists. There is another advantage of the point list—in MuPAD this is just a single object. Therefore point list uses less memory and plots faster than thousands of separate points.

We have two types of point lists—PointList2d and PointList3d. Objects declared as point lists are shown in the objects browser as single objects. There is a limitation, however, as all points in a point list have the same size and all other properties with one exception—each point may have a different color. Let us see what we can do with PointList2d and PointList3d.

Here is the syntax of the commands used to declare a point list:

plot::PointList2d(*Point1*, *Point2*, ..., *Point_n*, *options*)

plot::PointList3d(*Point1*, *Point2*, ..., *Point_n*, *options*)

Example 9.8

In MuPAD, we have a few ways to declare random data. One of the procedures that we can use to create random numbers is the function frandom(), which creates a random number from the interval $[0, 1]$. A good random function should cover the whole interval entirely, in this case $[0, 1]$, if we use it repeatedly enough. Let us see how good the frandom function is in MuPAD. We will create 5000 points from the square $[0, 1] \times [0, 1]$. These points should cover the square densely and uniformly. Here is the complete code for this experiment.

```
• NR := 5000:
  RandomData := [
     [frandom(), frandom(), // point coordinates
        [frandom(), frandom(), frandom()] //color
     ] $ i = 1..NR]:

  RandomPoints := plot::PointList2d(
     RandomData,
     PointStyle = FilledCircles,
     PointSize = 1.5*unit::mm
  ):

  Square := plot::Rectangle(0..1, 0..1,
     LineColor = RGB::Black
  ):

  plot(RandomPoints, Square, Scaling = Constrained)
```

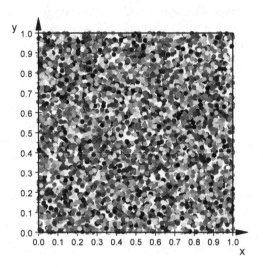

Try to experiment with this code using different values for the constant NR. As you can easily see, the function frandom() represents the concept of random data quite well.

Let me explain a few of the things that I used in this example. I used a very unusual way of declaring points in the point list here. It would be enough to declare a list of random points like this,

```
[[frandom(), frandom()] $ i = 1..NR ]
```

This would create a list of NR pairs of random numbers from the interval $[0, 1]$. The role of the additional triplet in my declaration is to declare a separate color for each point. So, for example, this way

```
    [[1,2,RGB::Yellow], [3,4,RGB::Blue]]
```

we can declare a list with two points where one point is yellow and the other is blue.

In our case we declare 5000 points and each of them with its own color. Note that the function frandom() is very useful for declaring colors, as it always produces numbers from the interval $[0, 1]$.

I also used the class Rectangle in this example. Its declaration is very intuitive. We will talk more about rectangles and similar primitives later on. ∎

Line2d and Line3d are other useful classes. We may declare a line by defining the coordinates of the two ending points of the line. The syntax of a line objects is as follows:

> plot::Line2d([$x1,y1$],[$x2,y2$], *options*)
>
> plot::Line3d([$x1,y1,z1$],[$x2,y2,z2$], *options*)

We can apply all the options here that are relevant to lines, such as LineWidth, LineColor, etc. You can also create a more complicated object using lines. For example, you could create a sequence of lines and then plot the entire sequence.

Example 9.9

Let A, B, P and Q be four different points in 3D. Divide the segment of the line connecting points A and B into n equal pieces. Then, join the end of each piece with the two given points P and Q. This is just the general form of a well-known geometry exercise where you need to connect the center of a segment with another point. Here, the major mathematical difficulty is to split a segment into n equal segments. From the MuPAD programming point of view, this is an interesting problem, as you can practice the ability to create complex geometric objects. Let us start with the mathematics and develop the formulae for the points of division.

It must be noted that each coordinate can be considered separately as a one-dimensional axis. Thus, when we have two points a and b on it, we have a segment (a, b). Then, $1/n$ of the segment is equal to $(b - a)/n$ and the coordinates of the division points can be expressed by the formula $c_i = a + i(b - a)/n$. From here, it is easy to conclude that the points dividing segments A and B into n equal pieces will have these coordinates:

$$C_i = [a_x + i\frac{(b_x-a_x)}{n}, a_y + i\frac{(b_y-a_y)}{n}, a_z + i\frac{(b_z-a_z)}{n}], \ i = 0,..,n,$$

where $A = [a_x, a_y, a_z]$ and $B = [b_x, b_y, b_z]$. You have probably already guessed that this is the end of the mathematical part of this exercise. Now it is time to go back to MuPAD. We shall start with the declarations of the points A, B, P, Q and the number n. Of course, we shouldn't forget to export the Point3d and Line3d procedures.

- export(plot, Point3d, Line3d):
- ax := 10:
  ay := 0:
  az := 10:
  bx := 0:
  by := 10:
  bz := 10:
  A := [ax,ay,az]:
  B := [bx,by,bz]:
  P := Point3d([0,0,0], PointSize = 2.60):
  Q := Point3d([10,10,10], PointSize = 2.60):
  n := 7:

The key construction in this example will be the creation of a sequence of points and two sequences of lines. We will use exactly the same scheme as in the previous example. Thus, we have this sequence of points from the segment $[A, B]$:

- points := Point3d(
    [ax+i*(bx-ax)/n, ay+i*(by-ay)/n, az+i*(bz-az)/n],
    PointColor = [i/n, 1-i/n, i/n],
    PointSize = 1.5,
    PointStyle = FilledCircles
) $ i = 0..n:

a sequence of lines connecting the point P and the points on $[A, B]$,

- lines1 := Line3d([0,0,0],
    [ax+i*(bx-ax)/n, ay+i*(by-ay)/n, az+i*(bz-az)/n],
    LineColor = [i/n, 1-i/n, i/n]
)$ i = 0..n:

and a sequence of lines connecting the point Q and the points on $[A, B]$:

- lines2 := Line3d([10,10,10],
    [ax+i*(bx-ax)/n, ay+i*(by-ay)/n, az+i*(bz-az)/n],
    LineColor = [i/n, 1-i/n, i/n]
) $ i = 0..n:

Finally, we can add the line connecting the points A and B, and plot the entire scene.

```
line0 := Line3d(A, B, LineWidth = 0.15):
plot(P, Q, points,line0, lines1, lines2)
```

Here is the final result:

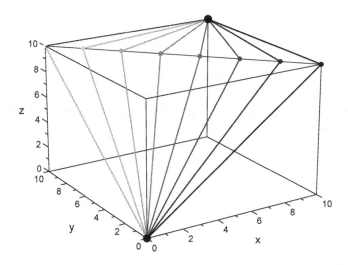

The above plot gives you some idea of what interesting geometric constructions you can develop using sequences of points and lines. However, there is still a bit more to learn.   ∎

One of the most useful classes in the plot library are the classes Polygon2d and Polygon3d. Before we go on to the next example, let me quickly explain how to declare polygons. As with other geometry objects, the declaration of a polygon follows the standard format.

plot::Polygon2d(*Point1*, *Point2*, *Point3*, ..., *Point_n*, *options*)

plot::Polygon3d(*Point1*, *Point2*, *Point3*, ..., *Point_n*, *options*)

where *Point1*, *Point2*, ..., *Point_n* are coordinates of points on the plane or in 3D space. You can use a number of options that you have already learned in this book. However, the most useful options are Closed, Filled, LineWidth and LineColor. You know all these options already, so I do not need to remind you about them. There is one thing here that is important, though. The option Filled=TRUE   on a 2D plane can be applied to all polygons. However, in 3D it can only be applied to triangles. Thus, if you wish to develop a filled polygon in 3D you will have to build it out of filled triangles. Now we are ready to look at a nice example where we use Polygon3d.

Example 9.10

In geometry, we often deal with various pyramids. You have already noticed perhaps that there are no classes to declare a pyramid in the MuPAD plot library (version 3.0). Well, this is good, as you can make one on your own. We will develop a regular pyramid — that is, a pyramid where the base is a equilateral polygon and its top point is located exactly over the center of the base.

Every time we draw a pyramid, we need to determine its size and the number of sides it has. The size can be defined, for example, by giving the height of the pyramid and the radius of the circle bounding the base. Thus, we will start our code with a declaration of all the necessary constants:

```
• r := 1:
  height := 2:
  n := 11:
  angle := 2*PI/n:
  topPoint := [0,0,height]:
  bottomPoint := [0,0,0];
```

In the next step, we can build the base of the pyramid.

```
base := plot::Polygon3d (
    [bottomPoint,
     [r*cos(i*angle), r*sin(i*angle), 0],
     [r*cos((i+1)*angle),r*sin((i+1)*angle),0]],
    Closed = TRUE,
    Filled = TRUE
) $ i = 0..n-1:
```

Finally, we will have to develop the sides of the pyramid. We will make them exactly the same way we made the base. We should also add some colors to the side triangles, otherwise everything will be plotted in the default colors.

```
sides := plot::Polygon3d ([
    topPoint,
    [r*cos(i*angle), r*sin(i*angle), 0],
    [r*cos((i+1)*angle), r*sin((i+1)*angle), 0]],
    Closed = TRUE,
    Filled = TRUE,
    FillColor = [i/n, 1-i/n, i/n],
    LineColor = RGB::Black80
) $ i = 0..n-1:
```

Now we are ready to see what we have developed. Let us type in the plot command. Do not forget, the picture of the pyramid should be plotted in the right proportions. So, we will use Scaling= Constrained. We can also remove the axes from the plot.

- plot(base, sides, Axes = None, Scaling = Constrained)

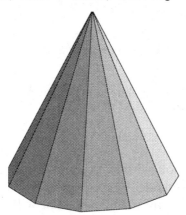

The resulting picture looks quite nice. You could develop a number of other geometry solids this way. However, before you start doing this, you might want to consider making some useful changes to our program.

It would be very good if we could have a procedure Pyramid(...), which we would call with some parameters. From our current program, it's only one small step to such a procedure. All we need to do is to move some variables to the header of the procedure, declare some other variables used inside our program as local variables, and finally take care to ensure the proper output of data from the procedure. The final form of the procedure might look like this.

```
• Pyramid := proc(r,height,n)
    local i, angle, topPoint, bottomPoint, base, sides;
  begin
    angle := 2*PI/n:
    topPoint := [0,0,height]:
    bottomPoint := [0,0,0];

    // ------ base declaration
    base := plot::Polygon3d (
        [bottomPoint,
        [r*cos(i*angle), r*sin(i*angle), 0],
        [r*cos((i+1)*angle),r*sin((i+1)*angle),0]],
```

```
      Closed = TRUE,
      Filled = TRUE
   ) $ i = 0..n-1:

   // ------ sides declaration
   sides := plot::Polygon3d ([
      topPoint,
      [r*cos(i*angle), r*sin(i*angle), 0],
      [r*cos((i+1)*angle), r*sin((i+1)*angle), 0]],
      Closed = TRUE,
      Filled = TRUE,
      FillColor = [i/n, 1-i/n, i/n],
      LineColor = RGB::Black80
   ) $ i = 0..n-1:
   return(base, sides)
end:
```

You have perhaps noticed that in some of our examples, whenever I wish to simplify the code, I export the graphical classes that will be used in the example and then I use them without the plot:: prefix. However, when developing procedures to produce graphical objects, we should avoid exporting anything, and use the plot:: prefix instead. This way our procedures can work independently, like a black box. Of course we should also protect our procedure against some other errors that the users can make. In our example, you should add statements checking if the input data is correct, and print error messages if something is wrong. I will leave this task for your future improvements. Now it is time to see how the procedure works. Let's create a pyramid with 30 sides, with radius of the base equal 3 units and a height 2 units.

```
Pyr := Pyramid(3,2,30):
plot(Pyr, Scaling = Constrained, Axes = None)
```

You could try to develop a number of similar procedures. For example, how would you construct a prism? ■

One of the most interesting additions in the plot library that came with MuPAD 3.0 is the SurfaceSet class. This is incredibly useful graphical class that can be used to create various 3D shapes using a mesh of points. The syntax of the declaration is quite complex. Depending on some options, we can get different shapes from the same data. It uses a more sophisticated color definition than other plot objects. In the pages of this book, I will show you the simplest way to declare a SurfaceSet object in order to create something interesting. Here is the syntax to declare a SurfaceSet object,

> plot::SurfaceSet(*mesh_list*, MeshListType=*type*, *options*)

where *mesh_list* is a list of point coordinates in 3D and the MeshListType can be Triangles, TriangleStrip, TriangleFan, Quads, QuadStrip and ColorQuads.

Example 9.11 Stella octangula

Stella octangula is a nice 3D solid that can be created and animated using eight triangles in 3D space. In fact, we produce a *tetrahedron* and we make a copy of it and then rotate the copy 180° with respect to the original *tetrahedron*. We will start by producing a *tetrahedron*, and then we will add a color function; finally we will produce its copy and rotate it.

```
• meshList := [
  //First triangle
    1.0,-1.0,-1.0,-1.0,-1.0, 1.0, 1.0, 1.0, 1.0,

  //Second triangle
    -1.0, 1.0,-1.0,-1.0,-1.0, 1.0, 1.0, 1.0, 1.0,

  //Third triangle
    -1.0, 1.0,-1.0,-1.0,-1.0, 1.0, 1.0,-1.0,-1.0,

  //Fourth triangle
    -1.0, 1.0,-1.0, 1.0, 1.0, 1.0, 1.0,-1.0,-1.0
  ]:
```

```
FS := plot::SurfaceSet(meshList,
    MeshListType = Triangles
):

plot(FS, Scaling = Constrained):
```

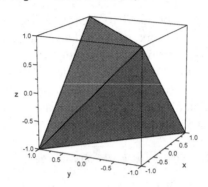

Before we proceed further, let us see how the object SurfaceSet is constructed. First, we declared a list of numbers. In this case, we treat each triplet in this list as the coordinates of a single point and three consecutive triplets as a single triangle. Then, in SurfaceSet, we use the option MeshListType=Triangles to be sure that MuPAD will interpret our data the same way as we do. However, instead of Triangles we can use other values that were mentioned earlier. In each case, the interpretation of data will be different. I suggest you experiment a bit with the different options to see what you can produce.

In the next stage, we will produce a rotated copy of the *tetrahedron*. For this purpose, we will use the plot::Rotate3d class. We have to define the angle of rotation one point and a vector on the axis of rotation. We will learn more about rotations in a minute. For now, let us return back to our example. Here is the complete code to produce our *stella octangula*.

```
• meshList := [
  //First triangle
      1.0,-1.0,-1.0,-1.0,-1.0, 1.0, 1.0, 1.0, 1.0,

  //Second triangle
      -1.0, 1.0,-1.0,-1.0,-1.0, 1.0, 1.0, 1.0, 1.0,
```

```
//Third triangle
  -1.0, 1.0,-1.0,-1.0,-1.0, 1.0, 1.0,-1.0,-1.0,

//Fourth triangle
  -1.0, 1.0,-1.0, 1.0, 1.0, 1.0, 1.0,-1.0,-1.0
]:

FS := plot::SurfaceSet(meshList,
   MeshListType = Triangles
):

FS1 := plot::copy(FS):
FS1 := plot::Rotate3d(PI/2, [0,0,0], [0,0,1], FS1):

plot(FS,FS2, Scaling = Constrained):
```

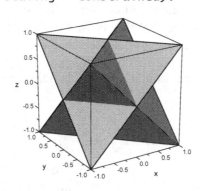

The final step of this creation is to add a color function to the code and change the code into a procedure. For this example, we will use a specific color function that depends not only on the variables x, y, z but also on n, where n is the number of a point in the mesh list. Here is the declaration of our color function:

```
rainbow := (n,x,y,z)->
   [[1,0,0],[0,1,0],[0,0,1],[1,1,0]][(n mod 4)+1]:
```

Note that we have defined here a list of four colors, and then we selected one of the colors from the list using an operation $(n \bmod 4 + 1)$.

Finally, we use our color function for the object FS.

```
FS := plot::SurfaceSet(meshList,
   MeshListType = Triangles,
   FillColorFunction = rainbow
):
```

I will not show you here the final picture that I got. It is not possible to represent rainbow colors properly in a gray picture. So, I suggest that you copy the code of this example from the book web site or type it on your own, add the color declaration and execute it in MuPAD. This way, you will find out more about color function in SurfaceSet than I can show here.   ■

The remaining geometry objects in the plot library are not as exciting as points, lines and polygons, but still useful. Let us see what is left. We will start with Rectangle.

plot::Rectangle(*x1..x2, y1..y2, options*)

where (*x1,y1*) is the bottom-left corner of the rectangle, and (*x2,y2*) is the opposite point. Note that this is only a 2D rectangle. In 3D space, we can declare a box using very similar syntax:

plot::Box(*x1..x2, y1..y2, z1..z2, options*)

Another class is Ellipse2d.

plot::Ellipse2d(*axis1, axis2, [x,y], options*)

Here, [*x,y*] is the center of the ellipse, while *axis1* and *axis2* are the lengths of its semi-axes. A 3D version of an ellipse is an Ellipsoid.

plot::Ellipsoid(*axis1, axis2, axis3, [x,y,z], options*)

In order to declare a circle or a sphere, we may either use an ellipse or ellipsoid with all of its semi-axes equal, or use the Circle2d, Circle3d or Sphere classes. Here is the appropriate syntax:

plot::Circle2d(*radius, [x,y], options*)

and

plot::Sphere(*radius, [x,y,z], options*)

Declaring a Circle3d is a bit more complicated. We have to give its radius, center and the normal vector for the plane of the circle.

> plot::Circle3d(*radius*, [*x,y,z*],[*xn,yn,zn*], *options*)

There are a few other plot classes in the plot library. Therefore, it is worthwhile to execute the command info(plot) in order to examine what's left. We may also expect that this library will expand in the future.

9.4 Transformations of Graphical Objects

In the previous section, we have already used the plot::Rotate3d class to rotate an existing object. Here are all the classes from the plot library that we can use to transform existing objects.

plot::Rotate2d(*angle, point, object*) - *rotations of objects on the 2D plane*
plot::Rotate3d(*angle, point, direction, object*) - *rotations of objects in 3D space*
plot::Scale2d([*sx, sy*], *objects*) - *scaling of objects on the 2D plane*
plot::Scale3d([*sx, sy, sz*], *objects*) - *scaling of objects in 3D space*
plot::Transform2d(*b, A, objects*) - *affine transformation on the 2D plane*
plot::Transform3d(*b, A, objects*) - *affine transformation in 3D space*
plot::Translate2d([*tx, ty*], *objects*) - *translation on the 2D plane*
plot::Translate3d([*tx, ty, tz*], *objects*) - *translation in 3D space*

Here, the parameter *angle* means the angle of the rotation and it must be given in radians; the parameter *point* represents the coordinates of the point in 2D or 3D that is the center of rotation; *direction* is a vector for the axis of rotation in 3D; [sx, sy], [sx, sy, sz], [tx, ty], [tx, ty, tz] are the vectors used for scaling and translation of objects. Finally A is a 2×2 or 3×3 matrix and b is a shift vector.

Note that the keywords for all transformations of geometry objects start with a capital letter. All these operations produce a new real object that contains a reference to an existing object. For this reason, it is more convenient to consider operations as classes of objects that were obtained by transforming existing objects.

In order to practice using some of these classes, we will try a simple exercise.

Example 9.12 Floor tiles

In many countries, the floors in old buildings are covered with tiles bearing geometric patterns. Many of them are created using rotations and translations. Let us examine such a design. The goal of this exercise will be to create a pattern like the one below.

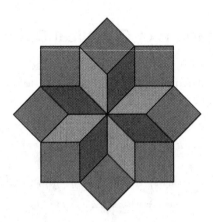

From the picture we can easily see that we will have to create three different objects — one square and two parallelograms.

Let us start with the basic code, and then I will explain its principles.

```
• X := cos(PI/4):
  Y := sin(PI/4):

  PBase1  := plot::Parallelogram2d(
     [X+1,Y], [X,Y],[1,0]
  ):

  PBase2 := plot::Parallelogram2d(
     [X,Y+1], [X,Y],[0,1]
  ):

  SqBase := plot::Rectangle(
     2*X..2*X+2, 2*Y..2*Y+2
  ):

  plot(PBase1, PBase2, SqBase, Scaling = Constrained)
```

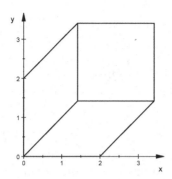

As you can see, we declared two parallelograms here — PBase1 and PBase2. When declaring the parallelograms, we use a point that will be the center of the figure, and two vectors pointing along each edge of the figure. Using simple trigonometric calculations, you can easily work out that the first vector for PBase1 is equal to [cos 45°, sin 45°], while its length and the length of the second vector are equal to 1. The length of each side of the parallelogram is 2. Now, we can add some color and pattern declarations to this code. Here is one of the many possible ways of doing this.

```
• X := cos(PI/4):
  Y := sin(PI/4):

  PBase1  := plot::Parallelogram2d(
     [X+1,Y], [X,Y],[1,0],
     Filled = TRUE,
     FillPattern = Solid,
     FillColor = RGB::WarmGray
  ):

  PBase2 := plot::Parallelogram2d(
     [X,Y+1], [X,Y],[0,1],
     Filled = TRUE,
     FillPattern = Solid,
     FillColor = RGB::Gold
  ):

  SqBase := plot::Rectangle(
     2*X..2*X+2, 2*Y..2*Y+2,
     Filled = TRUE,
     FillPattern = Solid,
     FillColor = RGB::Brick
```

```
):
```

We can now produce multiple copies of our basic objects and plot them all.

```
squares := plot::Rotate2d(i*PI/4, SqBase) $ i=1..8:
leaves := plot::Rotate2d(
   i*PI/2, PBase1, PBase2
) $ i = 1..4:

plot(squares, leaves, Scaling = Constrained)
```

The last statement will produce the pattern that was shown in the beginning of this example. If you wish to learn more about transforming plot objects, try to experiment with the enclosed code. For example, you can try to develop a floor design using multiple copies of the pattern that we created. You could also try to replace flat 2D objects with 3D objects and thus produce similar objects in 3D space. There are many other 2D and 3D patterns that can be created in a similar way.   ■

9.5 Turtle Graphics & L-systems

There are two plot classes in MuPAD graphics that you can treat either as mere toys, or very seriously. These are the two classes Turtle and Lsys. For someone dealing with calculus, these features are probably not especially important. However, for people who want to explore recursion or develop geometric patterns with interesting properties, these two classes are an invaluable source of interesting examples and activities. Using turtle graphics, you can generate visual representations of some less complicated algorithms. Let us see what we can get. First, I will need to introduce some basic concepts about MuPAD's turtle.

9.5.1 Turtle Graphics in MuPAD

The turtle class is implemented differently to other classes in MuPAD. In fact the turtle object that we create represents the path of movement of an imaginary turtle. There are two ways to create and plot such a path. We can build it at once by giving a list of orders to the turtle or create it step by step by appending lines to the existing turtle path. In each case the commands are slightly different. Let us see how both methods work, and most importantly, let us examine the differences between them.

Method 1 - building the turtle path using turtle commands

```
• A := plot::Turtle(
    [LineColor(RGB::Black),
     Forward(1), Right(PI/2),
     Forward(1), Right(PI/2),
     Forward(1), Right(PI/2),
     Forward(1)]
  ):
```

Method 2 - appending lines to the existing turtle path

```
T := Turtle():
T::setLineColor(RGB::Black):
T::forward(1): T::right(PI/2):
T::forward(1): T::right(PI/2):
T::forward(1): T::right(PI/2):
T::forward(1):
plot(T)
```

In both cases initially, the turtle waits in the center of the coordinate system and is oriented towards the positive direction of the y-axis. Subsequently, it moves according to our orders. Thus, in both cases we created exactly the same figure, a square.

Let us examine the differences in the syntax of both declarations. In the first case, all the commands inside the Turtle object begin with a capital letter. You can interpret them as orders given to an imaginary turtle. In the second case, all the commands start with lowercase characters; they are methods, applied to the created turtle object T. This is consistent with our notation where names of methods start with lowercase characters.

We can also work in a mixed mode, where we create a turtle object using the first method, and then we append elements to it using the second methods. However, I guess, using both methods at the same time could be highly confusing. Thus, it is best to stick to one of these methods. Note that the choice of method is quite important. Each method has its advantages and disadvantages. For example, using the first method, we cannot use any of the programming structures that we learned before. In the second method, we can fully take advantage of the MuPAD programming language. For instance, this construction will not work with the first method:

```
A := plot::Turtle(
   for i from 1 to 10 do
      Forward(1), Right(PI/5)
   end_for
):
plot(A)
```

but we can use it with the second method:

```
T := plot::Turtle():
for i from 1 to 10 do
   T::forward(1), T::right(PI/5)
end_for:
plot(T)
```

Therefore, with the second approach, we can build procedures and use them in multiple places.

Before we start working with the turtle, let us list all the turtle-related commands and methods.

Command	Method	Description
Left(*angle*)	left(*angle*)	*turn left given angle (in radians)*
Right(*angle*)	right(*angle*)	*turn right given angle (in radians)*
Forward(*angle*)	forward(*length*)	*draw forward a line of a given length*
Up	penUp()	*take the pen up*
Down	penDown()	*put the pen down*
Push	push()	*save the current stage of the turtle*
Pop	pop()	*move the turtle to the last saved stage*
LineColor(*color*)	setLineColor(*color*)	*change the current path color*

Important—the T::pop() command also deletes the last saved stage from memory. Therefore, using T::pop() again without first using

T:push() will produce an error.

After this short introduction to our virtual animal, we can start experimenting with it. As with many other graphical objects, we can develop a procedure that will perform various operations with our turtle. Such a procedure will be like a macro that forces the turtle to go through a given sequence of steps every time when the procedure is executed. For example, a procedure like the one shown below will force the turtle to draw an octagon with all sides having the same length and color, defined by the input parameter color.

```
Oct := proc(U, color)
    local i;
begin
    U::setLineColor(color);
    for i from 1 to 8 do
        U::right(PI/4);
        U::forward(1)
    end:
    return(U)
end:
```

Now, suppose that we want to draw five rows of octagons with 8 octagons in each row and each octagon drawn in a different color. In order to develop such a pattern, we have to find a way to move the turtle to a given point on the plane. First, we must save the current stage of the turtle. Then, we can perform some more operations with it.

```
dist := 1.4:
T := plot::Turtle():
T::push():
for i from 1 to 5 do
    T::pop();
    T::push();
    T::penUp();
    T::forward(i*dist);
    T::penDown();
    for j from 1 to 8 do
        Oct(T, [1/i, 1-1/j, 1/i]);
        T::right(PI/2);
        T::penUp();
        T::forward(dist);
        T::penDown();
        T::left(PI/2)
    end
end:
```

```
plot(T, LineWidth = 0.5)
```

I have already mentioned that every time we use the T::pop()
command, the last saved turtle position will be removed from the
memory. Thus, if you wish to preserve it for longer, you need to apply
this trick:

```
T::pop();
T::push()
```

The first of these commands will move the turtle to the last saved
position, while the second one will save this position again. Thus, if
you are careful with the pop() command, you will be able to keep a
specific turtle position in the turtle "memory" for a longer time. In our
example, we will save the initial position of the turtle, i.e. the point
(0,0). This will help us move the turtle, both along horizontal and
vertical rows. It is also easy to guess that the command

```
T::forward(i*dist):
```

will move the turtle forward vertically by dist=1.4 every time.
Finally, the construction

```
T::right(PI/2):
T::penUp():
T::forward(dist):
T::penDown();
T::left(PI/2)
```

will move the turtle forward horizontally by the dist=1.4, this time
without drawing a line. Note however, that before you can move the
turtle to the right, you must first turn it to face that direction, and then
move it forward. Here is the complete code for our program.

```
• Oct := proc(U, color)
     local i;
  begin
     U::setLineColor(color);
     for i from 1 to 8 do
         U::right(PI/4);
         U::forward(1)
     end:
     return(U)
  end:

  dist := 1.4:

  T := plot::Turtle():
  T::push():
```

```
for i from 1 to 5 do
   T::pop():
   T::push():
   T::penUp();
   T::forward(i*dist):
   T::penDown();
   for j from 1 to 8 do
      Oct(T, [1/i, 1-1/j, 1/i]);
      T::right(PI/2):
      T::penUp():
      T::forward(dist):
      T::penDown();
      T::left(PI/2)
   end
end:
```

```
plot(T, LineWidth = 0.5)
```

And here is the resulting pattern. For printing purpose, I drew it in black. It reminds me of the pattern inside the elevator in the house where I live.

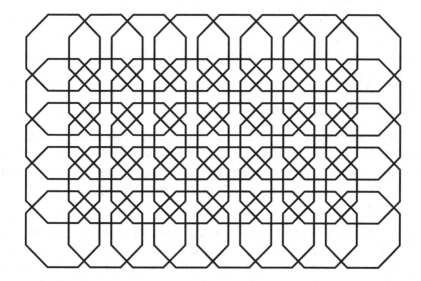

You can use the idea from this example to develop a number of interesting geometric patterns, tesselations, fractals, and so on. There is one hint that is worth remembering while experimenting with turtle graphics. Every time you execute some turtle-related code, you should execute it starting from the line where you initiate the turtle. Otherwise, your program will start developing the path of the turtle from its current position, without removing the previous pattern. You may thus end up with some rather unexpected drawings.

The turtle path can be animated, regardless of which method you used to produce it. We can animate both the length of the movement and the angle. Finally, we can use two or more turtles in one plot. Try out the animation that we can produce using simple code like this:

```
• T1 := plot::Turtle([
        LineColor(RGB::Green),
        Right(PI/4),
        Forward(1),
        Right(a),
        Forward(1+a)
     ], LineWidth = 0.8, a = 0..PI/2):

  U1 := plot::Turtle([
        LineColor(RGB::Red),
        Left(PI/4),
        Forward(1),
        Left(a),
        Forward(1+a)
     ], LineWidth = 0.5, a = 0..PI/2):

  plot(T1, U1)
```

9.5.2 Plotting L-systems

L-systems have origins in the biology of plants. The name *L-system* is short for *Lindenmayer system*, after *Aristid Lindenmayer*, who was one of the first people to use syntactic methods to model the growth of plants. Do not be afraid of the scientific jargon, though. I will explain the whole concept for you.

Imagine how some plants grow. Shortly after the seed was sown, you can observe a small, thin plant growing. There are no branches, and sometimes even no leaves. After some time, the plant transforms itself; it gets a few, say two, branches, its trunk grows a bit further, and its old part becomes woody. After another period of time, each branch develops its own small branches, and so on, and so on. You can observe how this looks in fig. 9.8.

Fig. 9.8 - L-system demonstrating growth of a plant

Of course, the above idea can be used in many situations. Plants or parts of plants are the most frequently seen examples, but you can see a similar process with sea-shells or snow flakes. Due to the similarity of the whole object to its individual parts, we consider L-systems as fractals.

In MuPAD we have a nice tool to develop L-systems and experiment with them. This tool is the class `plot::Lsys`, implemented as an application of the turtle graphics. Let us introduce the set of commands that will let us produce an L-system.

In order to declare an L-system and draw it on the screen, you need a two commands — a command to declare an L-system and a command to plot it on the screen. The resulting code can look like this.

```
MyLsys := plot::Lsys(
    2*PI/3, "F", "F"="F++F--F", Generations=7:
):

plot(MyLsys, Axes = None, Scaling = Constrained)
```

You have noticed perhaps that the most important information is hidden in the first statement. We declare several things here that will be important when creating the L-system. The very fist parameter, in this case 2*PI/3, is the angle that our turtle will turn every time we ask about this. The angle should be given in radians. The next expression, in this case "F", defines the shape that will be the starting object for our L-system. Here F is a default symbol for a line. So, you know that we will start with just a line. The third parameter is a rule that determines what transformations will occur between generations. In our example the rule is: replace each F by the turtle path—*draw a line 1 unit long, turn left 120 degrees, turn left again 120 degrees, again draw a line 1 unit long, turn right 120 degrees and draw a line 1 unit long.* In turtle language, the sequence F++F--F is equivalent to this sequence of commands:

```
T::forward(1):
T::left(2*PI/3):
T::left(2*PI/3):
T::forward(1):
T::right(2*PI/3):
T::forward(1)
```

Here is the meaning of all the symbols that you may use while developing the rules for the turtle drawing an L-system:

"F" - *draw a line, equivalent to* T::forward(1)

"f" - *move forward 1 unit, equivalent to* T::forward(1) with penUp

"+" - *turn left a given number of degrees, equivalent to* T::left(*angle*)

"-" - *turn right a given number of degrees, equivalent to* T::right(*angle*)

"[" - *save the turtle's position, same as* T::push()

"]" - *move the turtle to its last saved position, equivalent to* T::pop()

We may use more than one rule. For example, we could change the color of the plot, or add additional line elements. In order to do this, we have to define the additional element or color, like "B"=RGB::Red or "R"=Line, and use B and R in the L-system declaration. You will see how this can be done in later examples, but for now let us return to the MyLsys declarations. The rule "F"="F++F--F" means that every piece of the line F will be replaced by a shape similar to the letter Z. Thus, in the generation zero we only have F. In the first generation, the Z-like shape will replace F. In the second generation each occurrence of F will be replaced again by the Z-like shape, and so on. In each step, the shape of MyLsys gets more and more complicated. Can you imagine how complicated this shape will be after, say 8 or 9 generations? Let's produce it. Here is what we will get:

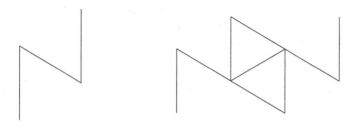

Fig. 9.9 MyLsys - first generation   *Fig. 9.10 MyLsys - second generation*

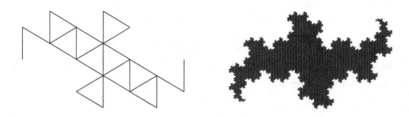

Fig. 9.11 MyLsys - third generation   Fig. 9.12 MyLsys - eight generation

Our example was quite simple, but I hope it gave you some idea about how the rules in Lsys declarations work. Now it is time to return to the primary concept of L-systems, and think about how plants grow.

Example 9.13

You can easily produce a hand-drawing showing how plants grow. By observing the pictures in fig. 9.8, you see that in every step, a single branch will be replaced by a shape resembling the Greek letter Ψ. So, the Lsys rule should produce exactly that shape. We can make our problem a bit more complicated, though. Imagine that the rule will make the old parts of the plant brown, and add those three young branches with a fresh green color. In order to do this, we shall introduce two colors, "B"=RGB::Brown, "H"=RGB::ForestGreen, and a new object representing the old branches R. Next, we must develop the rules for our system.

```
"F"="BR[+HF][-HF]HRF",
"R"="RR"
```

The first rule is easy to understand—make R brown, draw a green branch to the left, draw another to the right, and finally draw one straight branch. The second rule is applied to old branches. It makes them grow — you know that old branches also grow. Finally, having all these rules, you can write the complete code to develop and plot a plant. It will look like this:

```
• tree := plot::Lsys(PI/4, "F",
    "F"="BR[+HF][-HF]HRF",
    "R"="BRR",
    "F"=Line,
    "R"=Line,
    "B"=RGB::Brown,
    "H"=RGB::ForestGreen,
    Generations=6
):
```

```
plot(tree, Axes=None, Scaling=Constrained)
```

This makes a nice plot of a tree where the old branches are brown and all the new ones are green.

The angle I used here was 45 degrees given in radians, but you can try more or less to see what you will get.   ∎

This example concludes another, and the last, chapter about graphics in MuPAD. You will find a few more interesting examples in the programming exercises section.

9.6 Chapter Summary

In this chapter, we have explored a few applications of MuPAD graphics. You have learned about a few important functions that can be useful for calculus and geometry. In the last part of the chapter, we explored the turtle graphics and creating L-systems that are considered as fractals. Now, let us make a summary of the commands introduced in this chapter.

9.6.1 MuPAD Syntax Summary

Classes for calculus applications

plot::Density($f(x,y)$, $x = x1..x2$, $y = y1..y2$, *options*)

plot::VectorField2d([$u(x,y)$, $v(x,y)$], $x = x1..x2$, $y = y1..y2$, *options*)

plot::Inequality([$f1, f2, ..., fn$], $x = x1..x2, y = y1..y2$, *options*)

plot::XRotate(f, $x = x1..x2$, *other_options*)

plot::ZRotate(f, $x = x1..x2$, *other_options*)

Classes for presentation graphics

plot::PointList2d([[$n1, m1$], [$n2, m2$], ..., [nk, mk]], *options*)

plot::Polygon2d([[$n1, m1$], [$n2, m2$], ..., [nk, mk]], *options*)

plot::Bars2d(*data*, *options*)

plot::Bars3d(*data*, *options*)

plot::Piechart2d(*list_of_data*, *options*)

plot::Piechart3d(*list_of_data*, *options*)

Classes representing selected geometry objects

plot::Arc2d(*radius*, [x,y], *angle1..angle2*) - *2D arcs of a circle*

plot::Arrow2d([$x1,y1$], [$x2,y2$]) - *arrows in 2D*

plot::Arrow3d([$x1,y1,z1$], [$x2,y2,z2$], *options*) - *arrows in 3D*

plot::Box($x1..x2$, $y1..y2$, $z1..z2$, *options*) - *boxes in 3D*

plot::Circle2d(*radius*, [x,y], *options*) - *circles in 2D*

plot::Circle3d(*radius*, [x,y,z], [xn,yn,zn], *options*) - *circles in 3D*

plot::Cone(*radius1*, [$x1,y1,z1$], *radius2*, [$x2,y2,z2$], *options*) - *cones*

plot::Cylinder(*radius*, [$x1,y1,z1$], [$x2,y2,z2$], *options*) - *cylinders*

plot::Ellipse2d(*axis1*, *axis2*, [x,y], *options*) - *ellipses in 2D* plot::Ellipsoid(*axis1*, *axis2*, *axis3*, [*x,y,z*], *options*) - *ellipsoids*

plot::Line2d([*x1,y1*], [*x2,y2*], *options*) - *segment connecting two points in 2D*

plot::Line3d([*x1,y1,z1*],[*x2,y2,z2*], *options*) - *segment connecting two points in 3D*

plot::Parallelogram2d([*x,y*],[*v1,v2*],[*u1,u2*], *options*) - *parallelograms in 2D*

plot::Parallelogram3d([*x,y,z*],[*v1,v2,v3*],[*u1,u2,u3*], *options*) - *parallelograms in 3D*

plot::Point2d(*x, y, options*) - *points on a 2D plane*

plot::Point2d([*x, y*], *options*) - *points on a 2D plane*

plot::Point3d(*x, y, z, options*) - *points in 3D space*

plot::Point3d([*x, y, z*], *options*) - *points in 3D space*

plot::Pointlist2d(*Point1, Point2, ..., Point_n, options*) - *point list in 2D*

plot::Pointlist3d(*Point1, Point2, ..., Point_n, options*) - *point list in 3D*

plot::Polygon2d(*Point1, Point2, Point3, ..., Point_n, options*) - *open and closed polygons on a 2D plane*

plot::Polygon3d(*Point1, Point2, Point3, ..., Point_n, options*) - *open and closed polygons in 3D space*

plot::Rectangle(*x1..x2, y1..y2, options*) - *rectangles on a 2D plane*

plot::Sphere(*radius*, [*x,y,z*], *options*) - *spheres in 3D space*

plot::SurfaceSet(*mesh_list*, MeshListType=*type*, *options*) - *objects defined by mesh of points*

Platonic polyhedra

plot::Dodecahedron() - *dodecahedrons*

plot::Hexahedron() - *hexahedrons*

plot::Icosahedron() - *icosahedrons*

plot::Octahedron() - *octahedrons*

plot::Tetrahedron() - *tetrahedrons*

Transformations of graphics objects

plot::Rotate2d(*angle, point, object*) - *rotations of objects on a 2D plane*

plot::Rotate3d(*angle, point, direction, object*) - *rotations of objects in 3D space*

plot::Scale2d([*sx,sy*], *objects*) - *scaling of objects on a 2D plane*

plot::Scale3d([*sx,sy,sz*], *objects*) - *scaling of objects in 3D space*

plot::Transform2d(*b, A, objects*) - *affine transformation on a 2D plane*

plot::Transform3d(*b, A, objects*) - *affine transformation in 3D space*

plot::Translate2d([*tx,ty*], *objects*) - *translation on a 2D plane* plot::Translate3d([*tx, ty, tz*], *objects*) - *translation in 3D space*

Turtle graphics

T := Turtle() - *turtle initiation*

Turtle commands and methods

Command	Method	Description
Left(*angle*)	left(*angle*)	*turn left given angle (in radians)*
Right(*angle*)	right(*angle*)	*turn right given angle (in radians)*
Forward(*angle*)	forward(*length*)	*draw forward a line of a given length*
Up	penUp()	*take the pen up*
Down	penDown()	*put the pen down*
Push	push()	*save the current stage of the turtle*
Pop	pop()	*move the turtle to the last saved stage*
LineColor(*color*)	setLineColor(*color*)	*change the current path color*

Lindenmayer systems

MyLsys := plot::Lsys(*angle, init_statement, rules,* Generations=#):

L-system rules components

"F" - *draw a line (equivalent to* T::line(1))
"f" - *move forward 1 unit (equivalent to* T::move(1))
"+" - *turn left a given number of degrees (equivalent to* T::left(*degrees*))
"-" - *turn right a given number of degrees (equivalent to* T::right(*degrees*))
"[" - *save the turtle's position (same as* T::push())
"]" - *move the turtle to its last saved position (equivalent to* T::pop())

9.7 Programming Exercises

9.7.1 Calculus Applications

1. By hand, sketch the level curves of the function $z = x^3 - y^3$. Use $z = -2, -1, 0, 1, 2$. Develop a MuPAD program to obtain a graph of these level curves. Compare your graph with what you've obtained in MuPAD.

2. Develop a graph of the level curves for the function $f(x,y) = (x^2 - y^2)e^{-x^2-y^2}$.

3. Investigate the shape of the surface given by the equation $z = \frac{1}{1+x^2-y^2}$. Can you guess the shape of the surface, using only the level curves? Develop a plot of the level curves with 10 slices, from level 1.1 to level 10.1.

4. Use the density plot function to obtain the graphs of these functions. In each case, compare the density plot with a graph obtained using the Function3d class.

   a. $f(x,y) = x^3 - y^3$

   b. $f(x,y) = (x^2 - y^2)e^{-x^2-y^2}$

   c. $f(x,y) = \sin(x^2 + y^2)$

   d. $f(x,y) = xe^{-x^2-y^2}$

   e. $f(x,y) = \frac{xy}{x^2+y^2}$

5. Plot the vector fields of the following functions:

   a. $f(x,y) = [3x^2, -3y^2]$

   b. $f(x,y) = [2x\cos(x^2 - y^2), -2y\cos(x^2 - y^2)]$

6. Use the inequality plot class to draw the domains of these functions:

   a. $f(x,y) = \sqrt{y - x^2}$

   b. $f(x,y) = \frac{xy}{x^2 - y^2}$

c. $f(x,y) = \dfrac{1}{\sqrt{1 - \sin xy}}$

d. $f(x,y) = \dfrac{1}{\sqrt{(x-1)(y-1)}}$

e. $f(x,y) = \dfrac{1}{\sqrt{(x^2 - y^2)(2x - y^2 - 1)}}$

9.7.2 Geometry Creations

In 2D and 3D Cartesian geometry, there are a number of activities that can be a great inspiration for writing MuPAD procedures drawing complex geometry objects. The list of examples that I give you here is very limited. If you are fascinated by geometry or if you wish to find more applications of MuPAD in geometry, just open any textbook of analytic geometry and try to solve each problem by implementing it in the form of a MuPAD procedure.

Exercises on the plane

1. While developing a line passing through two given points, MuPAD produces a segment joining only these two points. For the given two points A and B, develop a procedure that will plot a line passing through these two points, and the points themselves. The line should go beyond both points.

2. As you know, the equation of a line can be given in many ways. Write the procedures that will draw the lines defined by:

   a. point A and a vector colinear with the line

   b. point A and a vector perpendicular to the line

   c. two points of intersection of the line l with the x-axis and y-axis.

3. Write a procedure that will plot a line parallel to the line l given by two its points and passing through the given point B.

4. Write a procedure that will draw a circle passing through the three given points A, B and C. The procedure should plot both the circle and the points.

5. Write a procedure that, for the three given points A, B and C,

will produce the line m, which is the bisection of the angle $\measuredangle(A, B, C)$

6. Write a procedure that, for a given triangle ABC will produce the medians of this triangle and the point of intersection of the medians.

7. Write a procedure that, for a given circle, will produce the line m, which is tangent to the circle and passes through the given point B.

8. Develop a procedure that, for a given triangle, will produce a circle inscribed in the triangle.

9. Develop a procedure that, for the three given points A, B, C, will generate the area of the triangle ABC.

10. Create a procedure that will check if three given points are colinear.

11. Develop a procedure that will produce the circumference of the triangle ABC.

12. Develop a procedure that will calculate the distance between two points.

Exercises for 3D space

1. Write a procedure that will calculate the distance between the given line m and the point A.

2. Develop a procedure that will produce, for three given points in 3D, a plane passing through these points.

3. Write a procedure that, for the plane Σ defined by the three points A, B, C, will produce the line m that passes through point D and is perpendicular to the plane.

4. Develop a procedure that will produce, for a given sequence of coplanar points in 3D, a filled polygon connecting all these points.

5. Create a procedure that will produce a cube with a given center and side lengths.

6. Write a procedure that will plot a sphere with a given center and

radius.

7. Develop procedures that will plot some of the well known polyhedra other than those existing in the MuPAD plot library.

9.7.3 Turtle Graphics and L-systems

Turtle graphics and L-systems can be useful tools for many interesting programming experiments with patterns that require recursive constructions or repeating tasks.

1. Develop a procedure using a turtle to produce n squares with a common center and the lengths of sides 1, 2, ..., n.

2. Tesselations are geometric patterns that have been known in mathematics and architecture for centuries. Develop procedures that will cover an area of a plane by tesselations of squares, triangles, hexagons, and octagons with a square.

3. The well-known *Koch snowflake* curve is a fractal curve. You start with a regular triangle. Then, you split each side of the triangle into three equal segments. Use the central segment as a base for a new regular triangle that points outside the main triangle. Repeat this construction for each side of the resulting figure. Write the MuPAD commands that will produce a *Koch snowflake* of the given order n.

4. The *Sierpinski triangle* can be produced by an iterative process that can be modelled using L-systems. Take a regular triangle, connect the centers of each side and remove the central triangle. Repeat this process for each resulting triangle. Write the MuPAD statements that will produce a *Sierpinski triangle* of the order n.

5. Use F as a starting shape, an angle of $\pi/2$ degrees and the generator rule "F"="-F+F-F-F+F" to produce an L-system. What will you get in the second, third and fourth generations? Can you explain why each two consecutive generations are so different?

6. In the previous example, change the rule to "F"="+F+F-F-F+F". What do you get this time? Are the shapes obtained in the previous example similar to the shapes from this one?

7. Start with a square F+F+F+F+, an angle of $\pi/2$ degrees, and then apply the rule "F"="F-F+F+F-F". How does the algorithm for the *Koch snowflake* differ from this one?

8. Try a new L-system, but this time use an angle of $\pi/3$ degrees and an equilateral triangle F++F++F++. Use the rule "F"="F+F--F+F" to produce a few generations of your object. What did you get this time? Observe that the triangles produced in each step point towards the center of the shape. How do you need to change the rule in order to get a *Koch snowflake*?

9. It is much easier to obtain an equilateral triangle using an angle of $2\pi/3$ degrees. You will only need F+F+F+. Now, try to use the rule F"="-F+F-F+F". Is this shape similar to those obtained in the previous examples? Draw the shape generated by your rule to see how the transformation that you applied looks this time.

Chapter 10 _____

Exploring Numbers with MuPAD

We have been dealing with numbers since the beginning of this book. Numbers are of course very convenient in explaining some of the MuPAD features. Now we will concentrate entirely on numbers, and I will show you what sort of interesting things you can do with numbers in MuPAD.

First of all, note that we have a number of domains in MuPAD with different kinds of numbers, but we also have many types that are useful for numbers type checking. Therefore, we should use Dom:: to declare our variables to represent objects from a given domain, and Type:: to declare the type of a given objects or to check its type.

While talking about numbers, it is important to distinguish between numbers and constants. Generally, MuPAD considers any object that does not contain symbolic identifiers as a constant. For example 5, sin(PI), exp(35) will be considered as constants, but the objects AB, sin(x+5) will be not considered as constants. There is also a special type Type::Constant to check if a given object is constant or not. We can easily see that in MuPAD objects PI, EULER, CATALAN, E, TRUE, FALSE, complexInfinity, I and a few other are considered as constants:

- testtype(EULER,Type::Constant)
  TRUE

- testtype(FALSE,Type::Constant)
  TRUE

In exactly the same way, we can also find out that any complex number or even quaternion is also treated as a constant.

10.1. Integers in MuPAD

MuPAD contains a number of predefined types for working with integer numbers. You can start off with a set of all integer numbers, or choose between different subsets of integers, such as negative, positive, non-negative, even or odd integers, or only zero. This gives you a lot of possibilities to concentrate on the specific properties of integers. Here is a list of the various types that are related to integer numbers.

Type::Integer - *integer numbers*

Type::NegInt - *all negative integer numbers*

Type::PosInt - *positive integers (do not contain 0)*

Type::NonNegInt - *non-negative integers (with zero)*

Type::Even - *even integers*

Type::Odd - *odd integers*

Type::Prime - *prime numbers*

A large set of functions for working with integer numbers is included in the standard library (stdlib). If you are fascinated with number theory, you can also find a separate library numlib, where a number of functions related directly to elementary number theory is defined.

Here are the most important functions from stdlib that you can use for working with integer numbers:

+, *, -, / - *arithmetical operations; note: / produces a rational number*

! - *factorial operation*

mod - *produces the remainder of integer division, e.g.* 25 mod 3 = 1

modp(n,m) - *produces* (n mod m)

div - *integer division of two integers, e.g.* 25 div 3 = 8

factor(n) - *factoring a given integer*

ifactor(n) - *factoring an integer into prime components*

igcd(n,m) - *greatest common divisor for integers*

igcdex(n,m) - *extended gcd for integers*

ilcm(n,m) - *the lowest common multiple for integers*

isprime(n) - *checks if the given integer is a prime number*

ithprime(i) - *produces i-th prime number*

max(n,m) - *maximum of two or more numbers*

min(n,m) - *minimum of two or more numbers*

random() - *generates a random integer number between 0 and* 10^{12}

random($n1..n2$) - *produces a function to generate random integers from the interval* $[n1, n2]$

Some of the above-mentioned functions will also work for other types of numbers. There are also a few functions that were designed for other types of objects but also produce satisfactory results for integers. For example, the function gcd, which produces the greatest common divisor for polynomials, will also work for integer numbers — a number can be considered a polynomial.

Most of the above-listed functions are quite common and you will not have any problems guessing how they work. For example,

- 12!

  479 001 600

- 25 div 3

  8

- 25 mod 3

  1

- factor(26403366438)

  $2 \cdot 3^2 \cdot 31 \cdot 47 \cdot 97^2 \cdot 107$

- isprime(26403366438)

  FALSE

- isprime(107)

  TRUE

- igcd(2640,3366)

  66

- igcdex(2640,3366)

  66, -14, 11

The igcdex function, however, produced a very puzzling result. You can certainly guess that 66 is the greatest common divisor of both numbers. However, what about −14 and 11? Here is a hint. Observe that,

$$-14 * 2640 + 11 * 3366 = 66$$

Thus, for two given numbers N and M, the igcdex function produces their greatest common divisor k as well as two numbers n and m so that $k = nN + mM$.

At this point, it is worth explaining what may be obtained from the random procedure. For example, the standard use of it will produce results like these:

- x := random()

  32062222085

- y := random()

  722974121768

However, when you use it with some parameters, you can get something very useful out of it. For example the command below will produce a function that can be used to obtain random integers from the interval 0, ... , 9

- myrand := random(10):
  x := myrand()

  9

  y := myrand()

  5

You can also use two integer numbers $n1 < n2$, to obtain random numbers from the interval $n1..n2$.

- myrand := random(5..6):
  x := myrand()

  5

  y := myrand()

  6

If you have already read the chapter about procedures, you will be able to go through a number of very interesting investigations with integers, and in particular with prime numbers. Just look back into the history of mathematics and see how many intriguing problems were considered by our ancestors. Here are some examples.

Example 10.1

We can easily produce a factorial of a given integer, for example 12! = 479 001 600. However, can you tell if a given integer n is the factorial of another integer? How do you check this? As in many other situations, there are a few solutions for this problem. For example, you can produce a list of all the factorial numbers in a given range and check if n belongs to this list. Another way to solve the problem is to trace all the consecutive integer divisors i of n. For each divisor i replace n by n div m, until (n div i = 1). Just take a look at the code below:

```
• isFactorial := proc(n:Type::PosInt)
      local i;
  begin
     i := 2;
     repeat
        if (n mod i = 0) and (n div i <>1) then
            n := n div i;
            i := i+1
        elif i=n then
            return(TRUE)
        else
            return(FALSE)
        end
     until FALSE end
  end:
```

Now you can try our procedure on some larger numbers.

```
• isFactorial(9!)
```

TRUE

You can even develop a small program checking which numbers in a given range are pointed out by the procedure as factorial numbers. You have keep in mind however, that for a larger n, the distance between $n!$ and $(n + 1)!$ will be very large. This means that searching a large interval of integers can be very slow and the results will not be very impressive. Here you have an example that required about 75 seconds to generate the final result.

```
• for n from 1 to 1000000 do
     if isFactorial(n) then
        print(Unquoted,expr2text(n)."- factorial number")
     end
  end:
```

2 - factorial number

6 - factorial number

24 - factorial number

120 - factorial number

720 - factorial number

5040 - factorial number

40320 - factorial number

362880 - factorial number

The fact that it took us so long to search for factorial numbers from a given interval is quite disturbing. However, these results can be obtained using a much faster approach. You simply need to produce the set of all factorials that are smaller or equal to n.

```
• factorials:= proc(n:Type::PosInt)
  local i, fa, setFact;
  begin
    i := 1;
    setFact := {};
    repeat
      fa := i!;
      setFact := setFact union {fa};
      i := i+1
    until fa >= n end;
    return(setFact)
  end;
```

And here is how this procedure works.

```
• factorials(10000000000)
```

$\{1, 2, 6, 24, 120, 720, 5040, 40320, 362880, 3628800, 39916800, 479001600, 6227020800, 87178291200\}$

This result was obtained in less than a second.   ∎

Factorial numbers, due to their rare distribution, are not especially exciting. However, there are many other types of numbers that are worth investigating. For example, ancient mathematicians were particularly intrigued by the so-called perfect numbers — numbers that are the sum of all the smaller numbers by which they can be divided. For example, 6 and 28 are prefect numbers.

$6 = 1 + 2 + 3$

$28 = 1 + 2 + 4 + 7 + 14$

Can you find out which positive integers are perfect numbers? This can be a nice programming exercise for both teachers and students.

What about the other families of numbers? Our predecessors were very inventive. They produced Bell numbers, Stirling cycle numbers, Ramanujan's numbers, Catalan numbers, Bernoulli numbers, Euler numbers, Fibonacci numbers, Lucas numbers and of course prime numbers. If you search a few mathematical texts, you will certainly find some more famous names that are associated with a specific family of numbers. This is almost a never-ending story. You could try to produce a library of MuPAD procedures to calculate elements from a few of the most important families of numbers. I will remind you about this later, at the end of the chapter. Of course, you do not need to write a procedure to find prime numbers or to check if a given number is a prime. These two procedure already exist in MuPAD and they work quite well. For example:

- `isprime(23)`

  TRUE

- `isprime(24)`

  FALSE

- `ithprime(5)`

  11

Having these two procedures, we can easily write a procedure that will produce a list of the first n prime numbers. This is a very easy job. First, we need to initiate an empty list, and then add all the primes to it using the `ithprime` procedure.

Your procedure should look like this one:

- ```
primes := proc (n : Type::PosInt)
  local lop, i;
  begin
    lop := [ithprime(i) $ i=1..n];
    return(lop)
  end:
```

Now we can use this procedure to produce a list of the first $n$ primes. For example—a list of the first 50 prime numbers will look like this:
```

- `primes(50)`

  [2, 3, 5, 7, 11, 13, 17, 19, 23, 29, 31, 37, 41, 43, 47, 53, 59, 61, 67, 71, 73, 79, 83, 89, 97, 101, 103, 107, 109, 113, 127, 131, 137, 139, 149, 151, 157, 163, 167, 173, 179, 181, 191, 193, 197, 199, 211, 223, 227, 229]

I will leave you to experiment with prime numbers and other types of numbers by yourself. This is really a long story, and it could fill another book. For now, let us move on to another interesting example.

Example 10.2 Arithmetic modulo m

In many textbooks of mathematics, you can find information about arithmetic modulo followed by a given number. For example, the arithmetic of integer numbers modulo 2, or modulo 7, and so on. In MuPAD, you can use the function modp. This function produces for the two given integer numbers n and m the remainder from integer division ($n \bmod m$). For example,

- `modp(34,7)`

  6

- `modp(7,3)`

  1

We can use this function to produce the results of various operations modulo a given integer. For example,

- `modp(3*4,7)`

  5

Now, you can think about how to develop a table of multiplication modulo any positive integer, for example 7. You only need to develop a matrix, say 7×7, and fill it with the appropriate results of the operation * taken modulo 7. Before you start writing such a procedure, I should remind you that while developing matrices, you can only use positive dimensions. This means that you cannot use 0,0 as the coordinates for the top-left element of the matrix. This slightly complicates our work, since numbers modulo 7 form the sequence 0, 1, 2, 3, 4, 5, 6 and the very first element of this sequence is zero. The code enclosed below shows how I solved this problem. Perhaps you can find another solution.

```
• range := 7:
  multmod := matrix(range,range):
  for i from 0 to range-1 do
     for j from 0 to range-1 do
        multmod[i+1,j+1] := modp(i*j,range)
     end:
  end:
```

```
• multmod;
```

Here is the resulting matrix:

$$
\begin{pmatrix}
0 & 0 & 0 & 0 & 0 & 0 & 0 \\
0 & 1 & 2 & 3 & 4 & 5 & 6 \\
0 & 2 & 4 & 6 & 1 & 5 & 6 \\
0 & 3 & 6 & 2 & 5 & 1 & 4 \\
0 & 4 & 1 & 5 & 2 & 6 & 3 \\
0 & 5 & 3 & 1 & 6 & 4 & 2 \\
0 & 6 & 5 & 4 & 3 & 2 & 1
\end{pmatrix}
$$

If you really want to make the result look like a multiplication table, you will need to add a row on top with the numbers 0, 1, 2, 3, 4, 5, 6; and on the left side, a similar column. The above-mentioned row and column will serve as a display for the elements that are multiplied while the remaining part of the matrix will show the results of the multiplication.

Finally, let me to mention that the matrix A can be created also like here:

```
• A := matrix(7,7,(i,j)->(i-1)*(j-1) mod 7)
```

Here, instead of filling cells of the matrix step-by-step we use a function to fill all cells.   ∎

Again, we have found a very interesting topic. Arithmetic modulo some integer numbers was investigated by many mathematicians in the past. You can find out a lot more about this in books about number theory.

10.2 Rational Numbers in MuPAD

We used to think about rational numbers as fractions $\frac{p}{q}$, where p and q are two integers. MuPAD contains a type Type::Rational that can be used while working with rational numbers. There are also a few other types like Type::PosRat, Type::NegRat, Type::NonNegRat, and of course Type::Zero, which can be used in relation to many sets of numbers.

Standard operations like +, −, /, * can all be used with rational numbers. These operations, when performed on rational numbers will always give you results that are rational, unless you force MuPAD to produce a different type of output. Thus, for example,

- 1/2 + 3/5

  $\frac{11}{10}$

- 2/3 * 6/7

  $\frac{4}{7}$

While trying to understand how the world is built, or just trying to sort out the inaccuracies that occur in science, we often have to compromise and use a rough approximation for a given number. Think about how one inch can be expressed in centimeters, or how a calendar year can be expressed by using moon months. Many such problems can be described using continued fractions. Here is an example showing how a continued fraction is created.

$$\frac{4131}{334} = 12 + \frac{123}{334} = 12 + \frac{1}{\frac{324}{123}} = 12 + \frac{1}{2 + \frac{88}{123}} = 12 + \frac{1}{2 + \frac{1}{1 + \frac{35}{123}}} = \ldots$$

We can continue to develop such fractions until the last one has the form $\frac{1}{m}$, where m is an integer number. Note that on each level, we have the expression $n + \frac{1}{x}$. Hence, we can express each continued fraction by a sequence of integer numbers. In our example these numbers are 12, 2, 1, ...

Now you can think about how to write a procedure to generate a continued fraction for a given rational or even real number. Well, you do not really need to do this, as it has been done already. However, the problem is interesting enough that you should at least try it.

I mentioned earlier that MuPAD contains a nice library numlib of procedures that are related to elementary number theory. One of these procedures is the numlib::contfrac procedure to produce continued fractions for a given real number. Generally, we do not need to import it. In most versions of MuPAD, there is also a more universal procedure contfrac that is automatically imported when MuPAD starts up. Here is a sample output from the contfrac procedure provided that we switched off the **Pretty Print Text** option in the **Format** menu.

- contfrac(4/5)

  $1/(1 + 1/(4 + 1/...))$

- myfrac := contfrac(41319/3340)

  $12 + 1/(2 + 1/(1 + 1/(2 + 1/(3 + 1/(2 + 1/(26 + 1/(2 + 1/...)))))))$

If the **Pretty Print Text** had not been switched off, then the output would be printed in text mode like on old text terminals. Because of the high complexity of some continued fractions, MuPAD doesn't typeset them.

If you create a continued fraction using the contfrac procedure, you can use the procedure nthcoeff to extract any coefficient of this continued fraction. For example,

- nthcoeff(myfrac,1)

  12

- nthcoeff(myfrac,2)

  2

- nthcoeff(myfrac,9)

  FAIL

The output FAIL shows that the 9th coefficient of our continued fraction doesn't exist.

While using procedures from MuPAD libraries, you should remember that some of these procedures were developed for more general purposes and quite often, we use them in a scaled-down problem. For example, the contfrac procedure was developed to produce the continued fraction approximation of a real number. The procedure nthcoeff was created to return the non-zero coefficients of a polynomial, even though it also works for continued fractions. Perhaps this would be the right time to develop something that will work explicitly for our purposes. Let us create a procedure that, for two given integers n and m, will produce coefficients of the continued fraction expansion for $\frac{n}{m}$.

Example 10.3 Continued fractions procedure

Take a look at how continued fractions develop. Suppose that you have the fraction 5/3. When converting it to a continued fraction, you perform the operations $\frac{5}{3} = 1 + \frac{2}{3} = 1 + \frac{1}{\frac{3}{2}}$. Now, you would apply the same transformation to the fraction $\frac{3}{2}$, and so on. In general, this means that the fraction $\frac{n}{m}$ is transformed into $\frac{n}{m} = (n \text{ div } m) + \frac{1}{\frac{m}{n \bmod m}}$.

Therefore, in the next step, you would replace n by m, and m by (n mod m). Finally, you can add an additional parameter that will limit calculating the coefficients of a continued fraction to a certain number of steps. Here is the procedure to calculate the coefficients of a continued fraction for the fraction $\frac{n}{m}$ represented by two given positive integer numbers n and m, with a finite number of *steps*. For the resulting output from the procedure, I have used a list. This will be convenient for further operations, for example to convert our continued fraction representation into MuPAD's representation.

```
• contFract := proc(n:Type::PosInt,
               m:Type::PosInt,
               steps:Type::PosInt)
  local list, i, result, remainder;
  begin
     list := [];
     i := 1;
     while i < steps do
         i := i+1;
         result := (n div m);
         remainder := (n mod m);
         list := list.[result];
         if remainder <> 0 and i< steps then
             n:= m;
             m:= remainder;
         else
             return(list)
         end;
```

```
      end;
   end:
```

Now, you can apply this procedure to a few pairs of integers and check if the produced results are the same as those from the numlib::contfrac procedure. Notice that the name of our procedure only slightly differs from the original MuPAD procedure — we used the name contFract, while the original one is called contfrac.

- contFract(41319,3340,10)

  [12, 2, 1, 2, 3, 2, 26, 2]

- numlib::contfrac(41319/3340)

  $12 + 1/(2 + 1/(1 + 1/(2 + 1/(3 + 1/(2 + 1/(26 + 1/(2 + 1/...)))))))$

Try using our procedure in a few more examples. Check if the results are properly calculated. Think about how you can improve it. ■

Once you have got some experience with continued fractions, you can go further. For example, you could develop a procedure that, for a given list of positive integer numbers, will produce a continued fraction and display it the same way that the numlib::contfrac does.

Note that in MuPAD, you can consider the algebra of continued fractions. Having two continued fractions, you can easily obtain their sum, difference, product or quotient. For example,

- a := contfrac(12/23)

  $1/(1 + 1/(1 + 1/(11 + 1/...)))$

- b := contfrac(23/34)

  $1/(1 + 1/(2 + 1/(11 + 1/...)))$

- a+b

  $1 + 1/(5 + 1/(22 + 1/(7 + 1/...)))$

- a*b

  $1/(2 + 1/(1 + 1/(5 + 1/...)))$

- a/b

  $1/(1 + 1/(3 + 1/(2 + 1/(1 + 1/(2 + 1/(4 + 1/(1 + 1/(2 + 1/...)))))))))$

- a-b

  $- 1 + 1/(1 + 1/(5 + 1/(2 + 1/(6 + 1/(4 + 1/(2 + 1/...))))))$

Analyze the above operations. Perhaps you can develop your own operations that will operate on lists of numbers, treating your lists as lists of coefficients of two continued fractions.

The story of continued fractions or arithmetic modulo could go on for a few more pages. However, I will leave this topic to your own investigations. Right now, we will move on to another important group of numbers.

10.3 Real Numbers in MuPAD

Real numbers have a significant place in numerical calculations. However, there is one thing that we must remember. In MuPAD, like in many other computing packages, any real number while converting it to the floating point notation will be represented in the form of a finite decimal fraction. This in fact makes it a rational number. There are still, however, a few important differences.

Let us check what MuPAD has to offer regarding real numbers.

There is, firstly, a number of predefined types that can be used with real numbers. Here are some of them:

Type::Real - *all real numbers* $\mathbb{R}$

Type::Positive - *all positive real numbers,* $\mathbb{R}^+$

Type::NonNegative - *all positive real numbers and zero,* $\mathbb{R}^+ \cup \{0\}$

Type::Negative - *all negative real numbers,* $\mathbb{R}^-$

Type::NonZero - *all real numbers non-equal to 0,* $\mathbb{R} - \{0\}$

Type::Zero - *only the number 0,* $\{0\}$

Type::Constant - *any constant*

Type::Interval - *interval of real numbers,* $[a, b]$

For real numbers, we have a number of operations at our disposal that we had already used for other types of numbers, as well as a lot of real number-specific operations.

Operations and relations for real numbers

$$+, -, *, /, <, <=, =, >, >=$$

Functions for real numbers

Here are some of the most important MuPAD functions for real numbers. You can find more information about the other functions in MuPAD help — just execute the statement ?functions.

abs(x) - *the absolute value of x*

sign(x) - *function sign*

log(base, x) - *function logarithm with a given base, e.g.* log(10,x)

ln(x) - *natural logarithm of x*

exp(x) - *exponential function of x*, exp(x) = e^x

sqrt(x) - *function* $\sqrt{x}$

sin(x), cos(x), tan(x), cot(x) - *trigonometric functions, x shall be in radians*

sec(x), csc(x) - *functions* sec(x) = $\frac{1}{\cos(x)}$ *and* csc(x) = $\frac{1}{\sin(x)}$

arcsin(x), arccos(x), arctan(x), arccot(x) - *inverse trigonometric functions*

arcsec(x), arccsc(x) - *inverse functions for* sec(x) *and* csc(x)

sinh(x), cosh(x), tanh(x), coth(x) - *hyperbolic functions*

sech(x), csch(x) - *hyperbolic functions* sech(x) = $\frac{1}{\cosh(x)}$ *and* csch(x) = $\frac{1}{\sinh(x)}$

floor(x), ceil(x), round(x), trunc(x) - *functions rounding real numbers, result is an integer number*

min(x,y,...,z), max(x,y,...,z) - *functions producing minimum and maximum*

Constants in real numbers

Finally, we have a few constants. Here are the most important ones.

PI - *the number* π

EULER - *the Euler-Mascheroni constant*

CATALAN - *Catalan constant*

E - *Napier's constant e, base for natural logarithms*

infinity - *constant representing infinity*

Most of these constants are irrational numbers. If you wish to calculate their decimal approximation, you will need to use the float function, like this:

- PI=float(PI)

  $\pi = 3.141592654$

- EULER=float(EULER)

  EULER = 0.5772156649

- CATALAN=float(CATALAN)

  CATALAN = 0.9159655942

- E=float(E)

  $e = 2.718281828$

- infinity

  ∞

Depending on the current value of the environmental variable DIGITS (no more than 2^{31}), you can use each of these constants with an appropriate number of decimal places. For example,

- DIGITS := 200:
- float(E)

  2.7182818284590452353602874713526624977572470936999595749669676277240766303535475945713821785251664274274663919320030599218174135966290435729003342952605956307381323286279434907632338298807531952510191

It is important to understand the role of the float procedure. MuPAD always tries to produce an exact value, and it will not produce a decimal number unless it is forced to. You can force MuPAD to produce the result in decimal form by using the float procedure like in the commands above, or by putting floating point numbers directly into your code. For example, this will force MuPAD to produce a decimal number:

- 1/3 + 0.0

  0.333333333

or

- float(1/3)

A number of very interesting opportunities are provided to us by the constant infinity. You can compare it with other numbers. For example:

- `bool(exp(23434567)<infinity)`

  TRUE

- `bool(-infinity < infinity+infinity)`

  TRUE

- `bool(infinity<infinity+infinity)`

  FALSE

As you see, by default infinity is a positive infinity. The last line provides us with some very interesting observations. Note that additionally,

- `bool(infinity=infinity+infinity)`

  TRUE

which, in fact, suggests that $\infty + \infty = \infty$. You can also easily determine that $\infty \cdot \infty = \infty$.

Of course, in mathematics there are many other specific constants that were discovered throughout the centuries. We have, for example, the very first known transcendental number discovered by Liouville in 1844, Gregory's numbers, Størmer's numbers, harmonic numbers, and so on, and so on. You can imagine how important some of these numbers were for our civilization. For example, how could one calculate many things about the circle without π, or how would our astronomy fare without logarithms and the number e? Let us pause for a moment and think about how some of these numbers can be calculated.

Example 10.4 To calculate an approximate value of $\sqrt{2}$

Before we do any coding, we must develop a method to calculate an approximate value of $\sqrt{2}$. Imagine that $q = \sqrt{2}$. Let us take the two numbers x_1 and y_1, such that $x_1 < q < y_1$. But how do we check if this inequality is true? Just take their squares. Supposing that $0 < a < b$, then, of course, $0 < a^2 < b^2$. This really means that $x_1^2 < q^2 = 2 < y_1^2$. Isn't it easy? We could take, for example, $x_1 = 1$ and $y_1 = 2$. Then of course, $1^2 = 1 < 2 < 2^2 = 4$. In this first step, we were able to locate $\sqrt{2}$ between the two numbers 1 and 2. Now, let us split this interval into two equal parts and see if the center of it, say p, lies to the left of q or to the right. In the first case, we will replace x_1 by p, and in the second case, we will replace y_1 by p. This will narrow down our interval and $q = \sqrt{2}$ will still be in the middle of the interval. Now, we will continue splitting the interval into two equal parts in the same way, until the length of the interval becomes quite small, say 0.0000001. This way, we can evaluate $q = \sqrt{2}$ with an accuracy equal to $\varepsilon = 0.0000001$.

We can now write a procedure that will produce $\sqrt{2}$ with a given accuracy. Here it is. I have made a few small improvements to it. See if you can find them and work out what they do.

- ```
sqrtNew := proc(n:Type::PosInt, m:Type::PosInt)
```

```
/* This is a procedure to calculate the square root
   of any poistive integer number n, with a given
   accuracy m. Number m must be a whole number.
   The procedure implements the bisection method */

local x, y, accuracy, middle ;

begin
   DIGITS := m+3; //number of digits in calculations
   x := 1;
   y := n; // right point > sqrt(n)
   accuracy := 0.1^m; // define length of interval
   repeat
      middle :=(x+y)/2;
      if middle^2<n then
         x := middle
      else
         y := middle
      end
   until (y-x)<accuracy end;
   return( float(x+y)/2)
end:
```

- ```
sqrtNew(2,200)
```
   1.4142135623730950488016887242096980785696718
   7537694807317667973799073247846210703885038753
   4327641572735013846230912297024924836055850737
   2126441214970999358314132226659275055927557999
   5050115278206057147 16

Now, we can check how good the obtained result is.

- %^2

  2.000000000000000000000000000000000000000000000
  000000000000000000000000000000000000000000000000
  000000000000000000000000000000000000000000000000
  000000000000000000000000000000000000000000000000
  00000000000000000043

As, you can see, the obtained result is quite good. Well, perhaps it could still be better for some very precise numerical calculations, but for school purposes, it is sufficient.

Did you notice that the above procedure can be applied to get the square root of any other positive integer? For example,

- `sqrtNew(13,10)`

  3.605551275497

- `(%)^2`

  13.00000000024

There are a few other interesting methods for obtaining the square root of any positive integer. However, this topic goes beyond the scope of this book. You can find a lot of information about these methods in any textbook of numerical methods or a good book about numbers. ■

## 10.4 Complex Numbers in MuPAD

Historically, complex numbers arose from the solutions of quadratic equations. Just imagine, how do we solve the quadratic equation $x^2 = -1$? This is quite easy, if we suppose that there is an imaginary number $i$, so that $i^2 = -1$. From this point, to solve any quadratic equation, we go through the well-known formulae to reach a solution that can be a real number or a complex number in the form $a + ib$, where $a$ and $b$ are two real numbers and $i$ is the so-called imaginary unit. We can represent complex numbers as points on a plane with two axes—the horizontal one being real and the vertical one being imaginary.

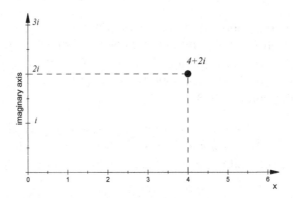

*Fig. 10.1 The plane of complex numbers*

Now, let us return to MuPAD and examine its functionality in relation to complex numbers. We have three types that are useful when working with complex numbers.

Type::Complex - *type for complex numbers*
Type::Imaginary - *type for numbers in the form a · i*
Type::Real - *the well-known type for real numbers*

A complex number in MuPAD can be written in the form a + b*I, where a and b are two real numbers. For example:

- 2 + 3*I

    $2 + 3i$

There are also a few operations that can be useful while dealing with complex numbers.

+, -, *, / - *arithmetic operations*
conjugate($a+b*I$) - *conjugate complex number, returns a-b\*I*
abs($a+b*I$) - *modulus of a complex number, returns* $\sqrt{a^2 + b^2}$
Re($a+b*I$) - *real part of a complex number, returns a*
Im($a+b*I$) - *imaginary part of a complex number, returns b*
rectform($z$) - *returns z in rectangular form a+b\*I*
sign($z$) - *returns z/abs(z)*
arg($z$) - *returns angle between vector* $\overrightarrow{0z}$ *and* x-axis.

Here are some examples showing some of the above functions in action:

- (2+4*I)+(4+5*I)

  $6 + 9 \cdot i$

- (2+4*I)*(4+5*I)

  $-12 + 26 \cdot i$

- (2+4*I)/(4+5*I)

  $\dfrac{28}{41} + \dfrac{6 \cdot i}{41}$

- conjugate(2+3*I)

  $2 - 3 \cdot i$

- abs(2+3*I)

  $\sqrt{13}$

- Im(2+3*I)

  3

- Re(2+3*I)

  2

- sign(2+3*I)

  $\left( \dfrac{2}{13} + \dfrac{3 \cdot i}{13} \right) \cdot \sqrt{13}$

- abs(%)

  $\sqrt{13} \cdot \sqrt{\dfrac{1}{13}}$

- C1:=1+I*exp(PI*(I/4+1/3))

  $i \cdot e^{\left( \frac{1}{3} + \frac{i}{4} \right) \cdot \pi} + 1$

- rectform(C1)

  $-\dfrac{e^{\frac{\pi}{3} \cdot \sqrt{2}}}{2} + 1 + i \cdot \dfrac{e^{\frac{\pi}{3} \cdot \sqrt{2}}}{2}$

The operations described above form a good foundation for many interesting investigations about complex numbers. For example, we can investigate the complex integers $a + bi$, where $a$ and $b$ are integer numbers. Today, we call them Gaussian integers. Then we can investigate Gaussian prime numbers. You can find a lot of information about these interesting facts in some of the older mathematical books, for example in the famous book by Courant and Robbins, *What is Mathematics?: An Elementary Approach to Ideas and Methods*, Oxford University Press.

Let us finish this chapter with an interesting example related to complex numbers.

### Example 10.5 Roots of complex numbers

For a given complex number, it is quite easy to obtain its power $z^n$, where $n$ is an integer number. For example,

- $(1/2+2/3*I)\wedge7$

$$\frac{25481}{93312} + \frac{4031}{69984}i$$

However, when you try to obtain the $n$-th root of a complex number in MuPAD, you will not get any result. This is because calculating the roots of complex numbers requires a different approach.

First of all, you should think about what you can call an $n$-th root of a complex number $a + bi$. This certainly will be the complex number $z$, such that $z^n = a + bi$. However, you have perhaps noticed that for a given number, there can be two or more such roots. For example, the numbers $-1$ and $1$ can both be considered as square roots of $1$. This is evident, because $(-1)^2 = 1$ and $1^2 = 1$. You may expect the same for complex numbers. This problem greatly interested the French mathematician Abraham DeMoivre (1667-1754). He discovered that for a given complex number $a + bi$ and a positive integer $n$, there are $n$ different complex numbers that can be called an $n$-th root of $a + bi$. He also discovered a formula for describing all these roots. Here is his formula:

$$\sqrt[n]{r}\left[\cos\left(\frac{\theta}{n} + \frac{2k\pi}{n}\right) + i\sin\left(\frac{\theta}{n} + \frac{2k\pi}{n}\right)\right]$$

where $r$ is the modulus of the complex number and $\theta$ is its argument. From this starting point, it is quite easy to calculate all the $n$-th roots of a complex number. Note that I am talking here about all the roots, not just a single one.

You should be able to easily determine what each instruction of the procedure below does. Note that the procedure will output all the $n$-th roots in the form of a list. You could instead use a set or a sequence. This depends on how you are going to use these roots in further calculations. Having organized them in the form of a list, you can access each root by its coordinate in the list.

- ```
  Moivre := proc(z:Type::Complex,n:Type::Integer)
  /* This procedure will calculate n complex
     roots of a given complex number z   */

  local r, k, roots, angle,
        a, newangle, newr, newRe, newIm, nthroot;
  begin
     r := abs(z);
     newr := r^(1/n);
     angle := arg(z);
     newangle := angle/n;
     a := 2*PI/n;
     roots := [];
     for k from 0 to (n-1) do
        newRe := cos(newangle + k*a);
        newIm := sin(newangle + k*a);
        nthroot := newr*(newRe+I*newIm);
        roots :=roots.[nthroot];
     end;
     return(roots)
  end:
  ```

Now you can try the procedure to see what output you will get. First, check some well-known cases. This will give you some idea about the nature of complex numbers. For example,

- `Moivre(1,2)`

  $[1, -1]$

- `Moivre(-1,2)`

  $[i, -i]$

- `Moivre(I,2)`

  $[(\frac{1}{2} + \frac{i}{2})\sqrt{2}, -(\frac{1}{2} + \frac{i}{2})\sqrt{2}]$

- `Moivre(1,3)`

  $[1, \frac{i}{2}\sqrt{3} - \frac{1}{2}, -\frac{i}{2}\sqrt{3} - \frac{1}{2}]$

Now, you can try something less obvious. For example, you could find all the cube roots of -8.
  ```

- Moivre(-8,3)

$$\left[i \cdot \sqrt{3} + 1, \; -2, \; 1 - i \cdot \sqrt{3} \right]$$

Well, I will leave all these experiments for you to try. This can be a time-consuming, but very interesting activity. By the way, do you know that for a given complex number z and a given integer n, the sum of all its n-th roots for $n > 1$ is equal to zero? Try to check it on our examples.

In this example we used an opportunity to practice programming and revise the concept of roots of complex numbers. However, in MuPAD you can produce roots of complex numbers even without knowing anything about complex numbers. For instance, you can always type in and execute the statement solve(z^n=a+b*I,z). However, solutions produced by our procedure are sometimes a bit simpler than those from the solve statement. Try, for example, solve(z^3=1+I,z) and Moivre(1+I,3).   ■

This was a very interesting chapter, and you learned many things here. Let us summarize all of them.

10.5 Chapter Summary

In this chapter, I introduced a large collection of functions and operations. Some of them are very specific to a given domain. For example, the function modp can only be applied to integer numbers or expressions representing integers. However, the function max(x,y) can be applied to integers, rational and real numbers or expressions representing these numbers. Generally, every function for real numbers can also be applied to integers as well as to rational numbers. However, the result may not belong to the same domain as the argument. For example, for the integer n, sqrt(n) may produce a result that is not an integer number.

Functions and operations for integer numbers

+, *, -, / - *arithmetical operations*

max(n, m) - *maximum of two or more numbers*

min(n,m) - *minimum of two or more numbers*

! - *factorial operation*

mod - *produces the remainder of integer division, e.g.* $(25 \bmod 3) = 1$

modp(n,m) - *produces* (n mod m)

div - *integer division of two integers, e.g.* $(25 \text{ div } 3) = 8$

factor(n) - *factoring a given integer*

ifactor(n) - *factoring integer into prime components*

igcd(n,m) - *greatest common divisor for integers*

igcdex(n,m) - *extended gcd for integers*

ilcm(n,m) - *the lowest common multiple for integers*

isprime(n) - *checks if the given integer is a prime number*

ithprime(i) - *produces i-th prime number*

random() - *generates a random integer number between* 0 *and* 10^{12}

random($n1..n2$) - *produces a function to generate random integers from the interval* $[n1, n2]$

Operations, functions and relations for real numbers

+, -, *, /, <, <=, =, >, >=

abs(x) - *function absolute value of x*

sign(x) - *function sign*

log($base, x$) - *function logarithm with a given base, e.g.* log(10,x)

ln(x) - *natural logarithm of x*

exp(x) - *exponential function of x,* exp(x) = e^x

sqrt(x) - *function* $\sqrt{x}$

sin(x), cos(x), tan(x), cot(x) - *trigonometric functions, x shall be given in radians*

sec(x), csc(x) - *functions* $\sec(x) = \frac{1}{\cosh(x)}$ *and* $\csc(x) = \frac{1}{\sinh(x)}$

arcsin(x), arccos(x), arctan(x), arccot(x) - *inverse trigonometric functions*

arcsec(x), arccsc(x) - *inverse functions for* sec(x) *and* csc(x)

sinh(x), cosh(x), tanh(x), coth(x) - *hyperbolic functions*

sech(x), csch(x) - *hyperbolic functions* $\text{sech}(x) = \frac{1}{\cos(x)}$ *and* $\text{csch}(x) = \frac{1}{\sin(x)}$

floor(x), ceil(x), round(x), trunc(x) - *functions rounding real numbers, result is an integer number*

min($x,y,...,z$), max($x,y,...,z$) - *functions producing minimum and maximum*

Constants

PI - *number* π

EULER - *the Euler-Mascheroni constant*

CATALAN - *Catalan constant*

E - *Napier's constant e, base for natural logarithms*

infinity - *constant representing infinity*

Operations and functions for complex numbers

+, -, *, / - *arithmetic operations*

conjugate(a+b*I) - *conjugate complex number, returns a-b*I*

abs(a+b*I) - *modulus of a complex number, returns* $\sqrt{a^2 + b^2}$

Re(a+b*I) - *real part of a complex number, returns a*

Im(a+b*I) - *imaginary part of a complex number, returns b*

rectform(z) - *returns z in rectangular form a+b*I*

sign(z) - *returns z/abs(z)*

arg(z) - *returns an angle between vector* $\overrightarrow{0z}$ *and* OX.

10.6 Programming Exercises

During this chapter, a number of interesting exercises were mentioned. You can easily find a large collections of problems that can be solved with MuPAD in books about number theory. Here, I will only mention a few of them.

1. A whole number is considered as a triangular number if it can be represented as a sum $\Delta_n = 1 + 2 + 3 + \ldots + n$. You can easily prove that $\Delta_n = \frac{n(n+1)}{2}$. Develop a procedure that will produce the triangular number Δ_n. Use this procedure to find some of the triangular numbers that are squares of some other integers.

2. Mathematicians have frequently considered the families of so-called polygonal numbers. We obtain different kinds of polygonal numbers by adding the first n terms of appropriate arithmetic progression, starting with 1. Thus:

    $1 + 1 + 1 + \ldots$ - gives counting numbers 1, 2, 3, 4, 5,..

    $1 + 2 + 3 + 4 + \ldots$ - gives triangular numbers 1, 3, 6, 10, 15,..

    $1 + 3 + 5 + 7 + 9 + \ldots$ - gives square numbers 1, 4, 9, 16, 25,..

    $1 + 4 + 7 + 10 + 13 + \ldots$ - gives pentagonal numbers 1, 5, 12, 22, 35,..

    Similarly, you can create hexagonal numbers, heptagonal, octagonal, and so on. Write a procedure that, for given whole number m, will produce the first n terms of a sequence of m-gonal numbers. Use this procedure to verify that every hexagonal number is a triangular number and that every pentagonal number is one-third of a triangular number. Can you find a geometric interpretation for these families of numbers?

3. The number of choices of r things from n can be expressed by the formula $\binom{n}{r} = \dfrac{n!}{r!(n-r)!}$. Write a procedure to obtain the choice number for two given whole numbers n and r. Use your procedure to verify that $\binom{n}{r} = \binom{n-1}{r} + \binom{n-1}{r-1}$.

4. Write down the middle numbers in the Pascal triangle. These are the numbers 1, 2, 6, 20, 70, 252, 924, ... Looking on them, you may think about the hypothesis that we can divide them by 1, 2, 3, 4, 5, 6,... respectively. A typical middle number Pascal's triangle has a formula $\binom{2n}{n}$. Using the mentioned idea we can write a formula for so called Catalan numbers:

$$c_n = \frac{1}{n+1}\binom{2n}{n}$$

Write a procedure that will produce the n-th Catalan number. Use your procedure to check if the above hypothesis is valid for the first 100 middle numbers in the Pascal triangle.

5. In one of our exercises, we had developed a procedure to create a table of multiplication modulo m. Write a procedure that will create a table of addition modulo m. What about the division of integers modulo m, or finding inverse numbers modulo m? You have perhaps noticed that for a given integer $0 < k < m$, there might exist another integer $0 < l < m$, such that $(kl \bmod m) = 1$. Thus, l can be considered as the inverse of k. Write a procedure that will check if for a given integer $0 < k < m$, there is an inverse modulo m. Try to find out when such an inverse always exists.

6. A whole number is considered perfect if it can be represented as the sum of all its divisors. For example, $6 = 1 + 2 + 3$, $28 = 1 + 2 + 4 + 7 + 14$. Write a procedure that will produce all the perfect numbers inside a given range, say $1..n$.

7. In 1640, Pierre Fermat hypothesized that all the numbers $2^{2^m} + 1$ are prime numbers. Today, these numbers are called Fermat's numbers. Write a procedure that will check if Fermat's numbers

from a given range are primes. Find as many Fermat numbers as possible that are not primes.

8. Write a procedure that will search all the integers $1..100$ for numbers that can be represented as a sum $n^2 + m^2$, where m, n are integers >0. For example, $2 = 1^2 + 1^2$, $17 = 4^2 + 1^2$. Which of these numbers are prime numbers?

9. The very famous Fibonacci sequence has a twin brother known as the Lucas sequence. Both of these sequences can be described by these quite amazing formulae:

$$F_n = \frac{1}{\sqrt{5}}\left\{\left(\frac{1+\sqrt{5}}{2}\right)^n - \left(\frac{1-\sqrt{5}}{2}\right)^n\right\}$$

$$L_n = \frac{1}{\sqrt{5}}\left\{\left(\frac{1+\sqrt{5}}{2}\right)^n + \left(\frac{1-\sqrt{5}}{2}\right)^n\right\}$$

I guess you might be surprised. These sequences do not look like sequences of integer numbers. Write a procedure that will calculate n terms of the Fibonacci sequence. Write another procedure that will calculate n terms of the Lucas sequence.

10. The tribonacci numbers are a generalization of the Fibonacci Numbers defined by $T_1 = 1$, $T_2 = 1$, $T_3 = 2$ and the formula $T_n = T_{n-1} + T_{n-2} + T_{n-3}$ for $n > 3$. Write a procedure that will produce n-th tribonacci number. Obtain the first ten tribonacci numbers.

11. In 1816, Farey proposed a spectacular way of arranging all the proper fractions $\frac{p}{q}$. He started with two fractions:

$$\frac{0}{1}, \frac{1}{1}$$

Then by inserting a median $\frac{a+c}{b+d}$ between the two consecutive fractions $\frac{a}{b}$ and $\frac{c}{d}$, he formed a second row, third row, and so on. There is one additional rule. You can insert a median fraction only if its nominator and denominator are no larger than the number of the row. Thus, rows of fractions in the Farey arrangement (we call them the Farey series) form a pyramid like this:

row 1: $\frac{0}{1}, \frac{1}{1}$

row 2: $\frac{0}{1}, \frac{1}{2}, \frac{1}{1}$

row 3: $\frac{0}{1}, \frac{1}{3}, \frac{1}{2}, \frac{2}{3}, \frac{1}{1}$

row 4: $\frac{0}{1}, \frac{1}{4}, \frac{1}{3}, \frac{1}{2}, \frac{2}{3}, \frac{3}{4}, \frac{1}{1}$

row 5: $\frac{0}{1}, \frac{1}{5}, \frac{1}{4}, \frac{1}{3}, \frac{2}{5}, \frac{1}{2}, \frac{3}{5}, \frac{2}{3}, \frac{3}{4}, \frac{4}{5}, \frac{1}{1}$

row 6: $\frac{0}{1}, \frac{1}{6}, \frac{1}{5}, \frac{1}{4}, \frac{1}{3}, \frac{2}{5}, \frac{1}{2}, \frac{3}{5}, \frac{2}{3}, \frac{3}{4}, \frac{4}{5}, \frac{5}{6}, \frac{1}{1}$

Write a procedure that will produce the n-th Farey series. You can also make a nice illustration of the Farey series. For each fraction $\frac{p}{q}$ from a given Farey series, draw a circle with the radius $r = \frac{1}{q^2}$ and the center $(\frac{p}{q}, r)$. You will thus obtain the so-called Ford circles.

12. It was proven by Lagrange that continued fractions are periodic for algebraic numbers of degree 2. This fact can be used to calculate some irrational numbers. Write a procedure that will calculate the approximate values of $\sqrt{2}$, $\sqrt{3}$, $\sqrt{5}$ by taking the first n terms of an appropriate continued fraction. You result should be obtained in the form of a rational number p/q. Use these formulae:

a. $\sqrt{2} = 1 + 1/(2 + 1/(2 + 1/(2 + 1/(2 + \ldots))))$

b. $\sqrt{3} = 1 + 1/(1 + 1/(2 + 1/(1 + 1/(2 + 1/(1 + 1/(2 + \ldots))))))$

c. $\sqrt{5} = 2 + 1/(4 + 1/(4 + 1/(4 + 1/(4 + 1/(4 + 1/(4 + \ldots))))))$

13. Write a procedure that will calculate a sequence of approximated values of Napier's constant e. Your results should be rational numbers in the form p/q. Use
$e = 2 + 1/(1 + 1/(2 + 1/(1 + 1/(1 + 1/(4 + 1/(1 + \ldots))))))$, where the sequence of consecutive denominators of the continued fraction forms the list

$$1, 2, 1, 1, 4, 1, 1, 6, 1, 1, 8, 1, 1, 10, \ldots$$

14. Write a program that will produce all the n-th roots of a given complex number and then plot them on a complex plane. The xy-plane will simulate the complex plane. Use the y-axis as the imaginary axis.

15. The story of complex numbers can go beyond two dimensions. You can investigate quaternions that are numbers of the form $q = a + bi + cj + dk$, where i, j, k are three imaginary units fulfilling Hamilton's rules $i^2 = j^2 = k^2 = -1$, $ij = k$, $jk = i$, $ki = j$, $ji = -k$, $kj = -i$, $ik = -j$. Quaternions do exist as objects in MuPAD — just look at Dom::Quaternion. However, you can always develop them in a slightly different way. Just think about a quaternion as a list of four numbers. For example, $2 + 3i + 4j + 7k$ can be represented as $[2, 3, 4, 7]$. From here, it is a simple step to produce the sum or a difference of two quaternions, for example $[2, 3, 4, 7] + [-1, 5, 4, 3] = [1, 8, 8, 10]$. Multiplication of quaternions is a bit more complicated, though. Write a procedure qprod to multiply two quaternions. Use the above rules to simplify the results.

Elementary Algebra and Trigonometry in MuPAD

A large part of undergraduate mathematics is devoted to working with algebraic expressions, simplifying and transforming them. In this chapter, you will learn how to define algebraic expressions, simplify or expand them, as well as many other things that are useful for solving problems in undergraduate mathematics. There will be not much programming in this chapter, though you will need to develop one or two simple procedures.

11.1 Polynomials

In order to declare a polynomial, you only need to type it in using MuPAD's specific notation. For example,

- `2 + 3*x + x^3 + x^2 - 4*x`
  $$x^2 - x + x^3 + 2$$

or assign it to a variable,

- `poly1 := 2 + 3*x + x^3 + x^2 - 4*x`
  $$x^2 - x + x^3 + 2$$

In order to make your commands more readable, remember to use spaces around the arithmetic operators + and −.

Now you can apply some of the standard MuPAD procedures to your polynomials:

- `factor(poly1)`
  $$(x + 2) \cdot (x^2 - x + 1)$$

- `simplify(%)`
  $$x^2 - x + x^3 + 2$$

- `solve(poly1 = 0, x)`

$$\left\{-2, \frac{1}{2} + \left(-\frac{i}{2}\right) \cdot \sqrt{3}, \frac{i}{2} \cdot \sqrt{3} + \frac{1}{2}\right\}$$

You can also apply arithmetic operations to your polynomials. For example, define two polynomials and add them, multiply them, etc.

- `poly1 := 5 - 6*x + 6*x^2 - 6*x^3 + x^4`
  $$6 \cdot x^2 - 6 \cdot x - 6 \cdot x^3 + x^4 + 5$$

- `poly2 := 15 + 2*x - 16*x^2 - 2*x^3 + x^4`
  $$2 \cdot x - 16 \cdot x^2 - 2 \cdot x^3 + x^4 + 15$$

- `poly1 + poly2`
  $$2 \cdot x^4 - 10 \cdot x^2 - 8 \cdot x^3 - 4 \cdot x + 20$$

- `poly1 - poly2`
  $$22 \cdot x^2 - 8 \cdot x - 4 \cdot x^3 - 10$$

- `poly1*poly2`
  $$(6 \cdot x^2 - 6 \cdot x - 6 \cdot x^3 + x^4 + 5) \cdot (2 \cdot x - 16 \cdot x^2 - 2 \cdot x^3 + x^4 + 15)$$

Note that the last expression was not simplified by MuPAD and it requires one more step of calculations:

- `expand(%)`
  $$8 \cdot x^3 - 2 \cdot x^2 - 80 \cdot x - 76 \cdot x^4 + 80 \cdot x^5 + 2 \cdot x^6 - 8 \cdot x^7 + x^8 + 75$$

However, if you feel that the result is too messy, you can try to get a clearer picture by factorizing your polynomial.

- `factor(%)`
  $$(x + 1) \cdot (x + 3) \cdot (x^2 + 1) \cdot (x - 5)^2 \cdot (x - 1)^2$$

You can build even more complicated expressions:

- `(poly1*poly2^2)^2`
  $$(6 \cdot x^2 - 6 \cdot x - 6 \cdot x^3 + x^4 + 5)^2 \cdot (2 \cdot x - 16 \cdot x^2 - 2 \cdot x^3 + x^4 + 15)^4$$

- `poly1/poly2`
  $$\frac{6 \cdot x^2 - 6 \cdot x - 6 \cdot x^3 + x^4 + 5}{2 \cdot x - 16 \cdot x^2 - 2 \cdot x^3 + x^4 + 15}$$

In this last case, the obtained result is a rational expression, but you can still apply many MuPAD procedures to it.

- `factor(poly1/poly2)`

$$\frac{x^2+1}{(x+1)\cdot(x+3)}$$

- `expand(%)`

$$\frac{1}{(x+1)\cdot(x+3)}+\frac{x^2}{(x+1)\cdot(x+3)}$$

We will talk about rational expressions a bit later.

In respect to polynomials, you can determine their greatest common divisor, least common multiple, or the degree of a polynomial. For example,

- `mylcd := gcd(poly1, poly2)`

$$x^2-6\cdot x+5$$

- `mylcm := lcm(poly1, poly2)`

$$2\cdot x-x^2-15\cdot x^4-2\cdot x^5+x^6+15$$

Again, if you are sometimes a bit lost with such large powers, you can factorize your polynomial expression in order to get its factors.

- `factor(mylcm)`

$$(x-1)\cdot(x+3)\cdot(x-5)\cdot(x+1)\cdot(x^2+1)$$

Finally, you can work out the degree of a more complex polynomial, or calculate it for a given value of the variable x.

- `degree(mylcd)`

2

- `evalp(mylcd, x=2.23)`

$$-3.4071$$

The names of operations that can be performed on polynomials are very natural and we have already gotten used to these names while working with polynomials at school. Thus, it is quite easy to remember all the names. You will find later that some of these operations can also be applied to other mathematical expressions with the expected output.

All the commands mentioned above are very convenient while solving problems in undergraduate mathematics. However, sometimes we need to go further beyond the informal concept of polynomials. For example, we may wish to consider polynomials with only rational coefficients, or polynomials with integer coefficients modulo some given number m. In MuPAD, we can also consider polynomials over a given ring. However, this concept is slightly beyond the scope of this book, so I will not talk too much about it here.

There are two ways of declaring polynomials. The first one is what you have already learned—I will refer to it as informal. There is also another method, which I will call a formal polynomial declaration.

In order to declare our polynomials in a formal way, we need to use the `poly` command with the appropriate parameters. We can also use the existing domains for polynomials:

```
Dom::Polynomial,
Dom::MultivariatePolynomial,
Dom::UnivariatePolynomial
```

or types

```
Type::PolyExpr,
Type::PolyOf.
```

Finally, MuPAD has a library called `polylib`, which contains a lot of advanced operations for polynomials.

Let us start with the basic commands using polynomial-specific procedures.

The simplest formal declaration of a polynomial needs to contain only the polynomial expression inside of the `poly()` statement.

- `poly(x^2 - x - 1)`
  $\text{poly}(x^2 - x - 1, [x])$

Quite often we specify also the variable used in the polynomial. This helps us to distinguish between variables and symbolic parameters. For example,

- `poly(a*x^2 + 2*x + 1, [x])`
  $\text{poly}(a \cdot x^2 + 2 \cdot x + 1, [x])$

Otherwise we may get,

- `poly(a*x^2 + 2*x + 1)`
  $\text{poly}(a \cdot x^2 + 2 \cdot x + 1, [a, x])$

We can also apply the `poly` command to an existing polynomial that was declared informally.

- `poly(mylcm,[x])`

  $poly(x^6 - 2 \cdot x^5 - 15 \cdot x^4 - x^2 + 2 \cdot x + 15, [x])$

Notice the first advantage of the formal declaration — the output contains all the powers of x printed out in descending order. This makes the output less messy and much easier to analyze.

The syntax of the polynomial declaration can be much more sophisticated. For example, you could try creating a polynomial with two variables x and y.

- `poly3 := poly(x*y + x + y + x^2 + x*y^2, [x,y])`

  $poly(x^2 + x \cdot y^2 + x \cdot y + x + y, [x,y])$

In the last case, the created polynomial contains two variables x and y and all the terms of the polynomial are sorted according to their powers in descending order. Examine the difference between the above polynomial, and a polynomial that looks identical, but where y is not considered as a variable.

- `poly4 := poly(x*y + x + y + x^2 + x*y^2, [x])`

  $poly(x^2 + (y + y^2 + 1) \cdot x + y, [x])$

In the latter case, y is treated as one of the coefficients of the polynomial. The difference between the two objects will become more evident if we use one of the polynomial-specific procedures that are available in MuPAD. For example, let us apply the procedure `poly2list` that changes a polynomial to a list displaying each term as a pair $[a, b]$ where the first element is the coefficient and the second is the power of the variable. Thus, $3x^2$ will be transformed to $[3, 2]$ and $3x^2y^5$ will be transformed to $[3, [2, 5]]$. Now observe how, for `poly3` and `poly4`, we get very different results.

- `poly2list(poly4)`

  $[[1, 2], [y + y^2 + 1, 1], [y, 0]]$

- `poly2list(poly3)`

  $[[1, [2, 0]], [1, [1, 2]], [1, [1, 1]], [1, [1, 0]], [1, [0, 1]]]$

Compare the obtained results closely, and try to work out how this can affect your calculations with polynomials.

The list notation for polynomials is a bit harder to read. However, it can be very convenient in calculations. Let us see what this really means.

Example 11.1

Suppose that we need to obtain all the polynomials of the form

$$x^n + \binom{n}{k} x^k y^{n-k} + y^n,$$

for fixed n and $k = 0..n$. Using the list notation for the polynomials, we can express each polynomial as a list:

$$\Big[[1, [1, 0]], \Big[\binom{n}{k}, [k, n-k] \Big], [1, [0, 1]] \Big].$$

Thus, we need to produce a list for each polynomial and then convert this list to the explicit polynomial form. Before we do this however, we are going to write a procedure to obtain $\binom{n}{k}$. This was one of the exercises for chapter 5. Therefore, you may use here the procedure that you had developed earlier. I will use a very simple procedure called nOverK.

- nOverK := proc(n: Type::NonNegInt,
                  k:Type::NonNegInt)

  ```
  local result;
  begin
     if k <= n then
        result := n!/(k!*(n-k)!);
        return(result)
     else
        return(0)
     end;
  end:
  ```

Now we can write the procedure to produce our polynomials. Note that I have used here the procedure poly to generate a polynomial from the list.

- shortPolynomials := proc(n: Type::NonNegInt)
  ```
  local allPoly, k, part1, part2,
        part3, mylist, polyresult;
  begin
  ```

```
allPoly := [];
for k from 0 to n do
    part1 := [1, [n,0]];
    part2 := [nOverK(n,k),[k,n-k]];
    part3 := [1,[0,n]];
    mylist := [part1,part2,part3];
    polyresult := poly(mylist,[x,y]);
    allPoly := allPoly.[polyresult]
end;
return(allPoly)
end:
```

You can easily understand every statement in this procedure. Most of the operations here are just assignments of values to the variables, and list declarations. Here are two results obtained by the procedure. The first one is for $n = 5$ and the second one is for $n = 10$. In both cases, I have reformatted the output in order to give you a better idea about the structure of the obtained polynomials.

- `shortPolynomials(5)`

  $[\,poly(x^5 + 2 \cdot y^5, [x,y]),$
  $poly(x^5 + 5 \cdot x \cdot y^4 + y^5, [x,y]),$
  $poly(x^5 + 10 \cdot x^2 \cdot y^3 + y^5, [x,y]),$
  $poly(x^5 + 10 \cdot x^3 \cdot y^2 + y^5, [x,y]),$
  $poly(x^5 + 5 \cdot x^4 \cdot y + y^5, [x,y]),$
  $poly(2 \cdot x^5 + y^5, [x,y])]$

- `shortPolynomials(10)`

  $[poly(x^{10} + 2 \cdot y^{10}, [x,y]),$
  $poly(x^{10} + 10 \cdot x \cdot y^9 + y^{10}, [x,y]),$
  $poly(x^{10} + 45 \cdot x^2 \cdot y^8 + y^{10}, [x,y]),$
  $poly(x^{10} + 120 \cdot x^3 \cdot y^7 + y^{10}, [x,y]),$
  $poly(x^{10} + 210 \cdot x^4 \cdot y^6 + y^{10}, [x,y]),$
  $poly(x^{10} + 252 \cdot x^5 \cdot y^5 + y^{10}, [x,y]),$
  $poly(x^{10} + 210 \cdot x^6 \cdot y^4 + y^{10}, [x,y]),$
  $poly(x^{10} + 120 \cdot x^7 \cdot y^3 + y^{10}, [x,y]),$
  $poly(x^{10} + 45 \cdot x^8 \cdot y^2 + y^{10}, [x,y]),$
  $poly(x^{10} + 10 \cdot x^9 \cdot y + y^{10}, [x,y]),$
  $poly(2 \cdot x^{10} + y^{10}, [x,y])]$

In a similar way, you can produce a number of other families of polynomials. ∎

Before we proceed to the next topic, I would like to mention something about working with polynomials over special domains. We can declare them in exactly the same manner as standard polynomials in a given variable or variables. The only difference is that we need to add the domain description. For example:

- `poly1 := poly(4*x^4 + 7*x^5*y^12,[x,y],`
  `    Dom::IntegerMod(3)`
  `)`

  $poly(x^5 \cdot y^{12} + x^4, [x,y],$ Dom::IntegerMod(3))

- `poly2 := poly(5*x^3 + 7*x^2*y^6,[x,y],`
  `    Dom::IntegerMod(3)`
  `)`

  $poly(2 \cdot x^3 + x^2 \cdot y^6, [x,y],$ Dom::IntegerMod(3))

- `poly1 + poly2`

  $poly(x^5 \cdot y^{12} + x^4 + 2 \cdot x^3 + x^2 \cdot y^6, [x,y],$ Dom::IntegerMod(3))

Note that in each case, all the coefficients were obtained using arithmetic modulo 3. However, some of the exponents are still represented by numbers larger than 2. This is not a bug. It's just that a rule like $x^5 = x^2$ is not valid in arithmetic modulo 3. For example,

- `modp(2^5,3)`

  2

- `modp(2^2,3)`

  1

Thus, in order to evaluate the expression $x^5 y^{12} + x^4 + 2x^3 + x^2 y^6$ modulo 3, we will have to wait until the final values have been substituted instead of x and y.

11.2 Rational Expressions

It is only a small jump from polynomials to rational expressions. You simply need to divide one polynomial by another. The world of rational expressions is very rich in interesting properties, theorems, as well as the names of the mathematicians that worked with them. On the side of operations, we can apply here most of the operations that we used with polynomials. There are also a few operations that are specific to rational expressions. We will talk about them in this chapter.

Let us see what interesting things related to rational expressions we can find in MuPAD. First of all, we have in MuPAD domains and types for rational expressions, for example `Dom::Fraction(R)` and `Type::RatExpr`. Therefore, a rational expression can be just a fraction, where the nominator and the denominator are representing a given domain. Let us show a simple example where we deal with rational expressions of polynomials with integer coefficients. We start by declaring a new domain,

- `F := Dom::Fraction(Dom::Polynomial(Dom::Integer)):`

Now, we can declare two elements from this domain.

- `a := F((1 - x)/(x + 2))`

  $-\dfrac{x-1}{x+2}$

- `b := F((-x + 1)/(2 - x))`

  $\dfrac{x-1}{x-2}$

On such rational expressions we can perform arithmetic and other suitable operations.

- `c:=a*b + a/b`

  $-\dfrac{2 \cdot x^2 - 6 \cdot x + 5}{x^2 - 4}$

- `factor(c)`

  $-\dfrac{2 \cdot x^2 - 6 \cdot x + 5}{(x - 2) \cdot (x + 2)}$

Note that you cannot use formally defined polynomials to obtain a rational expressions. For example,

- `poly1 := poly(x^2 + 1,[x])`

  $\text{poly}(x^2 + 1, [x])$

- `poly2 := poly(x^3 + 1,[x])`

  $\text{poly}(x^3 + 1, [x])$

- `poly1/poly2`

  FAIL

This last result suggests that if we wish to work with rational expressions, we will need to define them using informally declared polynomials or define our own domains.

MuPAD offers a number of procedures for working with rational expressions. First of all, let us see what our needs are. While working with rational expressions, we need to be able to perform some basic operations on them $(+,-,*,/)$, factorize or expand their components, and transform them to the form of partial fractions. Of course, you can think of a few more operations but the ones mentioned here are the essentials.

The operations $+,-,*$ and $/$ work in the same manner as they do for other arithmetic expressions. In the obtained results, each term is presented in the expanded form. For example,

- `rat1 := (x^2 - 2*x + 1)/(x^2 - 1)`
  $$\frac{x^2 - 2 \cdot x + 1}{x^2 - 1}$$

- `rat2 := (x^3 - 1)/(x^4 - 1)`
  $$\frac{x^3 - 1}{x^4 - 1}$$

- `rat1 + rat2`
  $$\frac{x^3 - 1}{x^4 - 1} + \frac{x^2 - 2 \cdot x + 1}{x^2 - 1}$$

- `rat1/rat2`
  $$\frac{(x^4 - 1) \cdot (x^2 - 2 \cdot x + 1)}{(x^2 - 1) \cdot (x^3 - 1)}$$

- `rat3 := rat1*rat2`
  $$\frac{(x^3 - 1) \cdot (x^2 - 2 \cdot x + 1)}{(x^2 - 1) \cdot (x^4 - 1)}$$

In order to get another form of the rational expression we may use some other procedures. Depending on what we wish to get, we may apply one of the procedures `simplify`, `factor`, or `expand`. The first procedure will produce the most simplified form of the numerator and the denominator of the rational expression. The second procedure will factorize its components. Finally, the last one will expand the numerator and the denominator of the fraction. For example,

- `rat4 := expand(rat3)`

$$\frac{x^2 - 2 \cdot x - x^3 + 2 \cdot x^4 - x^5 + 1}{x^2 + x^4 - x^6 - 1}$$

- `simplify(rat4)`

$$\frac{x^3 - 1}{(x^2 + 1) \cdot (x + 1)^2}$$

- `factor(rat4)`

$$\frac{(x + x^2 + 1) \cdot (x - 1)}{(x^2 + 1) \cdot (x + 1)^2}$$

Observe that in last two cases, all the common terms of the denominator and the numerator were removed. This makes the fraction simpler and easier to analyze.

The opposite operation is to expand our rational expressions into sums of simpler fractions. You can use the procedure expand, which simply expands the rational expression into a sum of fractions. Note that in some situations, the expand procedure has to be used twice before we get the sum of fractions. Like here:

- `rat5 := (x - 3)*(x - 2)/((x - 7)*(x - 5))`

$$\frac{(x - 2) \cdot (x - 3)}{(x - 5) \cdot (x - 7)}$$

- `expand(rat5)`

$$\frac{x^2 - 5 \cdot x + 6}{x^2 - 12 \cdot x + 35}$$

- `expand(%)`

$$\frac{x^2}{x^2 - 12 \cdot x + 35} + \frac{6}{x^2 - 12 \cdot x + 35} - \frac{5x}{x^2 - 12 \cdot x + 35}$$

If you wish to do a clean job and obtain the sum of the fractions in the simplest form, you should use partfrac. This is an operation that is extremely useful in calculus when integrating rational expressions. Here I will demonstrate how the procedure partfrac works.

- `partfrac(rat5)`

$$\frac{10}{x - 7} - \frac{3}{x - 5} + 1$$

Note how neat and short the result obtained by the `partfrac` procedure was, while the output from the `expand` procedure merely gave us an expanded expression. However, you will find many situations where an expanded rational expression will look much simpler than partial fractions. This depends mainly on the rational expression you are working with.

Finally, it is worth mentioning another function that can do a great job in simplifying rational expressions. This is the operation `normal`. It generates a new form of the fraction, where the common terms of the numerator and the denominator were removed. For example,

- `fraction := (x^3 - 1)/(x^2 - 1)`

$$\frac{x^3 - 1}{x^2 - 1}$$

- `normal(fraction)`

$$\frac{x + x^2 + 1}{x + 1}$$

11.3 Algebraic Expressions

By adding roots to rational expressions, you extend them to the class of so-called algebraic expressions. Here you have a number of tricky functions that are quite difficult to deal with as well as to systematize. These are some of them:

$$\frac{\sqrt{x} + x + \sqrt[5]{x^3}}{\sqrt{x} + x + x^2}, \quad \sqrt{x + \sqrt{x + 1}}, \quad \frac{1}{\sqrt{x} + x^2 + x^3}.$$

All the standard operations $+, -, *, /$ as well as $\sqrt[n]{}$, will produce algebraic expressions when applied to algebraic expressions. Also, note that you have the function `sqrt(x)` in MuPAD, which will produce $\sqrt[2]{x}$. However, if you wish to type in $x^{\frac{5}{3}}$, you will have to use `x^(5/3)`, and in general, `x^(m/n)` will produce $x^{\frac{m}{n}}$. For example:

- `alg1 := x^(3/4) + x^(2/3)`

$$x^{\frac{2}{3}} + x^{\frac{3}{4}}$$

- `alg2 := x^(4/5) + x^(5/6)`

$$x^{\frac{4}{5}} + x^{\frac{5}{6}}$$

- `alg1 + alg2`

$$x^{\frac{2}{3}} + x^{\frac{3}{4}} + x^{\frac{4}{5}} + x^{\frac{5}{6}}$$

For many algebraic expressions, the procedures mentioned in this chapter will not make any sense. We need to know exactly how we are going to transform such an expression and how to simplify it. For example, the standard `simplify` operation does not produce anything simpler in this case:

- `simplify(%)`

$$x^{\frac{2}{3}} + x^{\frac{3}{4}} + x^{\frac{4}{5}} + x^{\frac{5}{6}}$$

In many situations, it can be quite convenient to apply the procedure `radsimp` that looks at the radicals in a given algebraic expression and tries to simplify it according to the standard principles of simplifying mathematical formulae. For example, take a look at this useful transformation, which removes the radicals from the denominator of the fraction.

- `alg3 := (x - a)/(sqrt(x) - sqrt(a))`

$$\frac{a - x}{\sqrt{a} - \sqrt{x}}$$

- `radsimp(alg3)`

$$\sqrt{a} + \sqrt{x}$$

11.4 Trigonometric and Other Functions

In the previous chapter, we listed some of the trigonometric and hyperbolic functions. This makes a large group of functions that MuPAD understands. Before we do anything with trigonometric functions, we should note that MuPAD expects arguments for trigonometric functions to be given in radians—not in degrees. It is very important to understand this, as otherwise many situations might look like an error in the program. For example, `sin(1)` is considered as sine of one radian, not of one degree. Look at this:

- `sin(1)`

$$\sin(1)$$

- float(%)

  0.8414709848

Example 11.2

As you know, MuPAD expects that the arguments of trigonometric functions are given in radians. However, many of us are used to degrees, and we tend to write trigonometric functions with arguments given as degrees. I found that many of my students frequently fall into this trap. In order to help them, I have developed two simple procedures that convert between radians and degrees. The first procedure converts radians to degrees. I gave it the very simple name deg. Notice that I do not use the standard way of checking the type of the input variable here. The typical way,

```
deg := proc(x: Type::Real)
```

would not recognize π as a real number. It only syntactically checks if the input object looks like a real number. So, instead of this, I have used the procedure is(x, Type::Real) that checks whether x is really a real number, and not how it looks. This will work perfectly in our situation. Here is the deg procedure:

- ```
  deg := proc(x)
  begin
     if is(x, Type::Real) = TRUE then
        return(x*180/PI)
     else
        return(procname(args()))
     end;
  end:
  ```

I have also developed another procedure called rad to convert degrees to radians.

- ```
  rad := proc(x)
  begin
     if is(x, Type::Real) = TRUE then
        return(x*PI/180)
     else
        return(procname(args()))
     end;
  end:
  ```

Now we can check the results produced by the two procedures.

- deg(1)

  $\dfrac{180}{\pi}$

- deg(PI)

  180

- rad(180)

  π

- float(rad(1))

  0.01745329252

We will use these two procedures frequently in the next chapters of this book. You may also need them to demonstrate that $\sin(30°) = \frac{1}{2}$ and a few other basic facts. ■

Now, we will return to trigonometric expressions and transforming them in MuPAD.

First of all, you should notice that MuPAD already knows the most fundamental trigonometric identities. For example,

- trigexpr := (sin(x))^2 + (cos(x))^2

  $\cos^2 x + \sin^2 x$

- simplify(trigexpr)

  1

- trigexpr1 := sin(x)*cos(y) + cos(x)*sin(y)

  $\cos x \cdot \sin y + \cos y \cdot \sin x$

- simplify(%)

  $\sin(x + y)$

However, there are situations when MuPAD needs a bit of help in transforming trigonometric expressions. For example, the well-known identity $\tan(x + y) = \dfrac{\tan x + \tan y}{1 - \tan x \tan y}$ can be used to test MuPAD's abilities in simplifying trigonometric expressions. Let us try to simplify the right side of this identity. You might guess that we should get the expression that occurs on the left side.

- rside := (tan(x) + tan(y))/(1 - tan(x)*tan(y))
$$\frac{\tan x + \tan y}{1 - \tan x \cdot \tan y}$$

- simplify(rside)
$$\frac{\tan x + \tan y}{1 - \tan x \cdot \tan y}$$

As you can see, the result of the simplification is quite unsatisfactory. Let us try to make some changes in our formula that will help MuPAD in further transformations. First, let us replace all the $\tan(x)$ expressions with $\frac{\sin x}{\cos x}$. This will be done using the command rewrite:

- rewrite(%, sincos)
$$-\frac{\frac{\sin x}{\cos x} + \frac{\sin y}{\cos y}}{\frac{\sin x \cdot \sin y}{\cos x \cdot \cos y} - 1}$$

This expression still does not look good, but at least now you can simplify it:

- simplify(%)
$$\frac{\sin(x + y)}{\cos(x + y)}$$

The obtained result is the final one.

As you can see, in simplification of trigonometric expressions MuPAD sometimes needs a bit of help from our side, but we can easily obtain what we need. In the above transformation, I have used the procedure rewrite. I will explain it later in detail.

Since version 3.0 MuPAD has another, more powerful, tool to simplify expressions. This is the class Simplify (with the capital S), which is able to produce all the expressions equivalent to a given one, or even apply simplification rules defined by users. Describing this very sophisticated class in details is far beyond of the scope of this book. However, we can try to use some simple statements in order to see what we may get.

First of all, notice that without any special efforts from our side, we can in most cases produce the result that we want. For example,

- `Simplify(rside)`

$$\frac{\sin(x+y)}{\cos(x+y)}$$

The class `Simplify` is much more powerful than the procedure `simplify`. However, it can be slower when applied to some complex expressions. Let us see a more advanced use of this class. For example, we can use it with additional parameters. One of them is `All`. The statement `Simplify(rside, All)` will produce a list of all the expressions that are equivalent to the starting formula. Here I show only a small part of the output that we get in our example. Usually, the output is not typeset. So, I had to rewrite it.

$$[\frac{\sin(x+y)}{\cos(x+y)}, \frac{\tan x+\tan y}{\tan x \tan y - 1}, \frac{\tan x}{\tan x \tan y - 1} + \frac{\tan y}{\tan x \tan y - 1},$$

$$\frac{\cos x \sin y + \cos y \sin x}{\cos x \cos y - \sin x \sin y}, \ldots]$$

Transformations in another direction, that is, towards expanding trigonometric expressions, can be done using the expand procedure.

- `expand(sin(x+y))`

$$\cos x \cdot \sin y + \cos y \cdot \sin x$$

- `expand(tan(x+y))`

$$-\frac{\tan x + \tan y}{\tan x \cdot \tan y - 1}$$

- `expand(sin(6*x))`

$$6 \cdot \cos x \cdot \sin^5 x + 6 \cdot \cos^5 x \cdot \sin x - 20 \cdot \cos^3 x \cdot \sin^3 x$$

- `expand(sin(12*x))`

$$12 \cdot \cos^{11} x \cdot \sin x - 12 \cdot \cos x \cdot \cos^{11} x + 220 \cdot \cos^3 x \cdot \sin^9 x$$

$$-792 \cdot \cos^5 x \cdot 5 \sin^7 x + 792 \cdot \cos^7 x \cdot \sin^5 x - 220 \cdot \cos^9 x \cdot \sin^3 x$$

Finally, there is a large class of expressions with other functions, like e^x, a^x, $\ln x$, $\sinh x$, $\arctan x$, etc. The transformations for all these expressions go through the standard formulae for such functions. For example, $(ab)^n = a^n b^n$, $a^n a^m = a^{n+m}$, etc. The commands `simplify` and `expand` usually do a good job.

It is worthwhile to explain here a few procedures that are quite universal, and are a kind of bridge between the different classes of functions. We will start with the procedure `rewrite`, which was already mentioned in one of the trigonometric examples.

The procedure `rewrite` is useful for rewriting a formulae into a different notation. For example, we may wish to rewrite a trigonometric formula using $\tan x$ and $\cot x$ into a formula with $\sin x$ and $\cos x$, or rewrite a trigonometric formula into an expression with e^x. Here is the syntax of the command for the `rewrite` procedure.

$$\boxed{\text{rewrite}(formula,\ new\_component)}$$

Where *new_component* is one of the following parameters: `andor`, `arccos`, `arccot`, `arcsin`, `arctan`, `cos`, `cosh`, `cot`, `coth`, `diff`, `D`, `exp`, `fact`, `gamma`, `heaviside`, `ln`, `piecewise`, `sign`, `sin`, `sincos`, `sinh`, `sinhcosh`, `tan`, `tanh`. This way we can rewrite, for example, a formula with sine and cosine into the equivalent formula written using tangent, cotangent, or some other functions.

Here are a few examples showing how this transformation works.

- `expr1 := sin(x)/cos(x)`

$$\frac{\sin x}{\cos x}$$

- `rewrite(expr1, tan)`

$$\frac{2 \cdot \tan\left(\frac{x}{2}\right)}{1 - \tan\left(\frac{x}{2}\right)^2}$$

- `rewrite(expr1, exp)`

$$\frac{\frac{i}{2} \cdot e^{(-i)x} + \left(-\frac{i}{2}\right) \cdot e^{ix}}{\frac{e^{(-i)x}}{2} + \frac{e^{ix}}{2}}$$

- `rewrite(expr1, sinh)`

$$\frac{i \cdot \sinh((-i) \cdot x)}{2 \cdot \sinh((-\frac{i}{2}) \cdot x)^2 + 1}$$

- `rewrite(expr1, tanh)`

$$\frac{(-2 \cdot i) \cdot \tanh\left(\frac{i}{2} \cdot x\right)}{\tanh\left(\frac{i}{2} \cdot x\right)^2 + 1}$$

Now we can simplify each of obtained expressions like many others before.

- `Simplify(%)`

  $\tan x$

Another useful procedure is `combine`. We can use it to combine trigonometric, logarithmic or exponential expressions whenever this is possible. Note that as with `rewrite`, you must declare the form of the final result. For example, if you have a product of exponential expressions and you wish to combine it into a single exponential expression, you will need to specify `exp` as the second input parameter. Here are some examples showing how this can be done.

- `combine(exp(x)*exp(y), exp)`

  e^{x+y}

- `combine(exp(x)*exp(3*y)/exp(5*x), exp)`

  $e^{3 \cdot y - 4 \cdot x}$

11.5 Solving Equations and Inequalities

Solving equations is one of the most important activities, starting from primary schools, and up to very serious engineering applications. A computer package that is able to handle most of the special cases of equations and produce solutions for them is always in very high demand. In MuPAD equation solving procedures are among the most valuable ones. You can solve not only typical school equations, but also differential equations and many other types. You can solve equations in MuPAD by producing formulae representing solutions, or, if necessary, by obtaining numerical results. In this section, I will concentrate only on solving equations that are related to undergraduate mathematics. I will skip some of the most sophisticated methods, as they go beyond the scope of this book. You can find more information about MuPAD's solving possibilities in MuPAD's help.

Let us start by solving the most basic equations. As you have perhaps guessed, I mean polynomial equations.

- `solve(x^2 - 1 = 0,x)`

  $\{-1, 1\}$

This was quick and easy. You only have to remember that you need to issue the solve command with two parameters; the first one is your equation, and second one is the variable for which you wish to produce the results.

In the same way, you can use MuPAD's `solve` procedure to solve inequalities. For example,

- `solve(x^2 + 3*x > 0, x)`

  $(0, \infty) \cup (-\infty, -3)$

- `solve(35*x^2 - 50*x - 10*x^3 + x^4 + 24 >= 0, x)`

  $[2, 3] \cup (-\infty, 1] \cup [4, \infty)$

Here is the basic syntax of the `solve` command.

$$\text{solve}(formula,\ variable(s),\ options)$$

In the above command, *formula* denotes an equation, inequality or simply a mathematical formula. In this last case, if there is no relation operator =, <, >, <=, or => in the formula, MuPAD will solve the equation *formula=0*.

We will later discuss some of the possible options we could use. For now, let us explore a few simple examples.

- `solve(x^3 + 23 = 0,x)`

  $$\left\{ -\sqrt[3]{23}, -\frac{\left(i \cdot \sqrt{3} - 1\right) \cdot \sqrt[3]{23}}{2}, \frac{\left(i \cdot \sqrt{3} + 1\right) \cdot \sqrt[3]{23}}{2} \right\}$$

As you see above, MuPAD has produced three roots for a polynomial equation of order 3. Some of these roots are complex, and each of them is expressed by a nice formula. However, MuPAD is not always able to produce roots by giving their formulae. After all, you know quite well that it is not always possible to solve an equation by producing the formula for the root. Many polynomial equations of higher orders are like this. Take a look at this example.

- `solve(x^5 + 5*x + 25 = 0,x)`

  $RootOf(X8^5 + 5 \cdot X8 + 25, X8)$

It looks like MuPAD gave up solving the equation and returned what you put in as the input parameter. This is not exactly true. MuPAD has produced an object that represents a set of roots of the above equation. This object can be used to obtain numerical values of roots, or to perform other operations on roots. Let us see what the roots are in this case.

- `float(%)`

  $\{-1.74684711,$
  $-0.7142041288 + 1.73258888 \cdot i,$
  $-0.7142041288 - 1.73258888 \cdot i,$
  $1.587627684 + 1.246804871 \cdot i,$
  $1.587627684 - 1.246804871 \cdot i\}$

Note that we've now got five different roots expressed by complex numbers.

Solving polynomial equations of order 2, 3 and 4 is a bit easier. If you do not wish to give up on solving equations of order three or four, there is a way to get an exact solution using the Cardano formulae. However, the obtained formulae can be unpleasant and very long. For example, if in the `solve` command you would specify the option MaxDegree=3 or MaxDegree=3, you would get:

- `solve(x^3 + 3*x + 25=0, x, MaxDegree = 3)`

$$\left\{ \sqrt[3]{\frac{\sqrt{629}}{2} - \frac{25}{2}} - \frac{1}{\sqrt[3]{\frac{\sqrt{629}}{2} - \frac{25}{2}}} , \right.$$

$$\frac{1}{2\sqrt[3]{\frac{\sqrt{629}}{2} - \frac{25}{2}}} - \frac{\sqrt[3]{\frac{\sqrt{629}}{2} - \frac{25}{2}}}{2} - \frac{i}{2}\sqrt{3}\left(\left(\frac{\sqrt{629}}{2} - \frac{25}{2} \right)^{\frac{-1}{3}} + \sqrt[3]{\frac{\sqrt{629}}{2} - \frac{25}{2}} \right),$$

$$\left. \frac{1}{2\sqrt[3]{\frac{\sqrt{629}}{2} - \frac{25}{2}}} - \frac{\sqrt[3]{\frac{\sqrt{629}}{2} - \frac{25}{2}}}{2} + \frac{i}{2}\sqrt{3}\left(\left(\frac{\sqrt{629}}{2} - \frac{25}{2} \right)^{\frac{-1}{3}} + \sqrt[3]{\frac{\sqrt{629}}{2} - \frac{25}{2}} \right) \right\}$$

As you have seen before, MuPAD always organizes all the roots in the form of a set. We like sets in mathematics, but they also have some disadvantages. One of them is that multiple objects cannot be represented in a set. For example,

- {1,1,1}

  {1}

This feature of sets is quite important, as many equations have multiple roots. All these roots will be squeezed to a single root. Again, look at what you might get:

- solve((x - 1)^5 = 0,x)

  {1}

It doesn't look right, does it? We know quite well that the equation $(x - 1)^5 = 0$ should have a multiple root 1. So, we expected to get five identical roots. However, there is a way to get a more meaningful result. As I mentioned at the beginning of this section, we can use the solve command with various parameters. Let us try one of them now.

- solve(x^5+2*x^4-10*x^3-8*x^2+33*x-18, x, Multiple)

  $\{[1,2],[2,1],[-3,2]\}$

The Multiple option used in this example forced MuPAD to produce the roots of the equation and their multiplicity. Thus, you know that in the given example 1 is a double root, 2 is a single root and −3 is again a double root.

Finally, such output can be easily converted to a sequence or a list of all roots without removing multiple roots. We can use for this purpose the expand procedure.

- Roots:=solve(x^5+2*x^4-10*x^3-8*x^2+33*x-18,x,Multiple):
  expand(Roots)

  1, 1, 2, -3, -3

You may be wondering how you can represent roots in a more useful structure, which you would then be easily able to use in further calculations. For example, sometimes a list may be more useful. In MuPAD, there is a very useful procedure op that allows you to unwrap an object from external braces. This may change a set or a list into a sequence. Let us give it a try.

- roots := [op(solve(x^3 - 11 = 0,x))]

$$\left[-\frac{\left(i \cdot \sqrt{3} + 1\right) \cdot \sqrt[3]{11}}{2}, \frac{\left(i \cdot \sqrt{3} - 1\right) \cdot \sqrt[3]{11}}{2}, \sqrt[3]{11} \right]$$

Now you can use the structural format of roots and call each root separately.

- roots[1]

$$-\frac{\left(i \cdot \sqrt{3} + 1\right) \cdot \sqrt[3]{11}}{2}$$

- roots[2]

$$\frac{\left(i \cdot \sqrt{3} - 1\right) \cdot \sqrt[3]{11}}{2}$$

- roots[3]

$$\sqrt[3]{11}$$

Using the assume statement, we can narrow down solving equations to specific types of numbers. For example,

- assume(n, Type::Integer):
- assume(m, Type::Integer):
- solve(m^2 < 100, m)

$$\{-9,-8,-7,-6,-5,-4,-3,-2,-1,0,1,2,3,4,5,6,7,8,9\}$$

- solve(m-sqrt(n)=0, m)

$$\begin{cases} \{\sqrt{n}\} & \text{if} \quad \sqrt{n} \in \mathbb{Z} \\ \varnothing & \text{if} \quad \neg\sqrt{n} \in \mathbb{Z} \end{cases}$$

- assume(x, Type::Real):
- solve(sin(x)=2,x)

$$\varnothing$$

- assume(x, Type::Complex):
- solve(sin(x) = 2,x)

$$\{\pi - \arcsin(2) + 2 \cdot \pi \cdot k \mid k \in \mathbb{Z}\} \cup \{\arcsin(2) + 2 \cdot \pi \cdot k \mid k \in \mathbb{Z}\}$$

Certainly, you know that the equation $\sin x = 2$ does not have any real roots. Thus, MuPAD has produced an empty set. For complex numbers however, the situation changes and we are able to get complex solutions. This is what we got above. If you wish to get the numeric values of such solutions, you could try to apply the float command to the obtained results. This way, you can get:

- float(%)

  $\{6.283185307 \cdot k + 1.570796327 + 1.316957897 \cdot i) \mid k \in \mathbb{Z}\} \cup$
  $\{6.283185307 \cdot k + 1.570796327 - 1.316957897 \cdot i \mid k \in \mathbb{Z}\}$

Finally, we need to say something about solving equations with multiple variables or systems of equations. For example, how do we solve this system of equations?

- eq1 := x^2 + y^2 - 5 = 0:
- eq2 := x = 2*y:
- solve({eq1,eq2},{x,y})

  $\{[x = -2, y = -1], [x = 2, y = 1]\}$

As you have noticed, the only change in the statement is to put our equations and variables in a pair of curly braces. Of course, the number of equations and variables does not need to be the same.

One of the most interesting topics in solving equations are so-called recurrence equations. For example, the formula for the Fibonacci sequence can be given in the form of a recurrence equation:

$$f(0) = 0, f(1) = 1 \text{ and } f(n) = f(n-1) + f(n-2).$$

In one of the previous chapters of this book, I had shown you the explicit formula for Fibonacci numbers. It had looked quite complicated. Let us now try to solve the Fibonacci equation in MuPAD and check if the explicit formula was correct. In order to do this, we have to properly declare a recurrence equation.

The syntax for declaring recurrence equations in MuPAD is as follows:

rec(*equation*, *function*(*n*), *initial_conditions*)

where *equation* means a recurrence formula, for example $f(n) = f(n-1) + f(n-2)$; *function*(*n*) is the function for which we wish to solve the equation, e.g. $f(n)$. Finally, *initial_conditions* describe the first few elements of the sequence, e.g. $f(0) = 0$, $f(1) = 1$. Thus, the declaration of the Fibonacci sequence as a recurrence equation will have the following form:

- FibEq := rec(
      f(n) = f(n-1) + f(n-2), f(n), {f(0)=0, f(1)=1}
  ):

Now we can solve the Fibonacci equation by executing the statement,

- `FS:= op(solve(FibEq))`

$$\frac{\sqrt{5} \cdot \left(\frac{\sqrt{5}}{2} + \frac{1}{2}\right)^{n}}{5} - \frac{\sqrt{5} \cdot \left(\frac{1}{2} - \frac{\sqrt{5}}{2}\right)^{n}}{5}$$

We can now produce as many Fibonacci numbers as we wish. For example, the following loop will produce the first 20 Fibonacci numbers.

- ```
  for i from 1 to 20 do
     print(expand(subs(FS, n=i)))
  end;
  ```

Note, from the above solution can be used to define an explicit formula for the Fibonacci sequence. We define it as follows:

- `FibFun:= i->simplify(subs(FS, n=i)):`

Later we can use this formula to obtain any term of the Fibonacci sequence. For example:

- `FibFun(3), FibFun(15), FibFun(123)`

  2, 610, 22698374052006863956975682

You can find more examples of recurrence equations to solve at the end of the chapter, in the programming exercises section.

Before finishing this chapter, I would like to mention that in MuPAD you can find procedures to solve equations more sophisticated than the ones that I mentioned in this chapter. Here are some of them.

In the `linalg` library, there are tools to solve systems of linear equations. In the `numeric` library, you can find a number of tools to solve equations and differential equations using numerical methods; and in `detools`, there is a procedure to solve partial differential equations.

Finally, you may end up with an equation that is quite difficult to solve, and MuPAD produces nonsensical results, like here:

- `solve(x^2 - tan(x) = 0, x)`

  $\text{solve}(x^2 - \tan(x) = 0, x, \text{IgnoreProperties} = \text{TRUE})$
  ```

In this case, you might want to try the `numeric::solve` procedure:

- `numeric::solve(x^2 - tan(x) = 0,x)`

  {−4.756559406}

The `numeric::solve` procedure had produced only a single root. However, you may be expecting there to be more than one root. In that case you can use the `numeric::realroots` procedure to find the intervals where the roots exist. For instance, the equation in our example might have many solutions:

- `numeric::realroots(x^2-tan(x)=0,x=-10..10)`

  [[-7.87109375, -7.861328125], [-4.765625, -4.755859375],

  [-1.85546875, -1.845703125], [-0.009765625,
  1.875958165e-2525222],

  [-1.875958165e-2525222, 0.009765625], [4.658203125,
  4.66796875],

  [7.83203125, 7.841796875]]

You can now clearly see that the solution produced by the `numeric::solve` procedure was the one from the interval [-4.765625, -4.755859375].

11.6 Chapter Summary

In this chapter, we have explored several topics from elementary algebra and trigonometry. First, we spent some time talking about various algebraic expressions. Then, we explored various ways of transforming these expressions, expanding and simplifying them. Finally, in the last part of the chapter we learned about solving equations in MuPAD. Certainly, we have left a number of topics completely untouched. However, the purpose of this book is only to give you a first glance at the things you can use while working in mathematics with MuPAD. Now, let us revise some of the procedures that were introduced in this chapter.

factor(*expression*) - *factorize an algebraic expression*

simplify(*expression*) - *simplify an algebraic expression*

expand(*expression*) - *expand an algebraic expression*

gcd(*poly1*, *poly2*) - *obtain the* gcd *of two polynomials*

lcm(*poly1*, *poly2*) - *obtain the* lcm *of two polynomials*

degree(*polynomial*) - *produce the degree of a given polynomial*

evalp(*polynomial*, *x=value*) - *evaluate a polynomial for given value*

poly(*polynomial*,[*x*]) - *declare a polynomial object*

poly2list(*polynomial*) - *transform a polynomial to a list*

partfrac(*ratexpr*) - *produce partial fractions*

normal(*ratexpr*) - *produce the normal form of a given rational expression*

radsimp(*expression*) - *simplify the radicals in a given expression*

rewrite(*formula, new_components*) - *rewrite the formula using new components*

combine(*formula, component*) - *combine terms according to a component*

solve(*equation, variable(s), options*) - *solve an equation*

numeric::solve(*equation, variable(s)*) - *produces numerical solutions an equation*

numeric::realroots(*f(x),x* = *x1...x2*) - *produces intervals containing real roots of a real univariate function*

rec(*equation, function(n), initial_conditions*) - *declare a recurrence equation*

op(*expression*) - *isolate the operands of a given expression, list, set, etc.*

nops(*expression*) - *the number of terms in a given expression*

is(*x*, Type::*Type_expression*) - *checks the type of x*

11.7 Programming Exercises

1. Use the appropriate commands to factorize the given polynomials:

   a. $x^2 - y^2$

   b. $x^4 - y^4$

   c. $x^5 - y^5$

   d. $x^2 - 4xy - 12y^2$

   e. $x^3 - \frac{8}{3}x^2 - \frac{5}{3}x + 2$

2. Obtain all the roots of these polynomials:

   a. $x^3 + 33x + 196$

   b. $x^3 + 3x^2 + 3x + 1$

c. $x^5 + 4x^2 + 4$

d. $x^4 - 2x - 3$

e. $x^5 + x^4 + x^3 + x^2 + x + 1$

3. Solve the following equations with MuPAD. In each case, analyze how many solutions you got and what the conditions are for obtaining them.

   a. $\frac{1}{x} + \frac{1}{y} = 1$, in x

   b. $\frac{x}{y} + \frac{y}{z} + \frac{z}{x} = 1$, in x

   c. $\frac{x}{y} + \frac{y}{z} + \frac{z}{x} = 0$, in x

   d. $\sin x + \sin y = 1$, in x

   e. $\sin(1/x) + \sin y = 1$, in x

   f. $\tan x = \cot x$, in x

   g. $\sin x = \cos x$, in x

4. Solve these systems of equations and find the geometric interpretations of the obtained solutions.

   a. $x^2 - 3y = 7, 6x + 4y = 3$, in x and y

   b. $x^2 - y = 1, y^2 - x = 1$, in x and y

   c. $x = y^2, y = z^2, z = x^2$

   d. $x = \sin y, y = \cos x$, in x and y

   e. $x = \sinh y, y = \cosh x$, in x and y

5. A sequence is defined recursively as follows:
   $s_k = s_{k-2}$, for all integers $k \geq 2$,
   $s_0 = 1, s_1 = 2$.

   a. Calculate first 10 terms of the sequence and guess an explicit formula for the sequence.

   b. Define the sequence s_n in MuPAD and solve it. Compare your formula with the result produced by MuPAD. Use the solution produced by MuPAD to declare an explicit formula for the sequence.

6. A sequence is defined recursively as follows:
   $s_k = 3s_{k-2}$, for all integers $k \geq 2$,
   $s_0 = 1$, $s_1 = 2$.

   a. Calculate first 10 terms of the sequence and guess an
      explicit formula for the sequence.

   b. Define the sequence s_n in MuPAD and solve it. Compare
      your formula with the result produced by MuPAD. Use
      the solution produced by MuPAD to declare an explicit
      formula for the sequence.

7. A sequence is defined recursively as follows:
   $t_k = k - t_{k-1}$ for all integers $k \geq 1$,
   $t_0 = 0$.

   a. Calculate first 5 terms of the sequence and guess an
      explicit formula for the sequence.

   b. Define the sequence t_n in MuPAD and solve it. Compare
      your formula with the result produced by MuPAD. Use
      the solution produced by MuPAD to declare an explicit
      formula for the sequence.

8. A sequence is defined recursively as follows:
   $u_k = 2u_{k-2} - u_{k-1}$, for all integers $k \geq 2$,
   $u_0 = 1$, $u_1 = 2$.

   a. Calculate first 5 terms of the sequence and guess an
      explicit formula for the sequence.

   b. Define the sequence u_n in MuPAD and try to solve it.
      Compare your formula with the result produced by
      MuPAD. Use the solution produced by MuPAD to declare
      an explicit formula for the sequence.

9. Solve the following recurrence equations. Use the solution
   produced by MuPAD to declare an explicit formula for the
   sequence. Obtain the first 5 terms of the recurrence sequence.

   a. $y(n) = 2y(n-1) + n$, and $y(1) = 3$

   b. $y(n) = y(n-1) + 5$, and $y(1) = 3$

   c. $y(n+2) + 3y(n+1) + 2y(n) = 0$, and $y(0) = 1$, $y(1) = 1$

d. $y(n + 1) + 2y(n) = 0$, and $y(0) = 0$, $y(1) = 1$

e. $y(n) + 2y(n - 1) + y(n + 2) = 0$, and $y(0) = 1$, $y(1) = 1$

f. $y(n) = 3y(n - 1) - 2y(n - 2)$, and $y(1) = 1$, $y(2) = 3$

g. On a plane, two lines can intersect only at one point. Three
   lines may produce no more than three points of
   intersection. Add one more line and check how many
   points of intersection you may get. Develop a recurrence
   formula expressing the number of points at which n lines
   can intersect. Solve this recurrence equation.

Chapter 12 _____

Working with Sets and Sentences in MuPAD

Another two important parts of undergraduate mathematics are logic and set theory. In previous chapters, we had already talked about sets and logical expressions. We will now try to summarize and expand this knowledge.

12.1 Logic in MuPAD

In mathematics, like in all other areas of our life, we use sentences to express our ideas. Some of these sentences are true and some others are false. Well, reality is more complex than that, but in this book I do not wish to go into complicated logical investigations. So, for now, I will suppose that true and false are the only two logical values. In MuPAD true and false are represented by the two objects TRUE and FALSE. We can use them in many situations, for example for checking logical conditions in loops. We can operate on them using the standard logical operations and, or, not and xor. You already know those first three operations. The last one, xor, is called an exclusive or. We will talk about it later. We call the objects TRUE and FALSE Boolean constants, and the operations and, or, not and xor Boolean operations. In mathematical logic, we usually use the letters p, q and r to represent sentences, we call them propositions, and we construct complex Boolean expressions using these letters together with Boolean constants and Boolean operations. For example,

```
p and (not q)
(p or q) and (p or r)
not (p or q) and (not p and not q)
```

However, in mathematics textbooks, we use special symbols to denote logical operations: $\neg p$ for "not p", $p \wedge q$ for "p and q", and finally $p \vee q$ for "p or q". Thus, the above formulae are usually represented as follows: $p \wedge \neg q$, $(p \vee q) \wedge (p \vee r)$, and $\neg(p \vee q) \wedge (\neg p \wedge \neg q)$.

The very first step in introducing logic to undergraduate students is to produce the truth tables for all fundamental logical operations. This can also be done with the help of MuPAD. Here you have some very simple programs that produce the truth tables for not, and, or, and xor. Note that I used a new option NoNL in the print statement here. As you remember, in the past we used the option Unquoted to remove quotes from the output of the print command. The option NoNL works in a similar way to Unquoted, but additionally forces the print command to not put in any additional new line character after the current line. This makes the output more compact. Also, we can use at the very beginning the statement

```
PRETTYPRINT := FALSE
```

which will help us to produce the output in the form of tables.

I guess the output from the enclosed examples could be better formatted, but I wanted to keep my code as simple as possible. So, we have:

The truth table for the operation not

```
• PRETTYPRINT := FALSE:
• print(NoNL, "p "," not p ");
  print(NoNL,  "--------------");
  for i in {TRUE,FALSE} do
      print( i, not i)
  end;
```

p,   not p

FALSE, TRUE
TRUE, FALSE

The truth table for the operation and

```
• print(NoNL, "p ", "q ","p and q ");
  print(NoNL,"----------------");
  for p in {TRUE,FALSE} do
      for q in {TRUE, FALSE} do
          print(p, q, p and q)
      end;
  end;
```

```
p ,      q ,     p and q
-----------------------
FALSE, FALSE, FALSE
FALSE, TRUE,  FALSE
TRUE,  FALSE, FALSE
TRUE,  TRUE,  TRUE
```

The truth table for the operation or

```
• print(NoNL, "p ", "q ","p or q ");
  print(NoNL,"----------------");
  for p in {TRUE,FALSE} do
      for q in {TRUE, FALSE} do
      print(p, q, p or q)
      end;
  end;
```

```
p ,      q ,     p or q
--------------------
FALSE, FALSE, FALSE
FALSE, TRUE,  TRUE
TRUE,  FALSE, TRUE
TRUE,  TRUE,  TRUE
```

The truth table for the operation xor

```
• print(NoNL, "p ", "q ","p xor q");
  print(NoNL,"----------------");
  for p in {TRUE,FALSE} do
      for q in {TRUE, FALSE} do
          print(p, q, p xor q)
      end;
  end;
```

```
p ,      q ,       p xor q
-------------------------
FALSE, FALSE,  FALSE
FALSE, TRUE,   TRUE
TRUE,  FALSE,  TRUE
TRUE,  TRUE,   FALSE
```

You certainly know that there are a few other logical operations, which have not been mentioned until now in this book. We will introduce some of them later.

You can use the above method to obtain truth tables for more complex logical expressions. Here is an example.

Example 12.1

Let us build the truth table for the logical expression "p or (not p)". I guess the expression may remind you of the well-known saying from Shakespeare *"to be or not to be"*. You can easily see that in mathematical logic this expression is always true. Here you have the proof—just check the last column of the output.

```
• print(NoNL, "p ","not p ","p or (not p)");
  print(NoNL,"-----------------");
  for p in {TRUE,FALSE} do
     print(p, not p, p or not p)
  end;

  p ,      not p ,    p or (not p)
  --------------------
  FALSE, TRUE,     TRUE
  TRUE,  FALSE,    TRUE
```

In mathematics, sentences like the above are called tautologies. Tautologies are always true for any possible values of their terms. ∎

12.1.1 Polish Notation

Until now, while talking about Boolean expressions, we have used the conventional, so-called infix notation that resembles the way in which we write algebraic formulae. However, infix notation is not convenient for many applications. For example, in order to write a complex formula, you need to use brackets to specify the order of operations. Parsing formulae written in this notation is quite complicated. In many mathematical packages, defining new operations in infix notation is impossible without developing a special parser of expressions. For example, how could we define a new logical operation iff that will require the syntax "p iff q"? The most natural way would be to write a procedure iff(p,q). This is the point where the notation introduced by the Polish mathematician Jan Lukasiewicz can be very useful. In mathematics textbooks as well as in computer science, this notation is called the Polish notation.

Just suppose that, instead of not, and, or you can use these prefix forms:

N for negation ("not"),

K for conjunction ("and"),

A for alternation ("or"),

and no brackets at all. This is how the Polish notation works. For example, the expression "not p" shall be written as Np, the expression "p and q" as Kpq and the expression "p or q" as Apq. This way, most of the complex logical expressions may be expressed in a very compact form. For example:

"(p and q) or (not p or not q)" will be replaced by AKpqANpNq.

Well, I know what you will tell me — the Polish notation is shorter, but more difficult to understand than the standard one. You are right, but there is a trade off. We can add brackets to emphasize the structure of the formula and commas in order to separate multiple arguments. So, instead of AKpqANpNq we can write A(K(p,q),A(N(p),N(q))). Now it is much easier to understand, and, more importantly, it is possible to implement new logical operations in MuPAD.

As I said earlier, there are a few logical operations that have not yet been introduced in this book. Let us concentrate on the most important two — implication and equivalence. Both operations are implemented in MuPAD as ==> and <=> respectively. However, I will use them to show how new Boolean operations can be implemented using Polish notation.

We say that the sentence *p implies q* is true if, every time when *p* is true, *q* is also true. When *p* is false the expression *p implies q* is always considered as true. In mathematics, we use the arrow symbol to represented the implication *p implies q*, like this: $p \rightarrow q$. You can easily see that implication, as a logical operation on two propositions, can be expressed be the formula *(not p) or q*. Thus, it can be easily implemented as a MuPAD function:

• C := (p,q) -> (not p) or q;

Now, like for other Boolean operations, we can produce a truth table for implication.

- PRETTYPRINT := FALSE:
- print(NoNL, "p ", "q ","C(p,q)");
  print(NoNL,"----------------");
  for p in {TRUE,FALSE} do
     for q in {TRUE, FALSE} do
        print(p, q, C(p,q))
     end;
  end;

```
p,        q,        C(p,q)
--------------------
FALSE,   FALSE,   TRUE
FALSE,   TRUE,    TRUE
TRUE,    FALSE,   FALSE
TRUE,    TRUE,    TRUE
```

Another important Boolean operation missing from our investigations is equivalence. Two logical expressions are equivalent if they have the same logical value for the same values of their components. For example, you will find that *not*(*not p*) is equivalent to *p*.

In mathematics, equivalence is represented by the symbol ↔ or ⇔, and sometimes also by ≡. For our purposes, we will denote it with the letter E. Thus, we can now produce a definition of equivalence as a MuPAD function. However, before we do this, we need to tell MuPAD to forget the previous meaning of the character E. This is an important step, as otherwise MuPAD will not allow you to use E as a name for your new function. You will still have access the former E constant by using exp(1).

- unprotect(E):
- E := (p,q) -> C(p,q) and C(q,p)

Here is the MuPAD code to produce a truth table for equivalence.

- print(NoNL, "p ", "q ","E(p,q)");
  print(NoNL,"----------------");
  for p in {TRUE,FALSE} do
     for q in {TRUE, FALSE} do
        print(p, q, E(p,q))
     end;
  end;

```
p ,      q ,         E(p,q)
--------------------
FALSE, FALSE,  TRUE
FALSE, TRUE,   FALSE
TRUE,  FALSE,  FALSE
TRUE,  TRUE,   TRUE
```

Now, in order to have all the logical operations defined in prefix notation, we must redefine the three standard logical operations:

- N := p -> not p:
  K := (p,q) -> p and q:
  A := (p,q) -> p or q:

In MuPAD, there is a useful procedure, bool, which is used to check if a given expression is true. Note that the command bool doesn't check if the given expression is a tautology. It checks if the expression is true in the given context. However, the procedure bool is limited to Boolean expressions constructed from objects only of the syntactical Type::Real. This means that you can use,

- bool(2^2 < 3)

but you cannot use,

- bool(sqrt(2) < exp(2))

because the expressions sqr(2) and exp(2) are not considered as Type::Real (see comments on syntax and value-checking in chapter 5). Thus, when executing the bool procedure, MuPAD does not calculate the values of expressions other than those with Type::Real. In our example, sin(PI) and exp(2) will not be calculated — they must already be of Type::real when we apply bool. If you wish to get a TRUE or FALSE value for complex expressions, you need to apply the is procedure. For example,

- is(sin(PI) < exp(2))
  TRUE

In most of our examples, this is the most convenient way of checking a condition inside of a loop. For example, a construction like the one below will work correctly:

```
• if is(sqrt(2)>3)
  then
     print("Yes")
  else
     print("No")
  end;
```

However, **neither** of these two constructions will work properly:

```
if sqrt(2)>3 //this will NOT work
then
   print("Yes")
else
   print("No")
end;
```

and

```
if bool(sqrt(2)>3) //This will NOT work
then
   print("Yes")
else
   print("No")
end;
```

Now, let us return to logical expressions.

Example 12.2 Checking tautologies in Polish notation

Well, now we've got a lot of tools for checking if a given logical expression is a tautology. You can even write down a procedure that will check for you. This seems to be a quick job. You might develop something like this:

```
• taut := proc(lexpr)
  local p, q, r, BV, bval;
  begin
     BV := {TRUE, FALSE};
     bval:=TRUE;
     for p in BV do
        for q in BV do
           for r in BV do
              bval:=bval and bool(lexpr(p,q,r))
           end;
        end;
     end;
     return(bval);
  end:
```

Note that I used three local variables here. This means that the procedure will work for logical expressions having no more than three parameters. If you have an expression with more than three parameters, you will have to modify my procedure.

In the above procedure, this line is very important:

```
bval := bval and bool(lexpr(p,q,r))
```

As you know, (TRUE and FALSE) is equivalent to FALSE. This means that if, for some values of the variables p and q, the value of lexpr(p,q,r) is FALSE, then the final output from the procedure will be also FALSE.

Now we can try a few examples. Let us remind ourselves about some of the most famous laws of mathematical logic.

$\neg(p \vee q) \Leftrightarrow \neg p \wedge \neg q$ - de Morgan Law

$\neg(p \wedge q) \Leftrightarrow \neg p \vee \neg q$ - another de Morgan Law

$p \wedge (p \vee q) \Leftrightarrow p$ - absorption law

$p \vee (p \wedge q) \Leftrightarrow p$ - another absorption law

You can find more such laws in textbooks of mathematical logic, or even in high school textbooks. We will use the above formulae to check how our procedure works.

- ```
  delete p, q;
  ```
- ```
  lexpr1 := (p,q)-> E(N(A(p,q)),K(N(q),N(p))):
  taut(lexpr1)
  ```
  TRUE

- ```
  lexpr2 := (p,q)-> E(N(K(p,q)),A(N(q),N(p))):
  taut(lexpr2)
  ```
  TRUE

- ```
  lexpr3 := (p,q)-> E(K(p,A(p,q)),p):
  taut(lexpr3)
  ```
  TRUE

- ```
  lexpr4 := (p,q)-> E(A(p,K(p,q)),p):
  taut(lexpr4)
  ```
  TRUE
  ```

Finally, let us try a formula that definitely is not a tautology, $p \lor (p \land q) \Leftrightarrow q$.

- ```
lexpr5 := (p,q)-> E(A(p,K(p,q)),q):
taut(lexpr5)
```
  FALSE

You can try to use the `taut` procedure to check a few other logical expressions with no more than three parameters.                                    ■

This exercise finishes the short section showing how we can use MuPAD in teaching topics related to mathematical logic. However, there are still many things to tell. For example, in mathematics quite often we use the term relation. In fact, relation is an expression that is true for some values of the parameters used in it, and false for other values. For example, the relation $x > 0$ is true for all the values of $x$ that are positive. From a mathematical as well as a MuPAD point of view, you can treat relation as a function from a given domain to the set $BV = \{TRUE, FALSE\}$. This makes it a logical expression. For example,

```
Less := x -> ( x < 0 )
```

is the MuPAD representation of the relation $x < 0$. We can easily check its logical value for any real number x.

- ```
is (Less(12.45))
```
FALSE

- ```
is (Less(-3.09))
```
TRUE

As you see here, we can use MuPAD to check if a given relation is true or false for elements from its domain. With some additional suppositions, we can also check if a relation is true for all values or for some values from its domain. However, in order to do it, we need to know more about sets. So, let us go on to the next section.

12.2 Working with Sets

We've already got significant experience when it comes to using sets in MuPAD calculations. In this section, we will summarize our knowledge about sets and add just a few small pieces of information.
```

As you know, we declare a set in MuPAD by listing its elements inside the curly brackets or by using a procedure that generates one or more elements. For example,

- {a,b,c}

  $\{a,b,c\}$

- solve(x^4 + 4*x + 3, x)

  $\left\{ -1, 1 - i \cdot \sqrt{2}, i \cdot \sqrt{2} + 1 \right\}$

You have already learned that the sequence operator $ can be a very useful tool for creating large sets. For example,

- A:={n^2/2 $ n=1..20};

  $\{ \frac{1}{2}, 2, \frac{9}{2}, 8, \frac{25}{2}, 18, \frac{49}{2}, 32, \frac{81}{2}, 50, \frac{121}{2}, 72, \frac{169}{2}, 98, \frac{225}{2}, 128, \frac{289}{2}, 162, \frac{361}{2}, 200 \}$

If you are going to generate a very large set, you should prevent it from being displayed on the computer screen by using the colon ":" symbol. This is especially useful if the set contains a few hundred or more elements and you really do not wish to see all of them. For example, a set of the first 10,000 primes, or a set of all the numbers that fulfill a given condition, etc. Just try this command without ending it with the colon symbol:

- B := {ithprime(n) $ n=1..10000}:

In the earlier chapters of this book, we also used a number of infinite sets. For example,

- P := Dom::Interval(5,30):
  Q := Dom::Interval(-10,10):

In MuPAD, we can use all the known basic operations on sets: union, intersect, minus.

- P union Q

  $(-10, 30)$

- P intersect Q

  $(5, 10)$

- (P union Q) intersect B

  $\{2, 3, 5, 7, 11, 13, 17, 19, 23, 29\}$

- solve(sin(x)=0,x) intersect P

  $\{2 \cdot \pi, 3 \cdot \pi, 4 \cdot \pi, 5 \cdot \pi, 6 \cdot \pi, 7 \cdot \pi, 8 \cdot \pi, 9 \cdot \pi\}$

Example 12.3 Define operations for set algebra

In MuPAD, the symbols +, * and - are not supposed to be used instead of the set operations ∪, ∩ and \, like it is done in many mathematical books. Therefore, if you do not wish to redefine them, you have to type in, A union B, A intersect B, and A minus B instead of using A+B, A*B and A-B. This is a bit inconvenient, especially when writing long expressions involving a few operations on sets. For example,

- (K intersect D) union (C intersect D)

  $(C \cap D) \cup (K \cap D)$

The good thing is that MuPAD produces nicely formatted formulae with symbols representing set theoretical operations. So, at least you will be able to check what you typed in. In order to get this kind of output, you should check the option **Typeset Expressions** in the menu **Format**. Otherwise, you may get a typical text-based output.

If you do not intend, at least for a while, to use the operations +, * and −, then you can redefine their symbols as set operations. In one of the early chapters of this book, I mentioned the operator procedure. Allow me to tell you more about this interesting procedure.

The statement,

operator("*name*", *function*, *type*, *priority*)

can be used to define our own operators. Here, *name* is a string representing the name of the new operator. This can be any string, even with spaces inside. For example, something like, "> <". The next argument is the function, which can be any function that is going to perform the role of the new operator. The type of the new operator can be Binary, Prefix, Postfix and Nary. Finally, the priority means the priority of the new operator in the given context. We define priority of the operator by using numbers from 1 to 1999.

Let us come back to our example.

Suppose that we wish to define + and * as the set theoretical operations union and intersect. Both operators should be declared as Binary, and the priority for union should be lower than for intersection. This is because in expressions like $A \cap B \cup C$, the operation $A \cap B$ will be done first, and C will be added second.

We now know all that we need in order to redefine +,− and ∗ to the new role. Here is the code defining the symbols +,−, and ∗ as set operations.

- ```
  reset():
  operator("+", _union, Binary, 1):
  operator("-", _minus, Binary,1):
  operator("*", _intersect, Binary, 2):
  ```

Observe that, in order to have the above declarations working properly, I used here the internal operations _union, _minus, _intersect, which are binary functions for sets. Note also that redefining some symbols may have very serious consequences. For example, I redefined here the symbol "−" to be the set theoretical operation minus. This means that we cannot use it in any other context. For instance, we could not define a new function using a declaration with the arrow symbol ->. Something like x:=x->x^2 will not work. Therefore, before declaring a new function, we will have to either delete the operator "−" or reset the whole session.

Now we can define a few sets and see how the new operators work.

- ```
  A := {1,2,3,4,5,6,7}:
  B := {4,5,6,7,8}:
  C := {6,7,8,9}:
  ```
- ```
  A+B
  ```

  $\{1, 2, 3, 4, 5, 6, 7, 8\}$

- ```
  A*B
  ```

  $\{4, 5, 6, 7\}$

- ```
  (A-B)-C
  ```

  $\{1, 2, 3\}$

- ```
  A*(B+C)
  ```

  $\{4, 5, 6, 7\}$

You can now check a few more examples with our new operations, or you can try defining your own operations for sets.

The operations declared in this example are very useful. You can use them for a while. This will significantly simplify your formulae. However, before going into another topic we have to return back the old meaning of the symbols +,−, and ∗. Therefore, let us reset the current MuPAD session.

• `reset():`

Now, we can use again +,−, and ∗ like we did this before.   ■

In set theory, there are two special sets—the empty set, which we denote by {}, and the `universe`. The empty set is just a set without any elements. On the other hand, the `universe` is something very big that contains all of our objects. This means that every set is a subset of the `universe`.

Of course, the universe set should be considered rather as a theoretical concept than a concrete set. This is reflected in all operations where the universe is used. For example,

• `U := universe`

  universe

• `B := {ithprime(n) $ n=1..10};`

  $\{2, 3, 5, 7, 11, 13, 17, 19, 23, 29\}$

• `U union B`

  universe

• `U intersect B`

  $\{2, 3, 5, 7, 11, 13, 17, 19, 23, 29\}$

• `U minus B`

  universe \ $\{2, 3, 5, 7, 11, 13, 17, 19, 23, 29\}$

• `B minus U`

  $\varnothing$

As you see above, the `universe` is a real universe containing all the sets that we are able to define in MuPAD.

You will easily notice that MuPAD knows the properties of operations on sets and applies selected well-known set theoretical identities to simplify or expand formulae where sets are involved. For example,

- X := (K union L) intersect (K union M)

$$(K \cup L) \cap (K \cup M)$$

- Y:=expand(X)

$$K \cup (K \cap L) \cup (K \cap M) \cup (L \cap M)$$

Still, not everything really works as you would expect. For example, you would expect that the simplify operation will transform the set Y to the form X, but you get nothing from the simplify command,

- simplify(Y)

$$K \cup (K \cap L) \cup (K \cap M) \cup (L \cap M)$$

Worse, you would expect MuPAD to treat X and Y as equal sets, but the bool operation on identity $Y = X$ produces false.

- bool(Y=X)

  FALSE

As you can see, the concept of sets in MuPAD needs a bit of expanding. In the next few paragraphs of this chapter, we will try to do this.

First, let us try to define some missing operations. We will start with the universe.

In many situations, the general concept of the universe set is quite useless. For example, in many concrete situations we often use a given set as a universe, and then we consider some of its subsets. For example, suppose that we deal with small sets that are subsets of the set $U = \{0, 1, 2, 3, 4, 5, 6, 7, 8, 9\}$ that contains all digits in decimal arithmetic. In such a case, it would be useful to consider the set of all digits as your universe. You can do this in two different ways:

1. By unprotecting the universe set and replacing it with U.

2. By ignoring the universe set and using U instead of it.

The first way is a bit tricky, and in fact we cannot produce a new universe with all of its specific properties, but the second one is quite acceptable, provided that we keep control of what we are doing.

Let us define a new local universe.

- U := {0,1,2,3,4,5,6,7,8,9}

From this moment, everything will go quite well if we remember to deal only with subsets of U. For example,

- ```
  A := {0,1,2,3,4};
  B := {4,5,6,7,8,9};
  C := {3,4,5}
  ```
  $\{0,1,2,3,4\}$

  $\{4,5,6,7,8,9\}$

  $\{3,4,5\}$

Now, you can try various operations on A, B and C. Of course, there are no surprises as all three sets were declared as subsets of U. So, the results must also be subsets of U. However, you may think, what about the complement of a given subset of U? This could be a critical point of this exercise. You noticed that in MuPAD, the complement of a given set is not defined. This leaves an open door for our inventions. You can declare the complement of a set in such way that it reflects the current universe set. Here is a simple solution.

- ```
  compl := X -> U minus X
  ```
  $X \rightarrow U \setminus X$

And now, you can try out our new operation:

- ```
  compl(A)
  ```
  $\{5, 6, 7, 8, 9\}$

- ```
  compl(B)
  ```
  $\{0, 1, 2, 3\}$

- ```
  compl(C minus B)
  ```
  $\{0, 1, 2, 4, 5, 6, 7, 8, 9\}$

As you can see, it works and works quite well. You can use the new operation to check some properties of the subsets of U.

- ```
  A union comp(A)
  ```
  $\{0,1,2,3,4,5,6,7,8,9\}$

- ```
  A intersect comp(A)
  ```
  $\varnothing$
  ```

- A intersect (A union B) = A

  $\{0,1,2,3,4\} = \{0,1,2,3,4\}$

Limiting the universe to a specific set allows us to produce a number of interesting exercises that are useful when we explore the concept of sets. However, you will agree that such examples do not give general proof of anything. They are useful for gathering experience. Later, you will need to refer to a general situation. So, you may be interested in seeing whether MuPAD understands some well-known identities for sets. For example,

$A \cap A = A$, $A \cup A = A$,

$A \cap (A \cup B) = A$, $A \cup (A \cap B) = A$.

You can try to ask MuPAD about them, but you will find that MuPAD knows only some of them. This part of MuPAD's knowledge still needs a lot of development.

In some applications, you may need to generate a large set of random elements with a random number of elements. This can be especially important for some exercises in statistics. MuPAD can easily be used to generate such sets and perform operations on them. For this purpose, you can use one of the random functions that are implemented in MuPAD. You know them already — we had talked about these functions in one of the earlier chapters. So, let us use them now.

Before we start doing anything, we will establish the limits for the size of our sets. For example, we may suppose that we are going to deal with sets that have less than 100,000 elements. This is still a lot.

First, declare a procedure that will produce random integers from the interval 0..100,000.

- `largeRand:=random(100000):`

Now, you can use this procedure to create your sets. For example, this will create two sets, each with a random number of elements.

- `Y := {largeRand() $ n=1..largeRand()}:`
  `X := {largeRand() $ n=1..largeRand()}:`

Just remember to put the colon at the end of each statement. Otherwise, you may end up with many screens of numbers. Now, you can check how many elements each set has, or perform some operations on them. In my example I got:

• nops(X)

  5207

• nops(Y)

  21314

• (X minus Y) union (Y minus X)

The last command may produce a very large set, so I do not show its output here.

Finally, using the subset relation that was introduced in MuPAD version 3.0, we can check if set X is a subset of set Y.

• X subset Y

  FALSE

• A := {1,2,3,4,5,6,7}: B:={4,5,6,7}: C:={6,7,8,9}:
• C subset A

  FALSE

• (C intersect B) subset A

  TRUE

Example 12.4 - Cartesian product

Cartesian products are very useful. We use them to define coordinates on a plane, in 3D, and in many other applications. In the latest version of MuPAD, you can find two slightly different implementations of the Cartesian product:

> combinat::cartesian(*set1*, *set2*, ..., *setN*)
>
> combinat::cartesianProduct(*set1*, *set2*, ..., *setN*)

The first procedure, considered as obsolete, produces a set of ordered N-tuples. The second one produces a list of N-tuples.

In this example, we will develop our own Cartesian product procedure in order to see what difficulties we may face while developing new operations on sets.

Let us see what we have to do. First, you know well that for two given sets, $A \times B = \{< a,b > \mid a \in A \text{ and } b \in B\}$. In the previous example, you learned how to go through all the elements of a given set. Finally, you know that an ordered pair < a, b > will be considered as a list of two elements $a \in A$ and $b \in B$. In lists, the order of elements is important, so there is no danger that a and b will be swapped like in the case of sets.

Now, we can write down the procedure CartProd. I used a capital C here, as Cart is a shortcut derived from the name of the famous mathematician Descartes.

- ```
CartProd := proc(A,B)
  local cp, a, b, x, y;
  begin
     if testtype(args(1),Type::Set) and
        testtype(args(2),Type::Set) then
        cp:=[];
        for x in A do
            for y in B do
                cp:=cp.[[x,y]]
            end;
        end;
        return({op(cp)})
     else
        return(procname(args()))
     end;
  end:
```

Now, you can apply this procedure to sets K and L that we declared in the previous example. The output will be quite long. So, perhaps it would be better if we use smaller sets.

- ```
C := {1,2,3,4,5};
F := {u,b,c,d,e,f}
```
$\{1,2,3,4,5\}$

$\{b,c,d,e,f,u\}$

- CartProd(C,F)

$\{[1,b],[2,b],[3,b],[4,b],[5,b],[1,c],[2,c],[3,c],[4,c],$

$[5,c],[1,d],[2,d],[3,d],[4,d],[5,d],[1,e],[2,e],[3,e],$

$[4,e],[5,e],[1,f],[2,f],[3,f],[4,f],[5,f],[1,u],[2,u],$

$[3,u],[4,u],[5,u]\}$

Now you can produce more examples and see what you get. Note that, when obtaining the Cartesian product of larger sets, the calculations take a lot of time. So, try to avoid large sets, say, sets with more than 1000 elements. You can also try to declare a procedure to obtain Cartesian product for three or more sets. ∎

In order to make the chapter about sets complete, I need to introduce two more MuPAD procedures. However, this is a very quick job.

As you perhaps remember, in undergraduate mathematics we sometimes use the so-called power set. It is a set of all the subsets of a given set A. The name power set is derived from the fact that if A has n elements, then its power set contains 2^n elements. Thus, quite often the power set of A is denoted as 2^A. In order to obtain a power set of a given set A, you can use the procedure `combinat::powerset` or a newer procedure `combinat::subsets`. In newer versions of MuPAD, the procedure `powerset` is considered obsolete. However, there is one and significant difference that is still worth noting. The procedure `powerset` produces a set of subsets, while `subsets` produces a list of subsets. And you know, sets and lists are quite different objects. Here I show how both procedures work:

- `combinat::powerset({x,y,z,t})`

  $\{\varnothing, \{t\}, \{x\}, \{y\}, \{z\}, \{t,x\}, \{t,y\}, \{t,z\}, \{x,y\}, \{x,z\}, \{y,z\},$

  $\{t,x,y\}, \{t,x,z\}, \{t,y,z\}, \{x,y,z\}, \{t,x,y,z\}\}$

- `combinat::subsets({a,b,c,d})`

  $[\varnothing, \{a\}, \{b\}, \{c\}, \{d\}, \{a,b\}, \{a,c\}, \{a,d\}, \{b,c\}, \{b,d\}, \{c,d\},$

  $\{a,b,c\}, \{a,b,d\}, \{a,c,d\}, \{b,c,d\}, \{a,b,c,d\}]$

This ends the chapter about sets and sentences. You can find a few more interesting examples in the programming exercises section.

12.3 Chapter Summary

In this chapter, we revised our knowledge about sentences, logical operations and sets in MuPAD. You learned a few new MuPAD procedures. Here are all the MuPAD procedures introduced in this chapter.

and, or, not - *logical operations*

union, intersect, minus - *operations on sets*

nops - *number of elements in a set, cardinality of a set*

operator("*name*", *function, type, priority*) - *declaration of a new operator*

combinat::cartesian(*set1*, *set2*, ..., *setN*)

combinat::cartesianProduct(*set1*, *set2*, ..., *setN*)

combinat::powerset - *power set of a given set, result is a set*

combinat::subsets - *power set of a given set, result is a list*

in - *checks if an object belongs to a set, e.g.* x in X, *result is* TRUE *or* FALSE

subset - *checks is a given set is a subset of another set*

12.4 Programming Exercises

1. The two Boolean operations — implication and equivalence — were introduced in MuPAD in the form of expressions p ==> q and p<=>q respectively. Use MuPAD to produce the truth tables for them. Check if they have the same truth tables as the operations C and E that we developed in the section about Polish notation.

2. As you have perhaps noticed, we defined the logical operations K and A as binary operations, i.e. working with two arguments. In reality, we prefer to use more general operators where the number of arguments is not fixed. Write down the two procedures K and A that replace our declarations and will work for any number of arguments.

3. Implement the procedure X that will implement the Boolean operation xor in Polish notation. It should be possible to use this procedure with any number of input parameters.

4. The *nand* operator, sometimes called the Sheffer's stroke and denoted by " | " is defined by $p \mid q \Leftrightarrow \neg(p \wedge q)$.

   a. Write a short program to produce a truth table for $p \mid q$.

   b. Show that $p \mid p$ is logically equivalent to $\neg p$.

   c. Write the procedure NAND to implement the operation $p \mid q \mid \ldots \mid r$ for any number of arguments.

   d. Find a few tautologies for the *nand* operator.

5. Find five laws of mathematical logic that were not mentioned in this chapter and check them by using the taut procedure.

6. Observe that each binary logical operation can be represented by four values of its output. For example, looking at last column of the truth table of the ∧ operation,

p	q	p and q
FALSE	FALSE	FALSE
FALSE	TRUE	FALSE
TRUE	FALSE	FALSE
TRUE	TRUE	TRUE

we may identify ∧ with the sequence FALSE, FALSE, FALSE, TRUE. Now, you certainly know that you can build $2^4 = 16$ such sequences.

a. Write down all such sequences, and identify which sequences represent logical operations introduced in this chapter.

b. Look at the remaining sequences. There is still a lot of them. Which of them may represent an interesting logical operation? Choose three of them and implement them as MuPAD functions or procedures.

c. Find a few non-trivial tautologies using these operations.

7. In computing, we use so-called bits and bytes. Bits are just the numbers 0 and 1. Bytes are sequences of bits, e.g. 01110011 (8-bit sequence). In MuPAD, it is much more convenient to represent bytes by lists, like [0,1,1,1,0,0,1,1].

a. Write in MuPAD procedures NOT, AND, OR, XOR that will act on bytes in a similar manner like the logical operations not, and, or, xor. Just remember that 1 should be understood as TRUE and 0 as FALSE. For example the operation AND should act like this,

$$AND([0, 1, 1, 1, 0, 0, 1, 1], [1, 1, 0, 1, 1, 0, 1, 0]) = [0, 1, 0, 1, 0, 0, 1, 0]$$

b. Use the function random to generate a few bytes represented as lists and check how your operations work.

8. If you look carefully into MuPAD help, you will find that MuPAD uses three logical values — TRUE, FALSE and UNKNOWN. Examine how the three logical connectors or, and, and not work in the three-valued MuPAD logic. Find a few new formulae using UNKNOWN that are tautologies in this logic.

9. Define in MuPAD the symmetric difference of two sets. You can use the formula $A \div B = (A - B) \cup (B - A)$. Use a finite domain to test your operation.

10. Declare a procedure to produce the union and intersection of an undetermined number of sets. Use these procedures to produce a union of three or more sets.

11. Check which of the well-known identities for sets MuPAD accepts. Try to also check these identities with MuPAD using some concrete sets.

12. Find a way to declare the following sets using the $ operator in MuPAD:

   a. $\{4, 8, 12, 16, 20\}$

   b. $\{101, 1001, 10001, \ldots, 100000001\}$

   c. $\{1, 4, 9, 16, 25\}$

13. A collection of non empty sets $\{A_1, A_2, \ldots, A_n\}$ is a partition of a set A, if and only if $A = A_1 \cup A_2 \cup \ldots \cup A_n$ and sets $A_1, A_2, \ldots, A_n$ are mutually disjoint. Write a procedure that will produce all partitions for the set $A = \{a, b, c, d\}$.

14. Write a procedure that for a given finite set A with even number of elements will produce all partitions containing sets with exactly 2 elements.

15. Write a procedure that for a given finite set A with $3k$ elements will produce all partitions containing sets with exactly 3 elements.

16. Write a procedure that will check if the given sets $A_1, A_2, \ldots, A_n$ are partition of a given set A.

Chapter 13 _____

Exploring Calculus with MuPAD

Quite often, when writing books about using computer technology in teaching mathematics, people concentrate on teaching calculus with CAS. This is the reason why in this book, I left the calculus topics for one of the last chapters. I tried to show you first how to use MuPAD for other mathematical topics that are not as popular as calculus. Now it is time to move on to calculus and explore it with MuPAD. Due to the limited size of this book, I will concentrate on the most important topics from undergraduate mathematics without going too deep into mathematical considerations.

13.1 Limits

Usually, a course of calculus or precalculus begins with the limits of functions and sequences. In MuPAD, we have the `limit` procedure available to check the limit of a function. Let us take a look at the syntax of this procedure. The simplest way of calling it is the command:

$$\text{limit}(f(x),\ x{=}x_0)$$

where $f(x)$ is a function of one variable declared before, or an expression placed inside the `limit` command. The point $x = x_0$ is the point for which we wish to obtain the limit. So, using this procedure you might get,

- `limit(cos(x)*sin(x)/x, x = 0)`

  1

- `F := x -> sin(x)/x;`

  $$x \rightarrow \frac{\sin(x)}{x}$$

- `limit(F(x), x = 0)`

  1

- `limit(abs(x)*(x-1)/(x^2+2), x = +infinity)`

  1

- `limit(abs(x)*(x-1)/(x^2+2), x = -infinity)`

  -1

As you can see, the `limit` procedure works quite well for most of basic examples. Note also that in last two examples, we used the limit for $x \to +\infty$ and $x \to -\infty$. This is very convenient for a number of situations where we wish to learn about the behavior of a function in infinity.

If you know a bit of calculus, then you will certainly remember a number of unpleasant examples where calculating the limit of a function is not so straightforward. Let us take a look at some of them now.

Let us define the function

- `h := x -> x/abs(x)`

  $$x \to \frac{x}{|x|}$$

The unpleasant thing about this function is that you cannot calculate the value of it for $x = 0$. Take a look at what I obtained:

- `h(0)`

  Error: Division by zero; during evaluation of 'h'

Now, let us try to see what will happen while trying to obtain the limit of the function for $x = 0$.

- `limit(h(x), x = 0)`

  undefined

We certainly know now that there is something wrong with our function for $x = 0$. So, we have to try to use other forms of the `limit` procedure. Here is their syntax.

limit($f(x)$), $x=x_0$, Left)
limit($f(x)$), $x=x_0$, Right)

These commands will produce a limit of the function for $x \to x_0$ from the left and right side respectively. Let us see what we get this time.

- `limit(h(x),x = 0, Left)`

   -1

- `limit(h(x),x = 0, Right)`

   1

Now you undoubtedly know what happened. Our function $f(x) = \frac{x}{|x|}$ has a discontinuity for $x = 0$, and neither the value of the function or the limit of the function exist for $x = 0$.

Let us take a look at another example. Here, our function is,

- `h := x->(x-1)/(x-3)`

   $x \to \dfrac{x-1}{x-3}$

You do not need to calculate anything in order to know that you cannot calculate the value $h(3)$. You will get division by zero, and while trying to obtain the limit of the function we will get:

- `limit(h(x), x = 3)`

   undefined

This case also makes the `Left` and `Right` options in the `limit` procedure very useful.

- `limit(h(x), x = 3, Left)`

   $-\infty$

- `limit(h(x), x = 3, Right)`

   ∞

Now we know how the function $h(x)$ behaves near $x = 0$. The left branch of the curve goes down to $-\infty$ as x approaches 3 from the left side, while the right branch of the curve goes up to $+\infty$ as x approaches 3 from the right side.

Before we go on to a slightly different topic, let us check one more example.

- limit(sin(1/x),x = 0)

  undefined

This time MuPAD was not able to calculate the limit. This is because our function has infinitely many oscillations between −1 and 1 in the nearest neighborhood of 0. Therefore, such a limit simply doesn't exist. This can easily be illustrated by the graph of the function.

- export(plot, Function2d):
  h := Function2d(sin(1/x), x = -1..1, Mesh = 10000):
  plot(h)

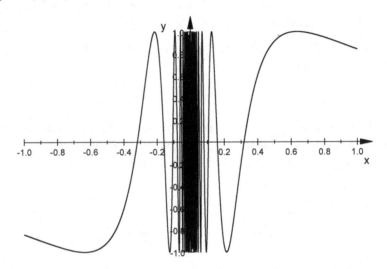

Now we can see the reason why the limit as well as the right and left limits do not exist.

In undergraduate mathematics, we also deal with infinite sequences. In reality, a sequence can be considered as a function, the domain of which is a subset of the natural numbers. For example, the sequence $a_n = \frac{1}{1+n}$, is a discrete function defined for all $n = 0, 1, 2, 3, \ldots$. For this reason, finding the limits for sequences in MuPAD uses the same procedure limit as for functions.

- a := n -> 1/(1-n)

$$n \rightarrow \frac{1}{1-n}$$

- limit(a(n), n = infinity)

  0

In reality we can ignore the fact that the domain of a sequence is only a set or a subset of natural numbers, and use the more general concept of the sequence as a function defined for integer numbers. This will allow us to calculate the limit of the sequence for $n \rightarrow -\infty$.

It is important to understand the behavior of the procedure limit in cases where a single limit does not exist. For example, let us analyze the sequence $b_n = \sin \frac{n\pi}{2}$. The values of this sequence are the numbers $1, 0, -1, 0, 1, 0, -1, \ldots$. It is difficult to say what we expect MuPAD to produce as a limit of this sequence. There is no limit at all. However, the whole sequence is built of just three values: -1, 0, and 1. So, it would be good if MuPAD could produce a set with these three values. In fact, MuPAD can produce an interval $[-1, 1]$ with the two boundary values for the possible limit or limits of all subsequences of the sequence b_n. Let us see how this can be done.

- b := n -> sin(n*PI/2)

  $n \rightarrow \sin\left(\dfrac{n\pi}{2}\right)$

- limit(b(n), n = infinity)

  undefined

We know that the limit of the sequence doesn't exist. However, let us use our limit procedure with the additional parameter Intervals. Just like here:

- limit(b(n), n = +infinity, Right, Intervals)

  $[-1, 1]$

This means that the limits for all possible subsequences of the sequence b_n are inside the interval $[-1, 1]$.

Note that MuPAD does not distinguish between the sequence a_n and the function $a(x)$ defined by the same formula. This sometimes leads to strange results or errors. Suppose that we take the sequence,

- a := n -> (-1)^(2*n)

  $n \rightarrow (-1)^{2n}$

Each term of this sequence is equal to $(-1)^{2n} = 1$. Therefore the limit of the sequence is also equal to 1. However, MuPAD produces a completely wrong result.

- limit(a(n), n = +infinity)
  undefined

We can easily work out what happened. MuPAD treated $a_n = (-1)^{2n}$ as a function $a(x) = (-1)^{2x}$ that of course does not have a limit when $x \to +\infty$. In a few situations, we can improve the performance of the limit procedure by declaring n as Type::Integer. Just like this:

- a := n -> (-1)^(2*n):
  limit(a(n), n = +infinity, Right) assuming(n,Type::Integer
  1

However, the above method may fail in many other situations. Check, for example, the sequence $b_n = (-1)^{2n+1}$.

Example 13.1 Plotting sequences

Before you can plot a sequence, you need to represent it as a PointList, Bars2d or even Polygon2d. However, in each case first you have to develop a list of points representing the sequence and then build the plot object.

For example, the sequence $b_n = (-1)^n \frac{n}{2n+1}$ will be represented as a sequence of points on a plane:

- b := n -> n*(-1)^n/(2*n+1):
  sequence := plot::PointList2d([[n, b(n)] $ n = 0..20])

You can use as many terms of the sequence as you wish, but you should think about how they will fit on your graph later on. The more points you have, the more difficult it will be to distinguish them on the graph. In fact, about 20–50 points can generally be enough to visualize the sequence and get some idea about its convergence or divergence. For our sequence, we get the plot:

- b := n -> n*(-1)^n/(2*n+1):
  sequence := plot::PointList2d([[n, b(n)] $ n = 0..50]):
  plot(sequence)

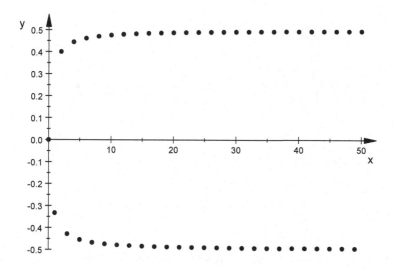

Looking only at the obtained picture, we can easily work out what is happening with the limit of our sequence.

Here is another graph of a sequence. Let us take the sequence $c_n = 2 - (-\frac{1}{2})^n$. By analyzing its formula, we may easily conclude that the number 2 is the limit of the sequence. This can be illustrated using the plot::Polygon2d object.

- ```
  c := n -> 2 - (-1/2)^n
  ```
  $$n \rightarrow 2 - \left(-\tfrac{1}{2}\right)^n$$

- ```
  sequence2 := plot::Polygon2d([[n, c(n)] $ n = 0..10]):
  plot(
      sequence2,
      GridVisible = TRUE,
      SubgridVisible = TRUE,
      PointsVisible = TRUE, PointSize = 2,
      LineColor = RGB::Black
  )
  ```

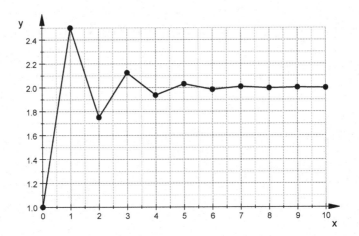

Now, we can try to visualize another fragment of the sequence. For example, for $n = 30..40$ we get:

- ```
  sequence2 := plot::Polygon2d([[n, c(n)] $ n = 30..40]):
  plot(
      sequence2,
      GridVisible = TRUE,
      SubgridVisible = TRUE,
      ViewingBoxYMax = 2.00000001,
      ViewingBoxYMin = 1.99999999,
      PointsVisible = TRUE, PointSize = 2,
      LineColor = RGB::Black
  )
  ```

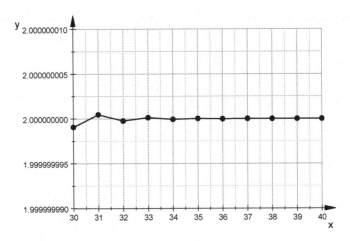
  ```

Note that this time, in order to produce a meaningful plot I had to narrow down the range of variable y. Otherwise, the points of the sequence would just form a horizontal line. ∎

13.2 Derivatives

After finding the limits of functions and sequences, calculating the derivative of a function or arithmetical expression is the second major activity in calculus. If you wish to illustrate the concept of a derivative, you can use MuPAD's limits to show what a derivative is, and how it is obtained. For example, if your function is $f(x) = ax^2 + b$, you can produce the limit of the difference quotient,

$$\lim_{h \to 0} \frac{f(x+h) - f(x)}{h}$$

and find its value. In MuPAD, you need only two commands for this purpose:

- `f := x -> a*x^2 + b`

  $x \longrightarrow a \cdot x^2 + b$

- `limit((f(x+h)-f(x))/h, h = 0)`

  $2 \cdot a \cdot x$

Now you can obtain the value of this derivative for any $x = x_0$. For example, for $x = 3$, you will get:

- `subs(%, x = 3);`

  $6 \cdot a$

You can repeat these steps for many examples of functions, and draw conclusions about the derivatives of various classes of functions. This is also the way to show why the derivative of a given function does not exist for some $x = x_0$. Here is such an example.

- `g := x -> abs(x)`

  $x \longrightarrow |x|$

Our function is the well-known function $|x|$, which has no derivative for $x = 0$. As you can see in the next two lines, MuPAD was not able to produce the limit of the difference quotient when $h \to 0$.

- limit((g(x+h) - g(x))/h, h = 0)

  undefined

You have perhaps already noticed that $x = 0$ is the point that is responsible for all this trouble. If, for example, you try to produce the above limit for $x = 0$, you will get the same result.

- limit((g(0+h) - g(0))/h, h = 0)

  undefined

You will be able to see where the is problem if you try to find the left and right limits for $h \to 0$.

- limit((g(0+h) - g(0))/h, h = 0, Left)

  -1

- limit((g(0+h) - g(0))/h, h = 0, Right)

  1

Now we finally know that the limit of the difference quotient does not exist for $x = 0$, and therefore, the derivative of the function $g(x)$ also does not exist for $x = 0$.

Experimenting with limits is invaluable when building knowledge about derivatives of various functions. However, in the later stages of working with calculus, you may need to calculate the derivatives of functions without going into the limits of the difference quotients. In MuPAD, there are several ways of producing the derivative of a function. The simplest method is the one that follows from standard mathematical notations where you use the derivative operators $'$, $''$, $'''$, etc. For example,

- h := x ->x*sin(x)*cos(x^5)

  $x \to x \sin(x) \cdot \cos(x^5)$

- h'(x)

  $\sin(x) \cdot \cos(x^5) + x \cdot \cos(x) \cdot \cos(x^5) - 5 \cdot x^5 \cdot \sin(x) \cdot \sin(x^5)$

- h"(x)

  $2 \cdot \cos(x) \cdot \cos(x^5) - x \cdot \sin(x) \cdot \cos(x^5) - 10 \cdot x^5 \cdot \cos(x) \cdot \sin(x^5)$
  $-30 \cdot x^4 \cdot \sin(x) \cdot \sin(x^5) - 25 \cdot x^9 \cdot \sin(x) \cdot \cos(x^5)$

Later, you can produce a few more derivatives using h"(x), h""(x), etc.

You can find another procedure in MuPAD to obtain a derivative. This is the procedure diff. However, you should be aware of the slight difference between applying the derivative operator ' and diff. The derivative operator produces the results only for functions. On the other hand, the diff procedure produces the derivatives for algebraic expressions and functions.

You can apply diff to an algebraic expression. For example,

- u := sin(x)*cos(x):
- diff(u, x)

$$\cos(x)^2 - \sin(x)^2$$

You can also apply it to a function, like v:=x->sin(x). Note the slight difference in the syntax of the command.

- v := x->sin(x)*cos(x):
  diff(v(x),x)

$$\cos(x)^2 - \sin(x)^2$$

Here is the general syntax for the diff procedure:

$$\text{diff}(f, x, y, \dots, z)$$

where f is a formula or function with variables, and x, y, z are the possible variables that may occur in f.

As you can see, the above procedure is more general than you would expect. First of all, using multiple variables you can produce partial derivatives. For example, if $f(x,y) = x^2 + xy + y^2$ then diff(f,x,y) will produce the mixed partial derivative $\dfrac{\partial^2 f}{\partial x \partial y}$. The appropriate MuPAD commands for this would be,

- f := x^2 + x*y + y^2

$$xy + x^2 + y^2$$

- diff(f,x)

$$2 \cdot x + y$$

- `diff(f,y)`

  $x + 2 \cdot y$

- `diff(f,x,y)`

  1

The `diff` procedure can also be used to calculate multiple derivatives in respect to the same variable. For example, here is how you would produce the third derivative of the expression $g(x) = \sin x \cos x$.

- `g := sin(x)*cos(x)`

  $\sin(x)\cos(x)$

- `diff(g,x,x,x)`

  $4\sin(x)^2 - 4\cos(x)^2$

You can use the sequence operator $ to produce derivatives of a higher order. For example, suppose you need to obtain the fifth derivative of the function $h(x) = \frac{\sin x}{x}$. To do this, you would use these commands,

- `h := sin(x)/x`

  $\dfrac{\sin(x)}{x}$

- `diff(h, x $ n = 1..5)`

  $\dfrac{\cos(x)}{x} - \dfrac{20\cos(x)}{x^3} + \dfrac{120\cos(x)}{x^5} - \dfrac{5\sin(x)}{x^2} +$
  $\dfrac{60\sin(x)}{x^4} - \dfrac{120\sin(x)}{x^6}$

This way, you can obtain derivatives of a very high order or even produce a sequence of derivatives.

Any calculus textbook is full of wonderful examples where you could apply MuPAD derivatives. So, I will limit my explorations here to just a few examples.

Example 13.2 Local minimum and maximum values

A typical undergraduate mathematics problem is to obtain the plot of a function and its derivative or derivatives on the same graph and to investigate the existence of local minimum and maximum values.

In this example, we will produce a graph of the function

$$z(x) = \frac{1 - x^3}{1 + x^4}$$

and its derivative; then, I will investigate the minima and maxima of $z(x)$.

We will begin by declaring the function $z(x)$.

- z := x -> (1 - x^3)/(1 + x^4)

$$\frac{1 - x^3}{x^4 + 1}$$

Now we can produce its derivative. We do not need to block the output from this command, as we may wish to see the final formula.

- u := x -> diff(z(x), x)

$$x \longrightarrow \frac{\partial}{\partial x} z(x)$$

Now, having both functions, we will produce the objects to be plotted. When declaring them, we will need to display their graphs in a slightly different way so that we can distinguish them. For this purpose, the original function $z(x)$ will be plotted using a blue line 0.5 mm thick. Its derivative will be plotted with the default parameters.

- z1 := plot::Function2d(z(x), x = -5..5,
      Color = RGB::Blue,
      LineWidth = 0.5
  ):

  u1 := plot::Function2d(u(x), x = -5..5):

  plot(z1, u1)

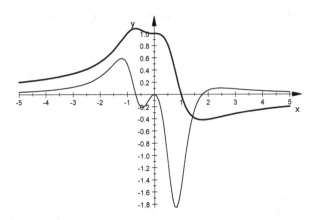

Looking at the above graph you can easily locate the critical points of the function $z(x)$. Here are three of them — one to the left of $x = 0$ (near $x = -1$), another one for $x = 0$, and finally, the last one near $x = 2$. You can calculate their exact values by solving the equation $u = 0$.

- ```
  solutions := solve(u(x) = 0, x)
  ```
  $\{0\} \cup \mathrm{RootOf}(X5^4 - 4 \cdot X5 - 3, X5)$

- ```
  solutions := float(solutions)
  ```
  $\{0.0\} \cup \{-0.6925048426, 1.784357981,$
  $-0.5459265692 + 1.45937795i, -0.5459265692 - 1.45937795i\}$

As you see, some of the solutions are complex numbers, which are useless for our further investigations. So, the best way is to select only real solutions or even to choose the ones that are inside the interval [-5,5]. Finally, if you wish to have access to each solution separately by its identifier, you may consider converting the set of the solutions into a list.

- ```
  rsol := solutions intersect Dom::Interval(-5,5)
  ```
  $\{-0.6925048426, 0.0, 1.784357981\}$

- ```
  rsol:= [op(rsol)]
  ```
  $[1.784357981, -0.6925048426, 0.0]$

Now, you can try to obtain the coordinates of these three points on $z(x)$ where the derivative of $z(x)$ turns to zero, and plot them on the same graph.

- ```
Z1 := [rsol[1], subs(z(x), x = rsol[1])];
Z2 := [rsol[2], subs(z(x), x = rsol[2])];
Z3 := [rsol[3], subs(z(x), x = rsol[3])];
```

$[1.784357981, -0.4203192453]$

$[-0.6925048426, 1.083024918]$

$[0.0, 1.0]$

Finally, here is the code to produce the complete graph.

- ```
P1 := plot::Point2d(Z1):
P2 := plot::Point2d(Z2):
P3 := plot::Point2d(Z3):
plot(z1, u1, P1, P2, P3,
    PointSize = 2*unit::mm,
    PointColor = RGB::Red
)
```

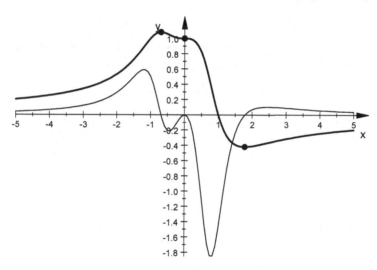

Looking at the graph of the derivative you can easily conclude that,

1. the point $Z2 = [-0.6925048426, 1.083024918]$ is the maximum point of $z(x)$ as the derivative of the function changes near $x = -0.6925048426$ from positive to negative,

2. the point $Z1 = [1.784357981, -0.4203192453]$ is the minimum point of $z(x)$ as the derivative of the function changes near $x = 1.784357981$ from negative to positive,

3. the point $Z3 = [0.0, 1.0]$ is not a local extremum point of $z(x)$ as the derivative of the function remains negative on both sides of $x = 0$.

Finally, by looking at the graph of the function, the number of solutions of $u(x) = 0$ and the graph of derivative, you may conclude that $Z2$ and $Z1$ are the global maximum and minimum points of $z(x)$ respectively. You can finish this example by producing $\lim\limits_{t \to \pm\infty} z(x)$. This will complete our investigations of the graph of $z(x)$. ∎

Example 13.3 Tangent lines and asymptotes of functions

When investigating the properties of functions, we also need to check the behavior of a function in the peripheral parts of the graph. This means developing and plotting the tangent lines and asymptotes of functions.

For the purpose of this exercise, we will use the function $h(x) = \dfrac{x^2 - 1}{|x - 2|}$. You can easily see that $h(x)$ may have a vertical asymptote for $x = 2$. A very basic plot of its graph may give you some idea about the other possible asymptotes.

- `h := x -> (x^2-1)/abs(x-2):`
- `R := 20:`
  `U := plot::Function2d(h(x), x = -R..R, y = -R..R):`
  `plot(U)`

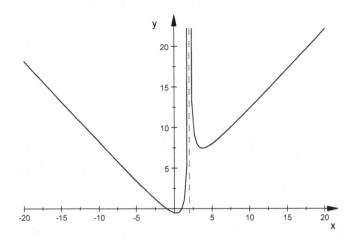

Certainly, you may also suspect the existence of two slant asymptotes.

Our goal will be to find out if these exist and to plot them.

First allow me to remind you the formula for a slant asymptote of the function $f(x)$. It will be represented by the equation of the straight line $y = ax + b$, where $a = \lim\limits_{x \to \infty} \frac{f(x)}{x}$ and $b = \lim\limits_{x \to \infty} (f(x) - ax)$.

For our function, we will consider two separate cases: $x \to +\infty$ and $x \to -\infty$. Thus, we have

- `a1 := limit(h(x)/x, x = +infinity);`

  1

- `b1 := limit(h(x) - a1*x, x = +infinity)`

  2

- `a2 := limit(h(x)/x, x = -infinity);`

  -1

- `b2 := limit(h(x) - a2*x, x = -infinity)`

  -2

And indeed, we've got two asymptotes, $y = x + 2$ and $y = -x - 2$. The first one is for $x \to +\infty$ and the second is for $x \to -\infty$. Here are the declarations for both the asymptotes as MuPAD graphical objects.

- ```
  asymptote1 := plot::Function2d(a1*x + b1, x = -R..R,
      Color = RGB::Blue
  ):
  asymptote2 := plot::Function2d(a2*x + b2, x = -R..R,
      Color = RGB::Black
  ):
  ```

Finally, you should remember about the vertical asymptote $x = 2$ that was already displayed on the graph.

You can check it out by calculating the limits $\lim\limits_{x \to 2^-} h(x) = +\infty$ and $\lim\limits_{x \to 2^+} h(x) = +\infty$. Thus, you have the declaration:

- `asymptote3 := plot::Implicit2d(x = 2, x = 0..3, y = -R..R):`

Note that I had to declare the vertical asymptote as an implicit plot. This is because the vertical line $x = 2$ cannot be considered as a function of the variable x.

Finally, we will collect all our components into one plot:
  ```

- ```
  plot(U,
      asymptote1,asymptote2,asymptote3,
      Scaling = UnConstrained
  )
  ```

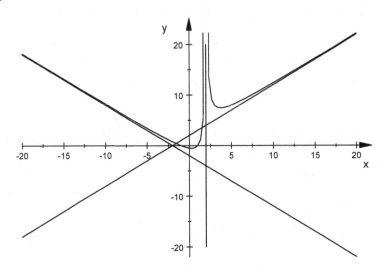

Sometimes, you may need to plot a tangent line for a function, given $x = x_0$. The equation of a tangent line to a given function $f(x)$ has the formula $y = f'(x_0)(x - x_0) + f(x_0)$. Thus, for example, with our function and $x_0 = 3$ we would get,

- ```
  A := float(subs(h(x), x = 3)):
  B := float(subs(h'(x), x = 3)):
  tangentLine := plot::Function2d(B*(x - 3) + A,
      x = -R..R, y = -R..R,
      Color = RGB::Brown
  ):

  plot(U,
      asymptote1, asymptote2, asymptote3, tangentLine,
      Scaling = Unconstrained
  )
  ```

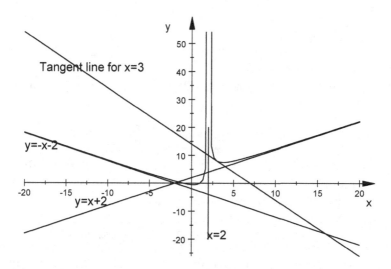

In the above output, I used VCam to add the equations of the asymptotes and a description of the tangent line.

Now you can try to work out on your own a few of the examples that are enclosed at the end of this chapter, or examples from any calculus textbook.   ■

13.3 Working with Integrals

Finding the derivatives of functions was quite a straightforward process. As you perhaps know, integration, which is the process opposite to differentiation, is much more difficult. You can find a number of functions where the indefinite integral does not exist at all, and many others where you need to look for special methods in order to calculate an integral. Before we start experimenting with integrals, let us take a look at the integration tools that are available in MuPAD.

The most basic procedure to produce the integral of a function is the procedure int. Here is its syntax:

int(*f*(*x*), *x*) - *returns the indefinite integral* $\int f(x)dx$

int(*f*(*x*), *x* = *a*..*b*, *options*) - *returns the definite integral* $\int_a^b f(x)dx$

In the second case, you can use two options—Continuous (do not look for discontinuities) and `PrincipalValue` (Cauchy principal value of the integral).

In many situations where finding the definite integral using the `int` procedure was unsuccessful, you can use the procedure `numeric::int` from the `numeric` library. Its syntax is similar to the standard `int`:

$$\text{numeric::int}(f(x), x = a..b)$$

There is also another numerical procedure designed to produce the definite integral. You can use it with the association of a few methods for numerical integration. Here is its syntax:

$$\text{numeric::quadrature}(f(x), x = a..b, \textit{options})$$

As this procedure goes deeper into numerical methods, I will not explain it here. If you are interested in numerical integration, you can refer to the MuPAD help files—just type in the command `?numeric::quadrature`.

You can find, in MuPAD, a small library called `intlib` containing the two procedures `byparts` and `changevar`. The first one uses the integration by parts formula to transform an integral, while the second procedure can be used to change a variable inside an integral. Here is the syntax of these procedures:

$$\text{intlib::byparts}(\textit{integral}, \textit{du})$$
$$\text{intlib::changevar}(\textit{integral}, \textit{equation})$$

I will explain how these two procedures are used later. For now, let us return to the beginning of integration. I will show you here step by step how to use each of the procedures mentioned above.

The simple `int` procedure can be used to integrate most of the functions from an undergraduate calculus textbook. For example, here are the commands to obtain the integrals $\int(x^3 + 5x^2 + 10)dx$, $\int \sin x \cos x dx$, and $\int \frac{5}{3+2x^2}dx$,

- `int(x^3 + 5*x^2 + 10, x)`

$$10 \cdot x + \frac{5 \cdot x^3}{3} + \frac{x^4}{4}$$

- `int(sin(x)*cos(x), x)`

$$-\frac{\cos(2 \cdot x)}{4}$$

- `int(5/(3 + 2*x^2), x)`

$$\frac{5 \cdot \sqrt{6} \cdot \left(2 \cdot \arctan\left(\dfrac{\sqrt{6} \cdot x}{3}\right) - \pi\right)}{12}$$

Note that MuPAD had only produced one of the antiderivatives, not the general one. So, if you need a specific solution, you will need to add the constant C to the integral, and find for it a value satisfying your conditions.

You can produce the definite integrals $\int_0^1 (x^3 + 5x^2 + 10)dx$, $\int_0^{\frac{\pi}{2}} \sin x \cos x \, dx$, $\int_0^{10} \frac{5}{3+2x^2} dx$, and $\int_{-\infty}^{+\infty} \frac{5}{3+2x^2} dx$, in a similar way.

- `int(x^3 + 5*x^2 + 10, x = 0..1)`

$$\frac{143}{12}$$

- `int(sin(x)*cos(x), x = 0..PI/2)`

$$\frac{1}{2}$$

- `int(5/(3 + 2*x^2), x = 0..10)`

$$\frac{5 \cdot \left(2 \cdot \arctan\left(\dfrac{10 \cdot \sqrt{6}}{3}\right) - \pi\right) \cdot \sqrt{6}}{12} + \frac{5 \cdot \pi \cdot \sqrt{6}}{12}$$

- `int(5/(3 + 2*x^2), x = -infinity..+infinity)`

$$\frac{5 \cdot \pi \cdot \sqrt{2} \cdot \sqrt{3}}{6}$$

As you can see, MuPAD was able to deal with definite integrals where the limits for integration are infinite. Of course, you can still produce an integral where the limit for integration is a finite number, and then calculate the limit of such an integral. For example, let us take the function $f(x) = e^{-x^2}$ and see how we can obtain its integral for the interval $[-\infty, +\infty]$

- `f := x->exp(-x^2)`

$$x \rightarrow e^{-x^2}$$

- U := int(f(x),x = -A..A)

  $\sqrt{\pi} \cdot \text{erf}(A)$

We've obtained quite an unusual result, but remember that the indefinite integral of $f(x) = e^{-x^2}$ cannot be expressed using elementary functions. So, next we will calculate the limit for $A \to \infty$.

- limit(U, A = infinity)

  $\sqrt{\pi}$

You can obtain exactly the same result just by integrating from $-\infty$ to $+\infty$,

- U := int(f(x),x = -infinity..infinity)

  $\sqrt{\pi}$

Using the int procedure many times, you can produce multiple integrals. For example, you can obtain $\int\int x\sin x\,dx\,dx$ by executing the statement

- int(int(x*sin(x),x),x)

  $-2 \cdot \cos(x) - x \cdot \sin(x)$

or produce the definite integral $\int_0^\pi \int_0^\pi x\sin x\,dx\,dy$. In our case, the first integration will produce a constant, and the second integration will produce the integral of a constant. Here you have the MuPAD command to produce this integral:

- int(int(x*sin(x), x = 0..PI), y = 0..PI)

  π^2

And in this way, we have come to multiple integrals of multivariate functions. This is quite an interesting topic, but it goes slightly beyond the scope of this book.

In the beginning of this section, I mentioned the two procedures intlib::byparts and intlib::changevar. Let us take a moment to check how useful they might be. First of all, it is important to note that in the byparts procedure, the first argument should be recognized by MuPAD as an integral. In order to obtain this effect, you must freeze the integral operator using the hold or freeze procedures. In command form, it might look like this:

- `intlib::byparts(hold(int)(x*exp(x),x), exp(x))`

$$x \cdot e^x - \int e^x dx$$

or

- `intlib::byparts(freeze(int)(x*exp(x),x),exp(x))`

$$x \cdot e^x - \int e^x dx$$

You can also freeze your integral in advance, and then apply the byparts procedure:

- `F := freeze(int)(x^2*exp(2*x),x);`

$$\int x^2 \cdot e^{2 \cdot x} dx$$

- `F := intlib::byparts(F,exp(2*x))`

$$\frac{x^2 \cdot e^{2 \cdot x}}{2} - \int x \cdot e^{2 \cdot x} dx$$

Now, you can finish calculating the integral:

- `eval(F)`

$$\frac{x^2 \cdot e^{2 \cdot x}}{2} + \frac{e^{2 \cdot x} \cdot (1 - 2x)}{4}$$

The byparts procedure can be quite useful in explaining how integration by parts works. You can probably also find a number of examples where this procedure may be used to help MuPAD obtain the final integral.

Observe that MuPAD is able to deal with most of the very difficult to integrate functions. For example, have you ever tried to obtain by hand the integral $\int x^2 e^{3x} \sin 2x dx$? I will show you here how MuPAD produces the final result for it. Note that I had to block the output from the int command as the result produced was too long to be printed here. I therefore had to simplify it, and get its simplest possible form. However, you can remove the output blocking colons placed at the end of each command and see what you get.

- `G := x^2*exp(3*x)*sin(2*x)`

$$x^2 \cdot e^{3x} \cdot \sin(2 \cdot x)$$

- `H := int(G,x):`
  `H := simplify(H):`
  `collect(H, exp(3*x))`

$$
-e^{3x} \cdot \left(\frac{92 \cdot \cos(2 \cdot x)}{2197} + \frac{18 \cdot \sin(2 \cdot x)}{2197} - \frac{24 \cdot x \cdot \cos(2 \cdot x)}{169} \right.
$$
$$
\left. + \frac{10 \cdot x \cdot \sin(2x)}{169} + \frac{2 \cdot x^2 \cdot \cos(2 \cdot x)}{13} - \frac{3 \cdot x^2 \cdot \sin(2 \cdot x)}{13} \right)
$$

The second procedure I mentioned, `intlib::changevar`, provides a very convenient way to change a variable inside of an integral. We already know that changing variables inside an integral can simplify it significantly. So, the `changevar` procedure is very important from a didactical point of view. Let us see how we can apply it.

Firstly, observe that like in the `byparts` procedure, you need to freeze the integral to which you wish to apply the `changevar` procedure. This is important, as MuPAD has to recognize the argument as an integral. So, in order to obtain an integral by substituting a new variable, you would have the following sequence of commands:

- `H := 2*(2*x + 1)^5 // declare the function`
  $2 \cdot (2 \cdot x + 1)^5$

- `IntH := freeze(int)(H,x) // freeze the integral`
  $\int 2 \cdot (2 \cdot x + 1)^5 dx$

- `// next substitute the new variable`
  `IntH := intlib::changevar(IntH, u = 2*x+1)`
  $\int u^5 du$

- `eval(IntH) // evaluate the new integral`
  $\dfrac{u^6}{6}$

- `subs(IntH, u = 2*x + 1)   // go back to original`
  variable
  $\dfrac{(2 \cdot x + 1)^6}{6}$

Note that you really do not need to stick to indefinite integrals. You can also use the same sequence of steps for definite integrals. In that case however, you will need to skip the very last command. Take a look at what you might get. Here, I will use the example of another function, but all the operations are the same.

- H := 2*x*sqrt(1 + x^2) // declare your function

$$2 \cdot x \cdot \sqrt{x^2 + 1}$$

- IntH := freeze(int)(H, x = 0..1)

$$\int_0^1 2 \cdot x \cdot \sqrt{x^2 + 1}\, dx$$

- IntH := intlib::changevar(IntH, u = x^2 + 1)

$$\int_1^2 \sqrt{u}\, du$$

- eval(IntH)

$$\frac{4 \cdot \sqrt{2}}{3} - \frac{2}{3}$$

You have certainly noticed that when changing a variable inside an integral, MuPAD also took care of the limits of integration.

Before finishing the section about integrals, I would like to point your attention to the student library. In this library, you will find a number of procedures that are useful for teaching mathematics with MuPAD. At the moment we are interested in the three procedures that produce numerical approximations of a definite integral using the Riemann, Simpson and trapezoid methods as well as in three other procedures that we can use to visualize the Riemann, Simpson and trapezoid approximations of definite integrals. Here, I will show you only the Riemann method. The statements using the other two methods are essentially similar. Here is a general syntax of these commands.

Procedures producing an object to be plotted by plot(object)

> student::plotRiemann(*f*(*x*)), *x* = *a..b*, *steps*, *method*, *plot_options*)
> student::plotSimpson(*f*(*x*)), *x* = *a..b*, *steps*, *plot_options*)
> student::plotTrapezoid(*f*(*x*)), *x* = *a..b*, *steps*, *plot_options*)

Procedures producing the numerical approximation of an integral

student::riemann(*f*(*x*), *x* = *a*..*b*, *steps*, *method*)

student::simpson(*f*(*x*), *x* = *a*..*b*, *steps*)

student::trapezoid(*f*(*x*), *x* = *a*..*b*, *steps*)

In all cases where $f(x)$ is a function of one variable or an expression of one variable, *a* and *b* are the limits for the integration and *plot_options* are 2D plot options. Note that both the Riemann procedures also use an additional option *method* that can be set to Left, Middle or Right. This way, you can simulate the integral using rectangles where the touching point with the curve goes through the top-left or top-right corners of the rectangle, or through the center of the top edge of the rectangle.

Let us start with the visualization of an integral using the Riemann method. Suppose that we wish to see how the integral of $f(x) = x^2$ would be calculated for $x = 0..1$. Here is the command, using 10 steps and the left approximation rectangles.

- `p := student::plotRiemann(x^2, x = 0..1, 10, Left):`
  `plot(p)`

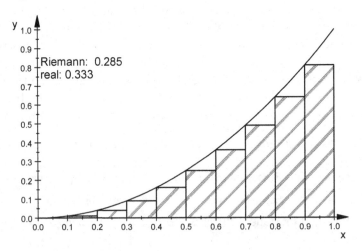

The same command with an accuracy equal to 20 and using the Middle option, produces a much tighter approximation.

* p := student::plotRiemann(x^2, x = 0..1, 20, Middle):
  plot(p)

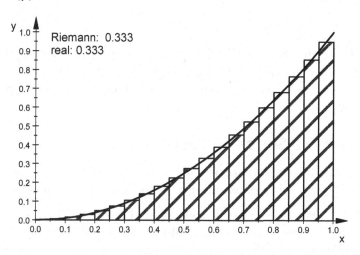

At the same time, the student::riemann procedure can be used to calculate the approximate value of the same integral. For example,

* M := student::riemann(x^2, x = 0..1, 20, Middle);
  L := student::riemann(x^2,x = 0..1, 20, Left);
  R := student::riemann(x^2,x = 0..1, 20, Right);

$$\frac{\sum_{i1=0}^{19}\left(\frac{i1}{20} + \frac{1}{40}\right)^2}{20}$$

$$\frac{\sum_{i2=0}^{19}\frac{i2^2}{400}}{20}$$

$$\frac{\sum_{i3=1}^{20}\frac{i3^2}{400}}{20}$$

* float(M), float(L), float(R)

0.333125, 0.30875, 0.35875

As you can see, all three approximations differ slightly from the exact value, which is equal to:

* float(int(x^2, x = 0..1))

0.3333333333333...

However, if you use more subdividing rectangles, for example about 200, you can get more accurate results:

0.33333125, 0.3308375, 0.3358375

I will leave the integrals at this point. You will certainly agree that there is much more to say about exploring integrals with MuPAD. However, we now need to explore another exciting calculus topic with MuPAD.

13.4 Infinite Series and Products

We have already talked about sequences in MuPAD. Now it is time to see what MuPAD can do with infinite sequences and their sums.

MuPAD contains a number of tools related to series. There is a special type Type::Series, and a small library Series containing the tools for series expansion. However, in this section we will concentrate on a few of the simplest procedures that can be applied in undergraduate mathematics. We will start with the sum procedure.

The point of the sum procedure is to obtain the sum of a given range of a series. For example, suppose we are dealing with the series $\sum_{n=2}^{\infty} \frac{1}{n^2 - 1}$. You can use the sum procedure to obtain an i-th partial sum of the series as follows:

* sum(1/(i^2 - 1), i)

$$-\frac{1}{2 \cdot (i-1)} - \frac{1}{2 \cdot i}$$

or the exact value of any i-th partial sum:

* sum(1/(i^2 - 1), i = 2..100)

$$\frac{14949}{20200}$$

If you are not satisfied with the quick output that you get from MuPAD, and you wish to see how MuPAD understood your statement, you can always use the hold operator to stop MuPAD from performing the final calculations. This will give the following results:

* hold(sum(1/(i^2 - 1), i = 2..100))

$$\sum_{i=2}^{100} \frac{1}{i^2 - 1}$$

And then you can evaluate the obtained expression to complete the calculations:

- eval(%)

$$\frac{14949}{20200}$$

Of course, you do not need to limit yourself to finite limits in your sum. You can have infinitely many terms calculated in the sum. For example,

- sum(1/(i^2 - 1), i = 2..infinity)

$$\frac{3}{4}$$

In a similar way, you can deal with infinite products in MuPAD. Suppose that you have the product of the fractions $\left(\frac{n}{n+1}\right)^n$, where n is changing from 1 to ∞. You can use the product procedure to express your product:

- H := hold(product((n/(n + 2))^n, n = 1..100))

$$\prod_{n=1}^{100} \left(\frac{n}{n+2}\right)^n$$

and then obtain its exact value,

- float(H)

  $4.489253488 \cdot 10^{-84}$

Using the sum and product procedures, you can do many exercises related to number series, checking their convergence, and so on.

Example 13.4 Checking the convergence of a number series

Show that the series

$$\sum_{n=1}^{\infty}\left(\frac{1}{2}\right)^n = \frac{1}{2} + \frac{1}{4} + \frac{1}{8} + \frac{1}{16} + \dots$$

converges and find its sum.

First, let us declare the above series as a MuPAD sum.

- h := hold(sum((1/2)^n, n = 1..k))

$$\sum_{n=1}^{k} (\tfrac{1}{2})^n$$

When working with sums and products, I prefer to use the command hold. This gives me a chance to check if I wrote the formula of my series properly. Then, in the next step, I can evaluate the frozen formula

- h;

$$1 - \left(\frac{1}{2}\right)^k$$

and calculate the limit.

- limit(%, k = +infinity)

  1

This way, you can explore many examples where you need to check if a given series is convergent or divergent. ∎

Number series are interesting but the real fun starts when we deal with power series and Taylor series. Let us start with the series procedure. Here is the most general syntax of the command for using this procedure:

series($f(x)$), $x = x_0$, *terms*, *dir*, NoWarning)

where,

1. $f(x)$ is a function of one variable, in this case variable x,

2. $x = x_0$ is a real number; if you leave only x, you will get the power series for $x_0 = 0$.

3. the word "terms" should be replaced by the number of terms of the power series expansion,

4. the word "dir" should be replaced by one of the keywords: Left, Right, Real or Undirected),

5. NoWarning means no warning messages.

The series command can also be used in simplified forms, where all the parameters starting from the declaration of x_0 can be optional. Let us see how it looks in real examples. Let $f(x) = \frac{x}{1+x^2}$. Hence, we end up with the commands:

- s := series(x/(1 + x^2), x)

$$x - x^3 + x^5 + O(x^7)$$

- s := series(x/(1 + x^2), x, 15)

$$x - x^3 + x^5 - x^7 + x^9 - x^{11} + x^{13} - x^{15} + O(x^{17})$$

- s := series(x/(1 + x^2), x = 1, 7)

$$\frac{1}{2} - \frac{(x-1)^2}{4} + \frac{(x-1)^3}{4} - \frac{(x-1)^4}{8} + \frac{(x-1)^6}{16} + O\left((x-1)^7\right)$$

- s(1.2)

0.491804

Having developed a series, you can use the `coeff` command to extract any of the already calculated coefficients. For our last example, you would get,

- coeff(s,6), coeff(s,12)

$\frac{1}{16}$, Fail

which simply means that the sixth coefficient is equal to $\frac{1}{16}$ but the 12th coefficient has not yet been calculated.

You can obtain similar results by applying the `taylor` procedure.

taylor($f(x)$), $x = x_0$, *terms*)

For example,

- taylor(exp(x), x, 10)

$$1 + x + \frac{x^2}{2} + \frac{x^3}{6} + \frac{x^4}{24} + \frac{x^5}{120} + \frac{x^6}{720} + \frac{x^7}{5040} + \frac{x^8}{40320} + \frac{x^9}{362880} + O(x^{10})$$

Example 13.5 Taylor polynomials

As you will have noticed, every time we produce the expansion of a function into a power series we get the $O(x^n)$ component, which represents a very small value near $x = 0$, or near $x = x_0$ in the case of an expansion for $x = x_0$.

In this example, we will develop a procedure to remove the $O(x^n)$ component and to obtain the Taylor polynomials approximating our function. The idea of the procedure is quite simple — to produce a Taylor series and copy all the terms from it, without the $O(x^n)$ component. We have already written similar procedures, so there is nothing new here.

- ```
  taylorPol := proc(expr, var, n)
  local u, i, newexpr;
  begin
      u := taylor(expr,var,n);
      newexpr := 0;
      for i from 0 to n-1 do
          newexpr := newexpr + coeff(u,i)*x^i
      end;
      return(newexpr)
  end:
  ```

Now we can try to see what this procedure gives us. For example, let us consider the function $f(x) = \sin x$, and produce a few Taylor polynomials approximating this function near $x = 0$.

- ```
  functions := (f.i := taylorPol(sin(x),x, n)) $ n=1..8
  ```

$$0, \; x, \; x, \; x - \frac{x^3}{6}, \; x - \frac{x^3}{6}, \; x - \frac{x^3}{6} + \frac{x^5}{120}, \; x - \frac{x^3}{6} + \frac{x^5}{120},$$
$$x - \frac{x^3}{6} + \frac{x^5}{120} - \frac{x^7}{5040}$$

Note that we obtained pairs of identical Taylor polynomials. This is because the coefficients of the terms x^{2k} for the function $f(x) = \sin x$ are always equal to 0.

It is always interesting to see how Taylor polynomials approximate the original function. Let us plot a graph of the function $f(x) = \sin x$ and a few of the Taylor polynomials approximating it. For this purpose, I will use a few higher degree polynomials. Let us declare the function and some of its Taylor polynomials as MuPAD plot objects:

- ```
  export(plot, Function2d):
  ```
- ```
  range := (x = 0..4*PI, y = -2..2):
  h6 := Function2d(taylorPol(sin(x), x, 6), range):
  h8 := Function2d(taylorPol(sin(x), x, 8), range):
  h14 := Function2d(taylorPol(sin(x), x, 14), range):
  h16 := Function2d(taylorPol(sin(x), x, 16), range):
  h24 := Function2d(taylorPol(sin(x), x, 24), range):
  h30 := Function2d(taylorPol(sin(x), x, 30), range):
  MyFun := Function2d(
      sin(x),range,
      Color = RGB::Blue,
  ```

```
    LineWidth = 0.5
):

plot(
    h6,h8,h14,h16,h24,h30,MyFun,
    ViewingBox = [0..12.5,-2..2]
)
```

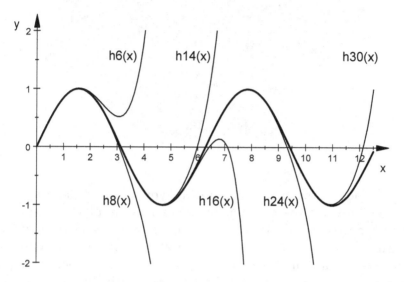

It would appear from the graph that the approximations are more accurate when n is larger and when x is closer to zero. ∎

You will find more explorations of series with MuPAD in the programming exercises section at the end of this chapter.

13.5 Calculus of Two Variables

I have always considered multivariable calculus as a great story, where I can let my imagination wander through amazing surfaces, curves and shapes. In this section, I will touch only a few topics related to multivariable calculus, and I will limit them to just two variables. The rest will have to wait until a better opportunity.

A two variable function can be represented as a surface in 3D space. For example, the graph of the well-known two variable function $f(x,y) = \left(x^2 - y^2\right)e^{-x^2-y^2}$ might look like fig.13.1.

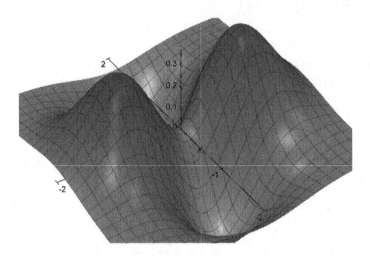

Fig. 13.1 $f(x,y) = (x^2 - y^2)e^{-x^2-y^2}$

You already know that with MuPAD, you can plot the graph of a function of two variables in many ways. You can plot it as a surface like the one shown in fig. 13.1, using contour or level curves, or as a density plot (check the earlier chapter about calculus graphics). You can also plot a domain of such a function using the inequality plot. Therefore, in this section we will concentrate on the things that we can calculate, for example the limits of functions of two variables, their derivatives, etc.

13.5.1 Limits of Functions of Two Variables

Producing the limits of functions with two variables can be slightly more complicated than for functions with just one variable. In the latter case, in the worst case we had a limit when x approaches x_0 from the right or left. For functions with two variables, however, you may have infinite directions pointing towards a point (x_0, y_0). In MuPAD, you can use the limit command with respect to each variable separately, or play a more serious game by considering all the possible directions. For example, here is shown how you can find the limit of the function $f(x,y) = \dfrac{(x+y)}{\sqrt{(1+x^2+y^2)}}$, for $x \to 0$ and then for $y \to 0$.

- limit(limit((x+y)/(sqrt(1 + x^2 + y^2)), x = 0), y = 0)

  0

You can easily see that it really doesn't matter from what direction the function (x,y) will approach the point $(0,0)$. We will always get $\lim\limits_{(x,y)\to(0,0)} \dfrac{(x+y)}{\sqrt{(1+x^2+y^2)}} = 0$.

Let us try another example, $h(x,y) = \dfrac{xy}{x^2+y^2}$. You could easily be mislead while trying MuPAD limits:

- limit(x*y/(x^2 + y^2),x = 0)

  0

- limit(x*y/(x^2 + y^2),y = 0)

  0

However, the real limit of this function does not exist when $(x,y) \to (0,0)$. Imagine that you are approaching the point $(0,0)$ from the direction $y = mx$. In this case, you can substitute $y = mx$ into the equation of the function. You will thus get $h(x) = \dfrac{xmx}{x^2 + m^2x^2}$

- h(x) := x*m*x/(x^2 + m^2*x^2)

  $\dfrac{m \cdot x^2}{x^2 + m^2 \cdot x^2}$

- limit(h(x), x = 0)

  $\dfrac{m}{m^2 + 1}$

It is evident from the last obtained result that a limit cannot exist, as it is different for each direction m. It could be a very good idea to plot a very accurate graph of the surface $\dfrac{xy}{x^2+y^2}$. You will be able to see what is really happening near the point $(x,y) = (0,0)$.

13.5.2 Partial Derivatives

The diff procedure can be used to obtain the derivative of a function with one variable, or a partial derivative of a function with two or more variables. Thus, we can consider $\dfrac{\partial f}{\partial x}$ as diff(f,x), $\dfrac{\partial^2 f}{\partial x\partial y}$ as diff(f,x,y), and so on.

Example 13.6 Plotting a tangent plane

Let $h(x,y) = 5 - 3x^2 - 4y^2$. We will use the `diff` procedure to obtain the tangent plane for the surface representing the function $h(x,y)$. Allow me to remind you the equation for a tangent plane to the surface given by the formula $h(x,y)$ for $x = a$ and $y = b$.

$$z = h(a,b) + \frac{\partial h}{\partial x}(a,b)(x - a) + \frac{\partial h}{\partial y}(a,b)(y - b)$$

Now we have all the necessary components to obtain a tangent plane and plot it together with the surface on the same picture. Let us start with the declarations for $h(x,y)$, the point (a,b) and the coefficients for the tangent plane,

```
• h := (x,y) -> 5 - 3*x^2 - 4*y^2:
  a := -0.3:
  b := -0.7:
  A := subs(diff(h(x,y),x),x = a, y = b):
  B := subs(diff(h(x,y),y),x = a, y = b):
  C := h(a,b):
```

Now we are ready to obtain the equation of the tangent plane:

```
• z := (x,y)->C+A*(x-a)+B*(y-b)
```

$$(x,y) \rightarrow C + A \cdot (x - a) + B \cdot (y - b)$$

Finally, we can export the `Function3d` and `Point` procedures and declare the objects to be plotted.

```
• export(plot, Function3d, Point):
• Surf := Function3d(h(x,y), x = -2..2, y = -2..2):
  Surf := Function3d(h(x,y), x = -2..2, y = -2..2):

  tplane := Function3d(
      z(x,y), x = a-1..a+1, y = b-1..b+1,
      FillColor = [0.5,0.5,0.5,0.7],
      Mesh = [2,2]
  ):
  tpoint := Point3d([a,b,C], PointSize = 2):
```

The very last step is to obtain the graph of these objects.

```
• plot(Surf,tplane,tpoint)
```

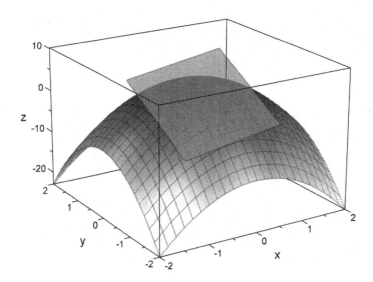

Now using the VCam tool, you can rotate the graph, zoom in and out, etc.   ■

Example 13.7 Finding extrema

Let us take the function $f(x,y) = \left(x^2 - y^2\right)e^{-x^2-y^2}$ (see fig. 13.1, pg. 434). We are going to calculate its extrema. The picture is of great help to us. We may expect two local minimum points and two local maximum points. However, we still need to obtain their exact values and find which of them might be the global maximum or minimum.

We will start with a declaration of the function, and by obtaining its first partial derivatives.

- `f := (x,y)->(x^2 - y^2)*exp(-x^2 - y^2)`

  $(x,y) \longrightarrow (x^2 - y^2) \cdot e^{-x^2-y^2}$

- `EQ1 := diff(f(x,y), x);`
  `EQ2 := diff(f(x,y), y);`

  $2 \cdot x \cdot e^{-x^2-y^2} - 2x(x^2 - y^2) \cdot e^{-x^2-y^2}$

  $2 \cdot y \cdot e^{-x^2-y^2} - 2y(x^2 - y^2) \cdot e^{-x^2-y^2}$

Now we will simplify the obtained partial derivatives and remove the expression $e^{-x^2-y^2}$ from them. This expression is always larger than zero and it will not have any influence on the obtained solutions, but it may cause a lot of trouble when solving equations in MuPAD.

- EQ1 := expand(EQ1*exp(x^2 + y^2)) = 0;
  EQ2 := expand(EQ2*exp(x^2 + y^2)) = 0;

  $2 \cdot x - 2 \cdot x^3 + 2 \cdot x \cdot y^2 = 0$

  $2 \cdot y^3 - 2 \cdot y - 2 \cdot x^2 \cdot y = 0$

- solve({EQ1,EQ2}, {x,y})

  ${[x = 0, y = 0], [x = 0, y = -1], [x = 0, y = 1], [x = -1, y = 0],}$
  ${[x = 1, y = 0]}$

Finally, you can easily obtain the values of the function for these points:

- f(0,0), f(0,-1), f(0,1), f(-1,0), f(1,0)

  $0, -e^{-1}, -e^{-1}, e^{-1}, e^{-1}$

- float(-exp(-1)), float(exp(-1))

  $-0.3678794412, 0.3678794412$

From the obtained values and the picture, you can conclude that $f(0,-1) = f(0,1) = -e^{-1} = -0.3678794412$ are the minimum values of the function, and $f(-1,0) = f(1,0) = e^{-1} = 0.3678794412$ are the maximum values of the function. Finally, the point $(0,0,0)$ is a saddle point for the surface. You can easily check this by producing a graph of the surface with a tangent plane passing through this point. ∎

13.6 Chapter Summary

limit($f(x)$), $x = x_0$) - *produces the limit of the function $f(x)$*

limit($f(x)$), $x = x_0$, Left) - *produces the limit of a function (left side)*

limit($f(x)$), $x = x_0$, Right) - *produces the limit of a function (right side)*

diff($f, x, y, ..., z$) - *produces a derivative of a function*

int($f(x)$), x) - *produces the indefinite integral of $f(x)$ with respect to variable x*

int($f(x)$), $x = a..b$, options) - *produces the definite integral $\int_a^b f(x)dx$*

hold(*object*) - *prevents the evaluation of an object*

freeze(*object*) - *creates an inactive copy of the given object*

numeric::int($f(x)$), $x = a..b$) - *produces a numerical approximation of definite integral* intlib::byparts(*integral*, *du*) - *integration by parts*

intlib::changevar(*integral*, *equation*) - *change of variable inside an integral*

student::plotRiemann(*f*(*x*), *x* = *a..b*, *steps*, *method*, *options*)

student::plotSimpson(*f*(*x*), *x* = *a..b*, *steps*, *options*)

student::plotTrapezoid(*f*(*x*), *x* = *a..b*, *steps*, *options*)

student::riemann(*f*(*x*), *x* = *a..b*, *steps*, *method*) - *numerical approximation of an integral*

student::simpson(*f*(*x*), *x* = *a..b*, *steps*) - *numerical approximation of an integral*

student::trapezoid(*f*(*x*), *x* = *a..b*, *steps*) - *numerical approximation of an integral*

series(*f*(*x*), *x* = x_0, *terms*, *dir*, NoWarning) - *produces a power series expansion*

taylor(*f*(*x*), *x* = x_0, *terms*) - *produces the Taylor series of a function*

13.7 Programming Exercises

1. In the problems below, find the points where the given function is not continuous. Use MuPAD to investigate the function limits for these points, and determine whether this discontinuity is removable.

   a. $f(x) = \dfrac{x}{(x+3)^3}$

   b. $f(x) = \dfrac{x-3}{x^2-9}$

   c. $f(x) = \dfrac{1}{1-|x|}$

   d. $f(x) = \dfrac{|x-1|}{(x-1)^3}$

   e. $f(x) = \begin{cases} -x & \text{if } x < 0 \\ x^2 & \text{if } x > 0 \end{cases}$]

   f. $f(x) = \begin{cases} 1+x^2 & \text{if } x < 0 \\ \dfrac{\sin x}{x} & \text{if } x > 0 \end{cases}$

   g. $f(x) = \begin{cases} \dfrac{1-\cos x}{x} & \text{if } x < 0 \\ x^2 & \text{if } x > 0 \end{cases}$

2. Use MuPAD to check if a given sequence is convergent or divergent. Try to use a PointList2d to visualize the behavior of the sequence.

   a. $a_n = \dfrac{2n}{5n - 3}$

   b. $a_n = \dfrac{1 - n^2}{2 + 3n^2}$

   c. $a_n = \dfrac{n^2 - n + 7}{2n^3 + n^2}$

   d. $a_n = \dfrac{n^3}{10n^2 + 1}$

   e. $a_n = \left(1 + \dfrac{1}{n}\right)^n$

   f. $a_n = \left(\dfrac{n - 1}{n + 1}\right)^n$

   g. $a_n = \left(\dfrac{2}{n}\right)^{3/n}$

   h. $a_n = \left(\dfrac{2 - n^2}{3 + n^2}\right)^n$

3. Investigate the given sequence by calculating its first 20 terms. Formulate a reasonable guess for the value of its limit, then apply the limit procedure to check if your guess was correct.

   a. $b_n = \sqrt{\dfrac{4n^2 + 7}{n^2 + 3n}}$

   b. $b_n = \left(\dfrac{n^3 - 5}{8n^3 + 7n}\right)^{1/3}$

   c. $b_n = e^{-1/\sqrt{n}}$

   d. $b_n = \dfrac{n^3}{e^{n/10}}$

4. Use MuPAD to investigate the following functions. Find their local minima and maxima, determine which of them is the global minimum or maximum. Try to find all the asymptotes of these functions. Finally, plot their graphs with the appropriate asymptotes. Emphasize the extreme points by plotting them as empty or filled circles.

   a. $f(x) = x^2 + 16/x$

   b. $f(x) = 3x^5 - 5x^3$

c. $f(x) = 2x + \dfrac{1}{2x}$

d. $f(x) = \dfrac{x}{1+x}$

e. $f(x) = \dfrac{x}{x^2 + 1}$

f. $f(x) = \dfrac{1-x}{x^2 + 3}$

g. $f(x) = 4x^4 - 11x^2 - 5x - 3$

5. Use MuPAD to obtain the equation of a line that is tangent to the given curve $y = f(x)$ at the point with the given coordinate. Plot the curve, its tangent line and the point on the same graph.

   a. $y = x\cos x,\ x = \pi$

   b. $y = \cos^2 x,\ x = \pi/4$

   c. $y = x - 2\cos x,\ x = 0$

6. For the given functions $y = f(x)$, find the lines that are tangent to the curve in infinitely many points. Plot the curve and its tangent lines in the same graph.

   a. $y = x - 2\cos x$

   b. $y = (5 + \sin x)/(3 + \sin x)$

   c. $y = x\sin\dfrac{1}{x}$

7. In the problems below, use the indicated substitution of a variable and the `intlib::changevar` procedure to evaluate the given integral.

   a. $\int (3x - 8)^{12} dx,\ u = 3x - 8$

   b. $\int 2x\sqrt{x^2 + 3}\, dx,\ u = x^2 + 3$

   c. $\int \dfrac{5x^2}{\sqrt[5]{2x^3 - 1}} dx,\ u = 2x^3 - 1$

   d. $\int \dfrac{\cos 3x}{\sqrt[3]{11 + 2\sin 3x}} dx,\ u = 11 + 2\sin 3x$

   e. $\int \dfrac{\cos \sqrt{x}}{\sqrt{x}} dx,\ u = \sqrt{x}$

f. $\int_0^{\pi/2} (1 + 3\sin x)^{3/2} \cos x\, dx, \quad u = 1 + 3\sin x$

8. In the following problems, use the `student::riemann`, `student::simpson`, and `student::trapezoid` procedures to obtain approximations of the given integrals. In each case, use the same number of steps equal to 20, 100 and 200. Compare the accuracy of the obtained results with the exact value.

   a. $\int_0^2 \frac{1}{x^2}\, dx$

   b. $\int_0^2 \sqrt{1 + x^3}\, dx$

   c. $\int_0^3 \frac{1}{1 + x^4}\, dx$

   d. $\int_{0.1}^1 \frac{\tan x}{x}\, dx$

   e. $\ln 2 = \int_1^2 \frac{1}{x}\, dx$

9. In following problems, use MuPAD to determine if a given series converges or diverges.

   a. $\displaystyle\sum_{n=1}^{\infty} \frac{2n + 1}{n^2(n + 1)^2}$

   b. $\displaystyle\sum_{n=1}^{\infty} \frac{1}{4n^2 - 1}$

   c. $\displaystyle\sum_{n=1}^{\infty} \frac{1}{n(n + 1)}$

   d. $\displaystyle\sum_{n=1}^{\infty} \frac{2}{n(n + 1)(n + 2)}$

   e. $\displaystyle\sum_{n=1}^{\infty} \frac{6}{n(n + 1)(n + 2)(n + 3)}$

   f. $\displaystyle\sum_{n=1}^{\infty} \frac{1 + 2^n + 5^n}{3^n}$

   g. $\displaystyle\sum_{n=1}^{\infty} \frac{7 \cdot 5^n + 3 \cdot 11^n}{13^n}$

10. Produce the n-th degree Taylor polynomials for $g(x) = e^x$ near $x = 0$. Plot the function $g(x)$ and its third and tenth order Taylor polynomials on the same graph.

11. Produce the n-th degree Taylor polynomials for the function $h(x) = \sinh x$, where $n = 2, 4, 6, 8, 10$ and 12. Plot $h(x)$ and its Taylor polynomials on one graph. For plotting, use the interval $x = -5..5$.

12. Produce the n-th degree Taylor polynomials for the function $h(x) = \cos x$, where $n = 6, 8, 14, 16, 22$ and 24. On one graph, plot $h(x)$ and its Taylor polynomials. Use the interval $x = -4\pi..4\pi$.

13. In the problems below, use MuPAD to find the limit or to show that it does not exist.

a. $\displaystyle \lim_{(x,y)\to(1,1)} \frac{1-xy}{1+xy}$

b. $\displaystyle \lim_{(x,y)\to(0,0)} \ln(1+x^2+y^2)$

c. $\displaystyle \lim_{(x,y)\to(0,0)} \frac{\sin xy}{xy}$

d. $\displaystyle \lim_{(x,y)\to(0,0)} \exp\left(-\frac{1}{x^2+y^2}\right)$

e. $\displaystyle \lim_{(x,y)\to(0,0)} \arctan\left(-\frac{1}{x^2+y^2}\right)$

14. Let $f(x,y) = \dfrac{x^2 y}{x^4 + y^2}$. Use MuPAD to show that the limit of the function for $(x,y) \to (0,0)$ from any direction $y = mx$ is equal to 0. Use MuPAD to show that this limit is equal to 1 when (x,y) approaches $(0,0)$ along the parabola $y = x^2$. Finally, use a MuPAD graph to show what is really happening around the point $(0,0)$.

15. For the enclosed functions of two variables, use MuPAD to find every point of the surface at which the tangent plane is horizontal. Determine whether it is a local minimum or maximum. On the same graph plot the surface and its horizontal tangent plane to check your conclusion.

a. $f(x,y) = x^2 + y^2 - 3x + 2y + 5$

b. $f(x,y) = -x^2 + y^2 - 3x + 2y + 3$

c. $f(x,y) = -x^2 - y^2 - 2x + 2y + 5$

d. $f(x,y) = (3x^2 - 2y^2)e^{-x^2-y^2}$

e. $f(x,y) = 2xye^{-2x^2-y^2}$

f. $f(x,y) = x^3 - y^3 + x^2 + y^2$

Chapter 14 _____

A Short Introduction to Linear Algebra

Depending on the country and the type of school, larger or smaller sections of linear algebra can be a part of the undergraduate curriculum. This makes the choice of topics for this chapter somewhat difficult. I will concentrate on the simplest and most common of topics.

14.1 Checking Resources

We will begin this chapter by discussing what linear algebra tools can be found in MuPAD. We have a number of procedures available, from simple to very sophisticated. Most important for us will be the linalg library, where you can find the most fundamental set of tools. However, you will also find some procedures for dealing with linear algebra problems in other libraries. For example, in the library numeric there are procedures like linsolve, eigenvectors, eigenvalues, expMatrix and a few others. These procedures are beyond of our interest, however, at least in this book. For now, we will limit our set of tools to certain selected procedures from the linalg library. MuPAD's linalg library is still too large to be covered in a single chapter of this book, and there will be many procedures that will not even be mentioned in this book.

You also need to remember some universal procedures from the stdlib that can be useful in linear algebra. The most important of these are solve and linsolve.

The linalg library contains a large collection of utilities. Here are some of them:

linalg::addCol,	linalg::addRow,
linalg::adjoint,	linalg::charmat,
linalg::charpoly,	linalg::col,
linalg::concatMatrix,	linalg::crossProduct,
linalg::delCol,	linalg::delRow,
linalg::det,	linalg::eigenvalues,
linalg::eigenvectors,	linalg::expr2Matrix,

```
linalg::gaussElim,              linalg::hilbert,
linalg::jacobian,               linalg::laplacian,
linalg::matdim,                 linalg::matlinsolve,
linalg::multCol,                linalg::multRow,
linalg::normalize,              linalg::randomMatrix,
linalg::swapCol,                linalg::swapRow,
linalg::vandermonde,
```

You can obtain a complete list of procedures that are enclosed in this library by executing the statement info(linalg). We will be using some of these procedures later and then there will be a good opportunity to learn more about them. Right now, let us see what else we can find in MuPAD.

In the Dom library, there are a few domains that can be useful in linear algebra; the most important is Dom::Matrix. We will use this domain quite frequently. We'll leave the remaining domains for later.

14.2 Solving Linear Equations

The well-known procedure solve can be used to solve any set of equations, including a set of linear equations. Like before, you need to list your equations and variables. For example, like this:

- solve({2*x + 3*y = 1, 4*x - 5*y = 2}, {x, y})

$$\{[x = \tfrac{1}{2}, y = 0]\}$$

If you wish to organize your work well, you can declare your equations in advance and then solve them, like here:

- equations := {
      x - 2*y - 3*z + t = 7,
      x + y + z + t = 1,
      5*x - 3*y - 3*z = 2
  }

$$\left\{ t + x + y + z = 1,\ t + x - 2 \cdot y - 3 \cdot z = 7,\ 5 \cdot x - 3 \cdot y - 3 \cdot z = 2 \right\}$$

- solve(equations, {x,y,z})

$$\left\{ \left[x = \frac{5}{8} - \frac{3 \cdot t}{8},\ y = \frac{15}{2} - \frac{5 \cdot t}{2},\ z = \frac{15 \cdot t}{8} - \frac{57}{8} \right] \right\}$$

With the solve procedure, you can also solve systems of equations that are not necessarily linear. For example,

- `solve({x^2 - y = 2, y^2 - x = 2},{x,y})`

$$\left\{ \begin{array}{c} \left[x = -\frac{\sqrt{5}}{2} - \frac{1}{2}, y = \frac{\sqrt{5}}{2} - \frac{1}{2} \right], \\ \left[x = \frac{\sqrt{5}}{2} - \frac{1}{2}, y = -\frac{\sqrt{5}}{2} - \frac{1}{2} \right], \\ [x = -1, y = -1], \\ [x = 2, y = 2] \end{array} \right\}$$

However, for solving linear equations, the `linsolve` function may be more useful, as it is prepared for dealing with the special cases that may occur when solving linear equations.

This procedure is used exactly like the `solve` procedure. For example,

- `linsolve({x - 2*y = 7, 5*x - 3*y = 2}, {x,y})`

$$\left[x = -\frac{17}{7}, y = -\frac{33}{7} \right]$$

Note that in both cases, you can organize the equations as well as the variables in the form of a list or set. Thus, these uses of the `linsolve` will all give the same or similar results. Note that {x,y} is a set and MuPAD is free to choose the most suitable order of variables.

- `linsolve({x - 2*y = 7, 5*x - 3*y = 2}, {x,y});`
  `linsolve([x - 2*y = 7, 5*x - 3*y = 2], [x,y]);`
  `linsolve({x - 2*y = 7, 5*x - 3*y = 2}, [x,y]);`
  `linsolve([x - 2*y = 7, 5*x - 3*y = 2], {x,y})`

Another interesting opportunity arises from the fact that you do not need to stick to variables of type x, y, z, etc. You can also use indexed variables, like this:

- `linsolve({x1 - 2*x2 = 7, 5*x1 - 3*x2 = 2}, {x1,x2})`

$$\left[x1 = -\frac{17}{7}, x2 = -\frac{33}{7} \right]$$

or even like this,

- `linsolve(`
  `    {x[1] - 2*x[2] = 7, 5*x[1] - 3*x[2] = 2},`
  `    {x[1], x[2]}`
  `)`

$$\left[x_1 = -\frac{17}{7}, x_2 = -\frac{33}{7} \right]$$

As you can see in the last output, MuPAD treats objects like x[1], x[2], etc., as indexed variables x_1, x_2. Of course, typing in the multiple square brackets is a bit inconvenient, but the obtained output will satisfy even the most pedantic person. Personally, I prefer to use just x1, x2, ..., xn. This is definitely less troublesome.

The linsolve command has another nice feature. You can use formulae instead of single variables in your equations. For example, you can solve a system of equations like this one:

$$5\cos x + 3e^x = 1$$
$$\cos x - 2e^x = 0$$

Here is the MuPAD code to solve this system. Note how the variables for which the system will be solved were declared.

- reset(): //reset the notebook
- linsolve(
      {5*cos(x) + 3*exp(x) = 1,cos(x) - 2*exp(x) = 0},
      {cos(x), exp(x)}
)

$$\left[\cos x = \frac{2}{13},\ e^x = \frac{1}{13} \right]$$

Another interesting feature of the linsolve procedure is that you can use it to solve linear equations in a given domain. This may lead to very interesting mathematical investigations.

Here is one of the most popular cases — solving an equation over the domain of integers modulo 7.

- linsolve(
      {3*x + 5*y = 1, -2*x - 5*y = 0}, {x,y},
      Domain = Dom::IntegerMod(7)
)

$$\left[x = 1 \bmod 7,\ y = 1 \bmod 7 \right]$$

You can easily find that a number of systems of linear equations may have solutions over a specific domain, and not have solutions over another domain. Solutions over two different domains can also be quite different. Here is an example demonstrating this fact:

```
linsolve(
    {x - 2*y = 7, 5*x - 3*y = 2}, {x,y},
    Domain = Dom::IntegerMod(7)
)
```

FAIL

```
linsolve(
    {x - 2*y = 7, 5*x - 3*y = 2}, {x,y},
    Domain = Dom::IntegerMod(11)
)
```

$[x = 7 \bmod 11, y = 0 \bmod 11]$

```
linsolve(
    {x - 2*y = 7, 5*x - 3*y = 2}, {x,y},
    Domain = Dom::IntegerMod(13)
)
```

$[x = 5 \bmod 13, y = 12 \bmod 13]$

A few interesting questions arise from this example. For instance, how would graphs of the lines $x - 2y = 7$, $5x - 3y = 2$ look over a domain of integer numbers modulo 7, 11 or 13, and why might the system of these two equations not have solutions in one domain and have solutions in another?

Here is a summary of the syntax for the linsolve procedure:

linsolve([*lin. equations*], [*variables or expressions*], *domain declaration*)

14.3 Matrices and Operations on Matrices

In this book, we have already talked two or three times about matrices. In linear algebra, we can see how important matrices are and for what we can use them. We will start this section by reminding ourselves how to declare a matrix, and later I will show you some operations on matrices that are available in MuPAD.

14.3.1 Declaring Matrices in MuPAD

The most common matrix declaration statement has the form

matrix(*n, m*, [[*a1,a2,...,am*], [*b1,b2,...,bm*], ..., [*d1,d2,...,dm*]])

where *n* is the number of rows, *m* is the number of columns, and then we have a list of lists. Each of the nested lists represents one row of the matrix. For example,

- matrix(3, 3, [[0,1,2], [1,0,3], [4,5,0]])

$$\begin{pmatrix} 0 & 1 & 2 \\ 1 & 0 & 3 \\ 4 & 5 & 0 \end{pmatrix}$$

The above command can be simplified. In fact, you do not need to declare the dimensions of the matrix, you may use a command like the one below.

- matrix([[0,1,2], [1,0,3], [4,5,0]])

$$\begin{pmatrix} 0 & 1 & 2 \\ 1 & 0 & 3 \\ 4 & 5 & 0 \end{pmatrix}$$

However, there is an advantage to declaring the dimensions of the matrix. By declaring them, you are passing to MuPAD additional, important information about your matrix. For instance, you may start with an incomplete matrix and MuPAD will add the missing information in the form of additional zeros. Look at the command

matrix(4, 5, [[1],[1,1], [1,1,1]])

Here, I declared a matrix with four rows and five columns, but each of my rows is incomplete and one row is completely missing. MuPAD will take care of all the missing information. Here is the result:

- matrix(4, 5, [[1],[1,1], [1,1,1]])

$$\begin{pmatrix} 1 & 0 & 0 & 0 & 0 \\ 1 & 1 & 0 & 0 & 0 \\ 1 & 1 & 1 & 0 & 0 \\ 0 & 0 & 0 & 0 & 0 \end{pmatrix}$$

This feature can be very useful when declaring large matrices with a number of zeros or with missing information that can be produced and added later.

Sometimes we have to declare matrices that are too large to be displayed on the screen. For example, a matrix like this

    A:=matrix(400, 400, [[1],[1,1], [1,1,1]])

will be very large and MuPAD will not be able to display it. However, you can still declare such matrix, access its elements and perform operations on it. When declaring such a large matrix, MuPAD will display only a warning message.

- A:=matrix(400, 400, [[1],[1,1], [1,1,1]])

  Warning: This matrix is too large for display. If you want to see all non-zero entries of large matrices, call doprint(..).
  [(Dom::Matrix(Dom::ExpressionField()))::print]

  $$\text{Dom::Matrix}()(400, 400, ["\dots"])$$

You can also create matrices with one row or one column only. These can be row-vector or column-vector. For example,

- matrix(1, 3, [5.32, 1.33, 2.56])

  $(5.32, 1.33, 2.56)$

- matrix(3, 1, [x, y, z])

  $$\begin{pmatrix} x \\ y \\ z \end{pmatrix}$$

While developing a new matrix, you can also provide a list of elements and use the option Diagonal to get a diagonal matrix where the elements from the list will be placed on the diagonal:

- matrix(5, 5, [5,3,3,5,8], Diagonal)

  $$\begin{pmatrix} 5 & 0 & 0 & 0 & 0 \\ 0 & 3 & 0 & 0 & 0 \\ 0 & 0 & 3 & 0 & 0 \\ 0 & 0 & 0 & 5 & 0 \\ 0 & 0 & 0 & 0 & 8 \end{pmatrix}$$

Another interesting option is Banded. Using this option, you will be able to produce a matrix were each element of the list will be placed on a slanted band parallel to the matrix diagonal top-left to bottom-right.

• matrix(5, 5, [1,2,3,4,5,6,7,8,9], Banded)

$$\begin{pmatrix} 5 & 6 & 7 & 8 & 9 \\ 4 & 5 & 6 & 7 & 8 \\ 3 & 4 & 5 & 6 & 7 \\ 2 & 3 & 4 & 5 & 6 \\ 1 & 2 & 3 & 4 & 5 \end{pmatrix}$$

If the list in a banded matrix is shorter than the number of bands, the remaining bands will be filled with zeros.

• matrix(5, 5, [5,3,3,5,8], Banded)

$$\begin{pmatrix} 3 & 5 & 8 & 0 & 0 \\ 3 & 3 & 5 & 8 & 0 \\ 5 & 3 & 3 & 5 & 8 \\ 0 & 5 & 3 & 3 & 5 \\ 0 & 0 & 5 & 3 & 3 \end{pmatrix}$$

Finally, you can use various functions to generate the contents of the matrix. Your function should be a function of two variables for a non-diagonal matrix, and a function of one variable for a diagonal matrix. For example,

• matrix(5, 5, (n,m) -> exp(n + m))

$$\begin{pmatrix} e^2 & e^3 & e^4 & e^5 & e^6 \\ e^3 & e^4 & e^5 & e^6 & e^7 \\ e^4 & e^5 & e^6 & e^7 & e^8 \\ e^5 & e^6 & e^7 & e^8 & e^9 \\ e^6 & e^7 & e^8 & e^9 & e^{10} \end{pmatrix}$$

- `matrix(5, 5, (n) -> exp(n), Diagonal)`

$$\begin{pmatrix} e & 0 & 0 & 0 & 0 \\ 0 & e^2 & 0 & 0 & 0 \\ 0 & 0 & e^3 & 0 & 0 \\ 0 & 0 & 0 & e^4 & 0 \\ 0 & 0 & 0 & 0 & e^5 \end{pmatrix}$$

Like in most similar situations, you can define your function earlier and then use it in the `matrix` command, like here,

```
UF := (n,m) -> exp(n * m):
U  := matrix(15, 15, UF)
```

You may consider matrices obtained using the Hilbert function $h(i,j) = (i+j-1)^{-1}$, the so-called Hilbert matrices, as a special case of matrices generated by a function. You can create them using the function $h(i,j)$ like in the above examples, or using two special procedures from the `linalg` library. Here are these two procedures.

linalg::hilbert(n)- *returns the n×n Hilbert matrix generated by the function*

$$h(i,j) = (i+j-1)^{-1}$$

linalg::invhilbert(n)- *returns the inverse of the n×n Hilbert matrix generated by the*

function $h(i,j) = (i+j-1)^{-1}$

An example of a Hilbert matrix is the matrix obtained by this command:

- `linalg::hilbert(5)`

$$\begin{pmatrix} 1 & \frac{1}{2} & \frac{1}{3} & \frac{1}{4} & \frac{1}{5} \\ \frac{1}{2} & \frac{1}{3} & \frac{1}{4} & \frac{1}{5} & \frac{1}{6} \\ \frac{1}{3} & \frac{1}{4} & \frac{1}{5} & \frac{1}{6} & \frac{1}{7} \\ \frac{1}{4} & \frac{1}{5} & \frac{1}{6} & \frac{1}{7} & \frac{1}{8} \\ \frac{1}{5} & \frac{1}{6} & \frac{1}{7} & \frac{1}{8} & \frac{1}{9} \end{pmatrix}$$

In the past, we have used the procedure `linalg::randomMatrix` to create a random matrix. Its syntax is as follows:

linalg::randomMatrix(*n*, *m*, *domain*, *bound*, *options*)

where *n* and *m* are the dimensions of our matrix, *n* is the number of rows, and *m* is the number of columns; *domain* is a domain from which we wish to generate the matrix elements; *bound* can be used to limit all the elements to a single interval; and finally, *options* could be Diagonal or Unimodular. This syntax is enough to create a number of interesting random matrices. Take a look at the examples below in order to see what you may get. We will start with an example where the elements of the matrix are taken from the domain of integer numbers.

- linalg::randomMatrix(5, 5, Dom::Integer)

$$
\begin{pmatrix}
337 & -220 & -502 & -702 & 109 \\
7 & -942 & 281 & -79 & 536 \\
-327 & 41 & 106 & -11 & -645 \\
-429 & 83 & -679 & -705 & -476 \\
-855 & -33 & 189 & 66 & 582
\end{pmatrix}
$$

You can limit your elements to a given interval; for example, to $[0,\ldots,9]$,

- linalg::randomMatrix(4, 4, Dom::Integer, 0..9)

$$
\begin{pmatrix}
5 & 3 & 9 & 1 \\
5 & 9 & 6 & 3 \\
3 & 6 & 7 & 3 \\
6 & 8 & 8 & 8
\end{pmatrix}
$$

- linalg::randomMatrix(4, 4,
    Dom::Integer, 0..100, Diagonal
)

$$
\begin{pmatrix}
84 & 0 & 0 & 0 \\
0 & 41 & 0 & 0 \\
0 & 0 & 30 & 0 \\
0 & 0 & 0 & 13
\end{pmatrix}
$$

The option Unimodular produces a matrix that has a determinant equal to 1 in a given domain. For example,

- linalg::randomMatrix(5, 5, Dom::Integer, Unimodular)

$$\begin{pmatrix} 0 & -4 & 6 & 3 & -3 \\ 0 & 0 & -1 & -3 & 3 \\ -3 & -1 & 3 & 3 & -4 \\ 1 & -7 & 6 & -3 & 3 \\ -6 & 5 & -8 & -10 & 8 \end{pmatrix}$$

- linalg::det(%)

  1

There is another, more sophisticated way of declaring matrices. For instance, suppose that you are going to deal with matrices of complex numbers. You can define your own domain of matrices of complex numbers. This is done as follows:

- CompMat := Dom::Matrix(Dom::Complex)

Now you have your own domain of matrices over complex numbers, and you can use it to create your own matrices. For example,

- A := CompMat([[I+3,1+I,3-I],[1,3+I,4],[1,2*I,5+3*I]])

$$\begin{pmatrix} 3+i & 1+i & 3-i \\ 1 & 3+i & 4 \\ 1 & 2 \cdot i & 5+3 \cdot i \end{pmatrix}$$

Of course, you can always develop and use a more sophisticated domain for your matrices. This depends only on your imagination and your needs.

14.3.2 Global Operations on Matrices

By now, you have learned a lot about creating matrices. Let us now see what operations on matrices can be performed with MuPAD.

Among the most important are some arithmetical operations performed on matrices—addition, subtraction, multiplication and finding the inverse matrix. These operations use matrices as a whole, and I will therefore call such operations global matrix operations.

Another group of important matrix operations are those used to manipulate rows and columns. For the rest of this chapter, I will call them local matrix operations.

Let us start with arithmetical operations on matrices. First, let us create a few matrices that we will need later to demonstrate matrix operations. In order to save space, for our exercises, I will use small matrices, with dimensions like 3×3. However, when you begin experimenting on your own, you can use matrices of any dimension.

- A := matrix(3,3,[[7,6,0],[8,0,3],[6,1,9]]);
  B := matrix(3,3,[[9,7,0],[0,6,1],[0,0,8]]);
  C := matrix(2,3,[[7,1,7],[0,8,4]]);
  F := matrix(3,2,[[1,8],[7,5],[7,0]]);

Here is our output,

$$A = \begin{pmatrix} 7 & 6 & 0 \\ 8 & 0 & 3 \\ 6 & 1 & 9 \end{pmatrix}, B = \begin{pmatrix} 9 & 7 & 0 \\ 0 & 6 & 1 \\ 0 & 0 & 8 \end{pmatrix},$$

$$C = \begin{pmatrix} 7 & 1 & 7 \\ 0 & 8 & 4 \end{pmatrix}, F = \begin{pmatrix} 1 & 8 \\ 7 & 5 \\ 7 & 0 \end{pmatrix}$$

Now, we are ready to experiment. So, we have,

- A + B

$$\begin{pmatrix} 16 & 13 & 0 \\ 8 & 6 & 4 \\ 6 & 1 & 17 \end{pmatrix}$$

- 3*A + 5*B

$$\begin{pmatrix} 66 & 53 & 0 \\ 24 & 30 & 14 \\ 18 & 3 & 67 \end{pmatrix}$$

You can also multiply matrices, provided they have the right dimensions.

- A*B

$$\begin{pmatrix} 63 & 85 & 6 \\ 72 & 56 & 24 \\ 54 & 48 & 73 \end{pmatrix}$$

Since version 3.0, the internal representation of matrices in MuPAD has been changed. This is reflected in the speed of operations on matrices, and especially matrices with many zeros. For example, try the following code and see how fast MuPAD produced U^{20}, where U is a random diagonal matrix.

- ```
  U := linalg::randomMatrix(20, 20, Dom::Integer, 0..2):
  V := U^20:
  V[14,15]
  ```
  2737253986969361892936228

Before we go on further with our investigations, we should revise some of the most important properties of matrix multiplication.

In order to multiply two matrices, their dimensions must match. The number of columns in the first matrix must be equal to the number of rows in the second matrix. Look how MuPAD reacts when this rule is not satisfied.

- B*F   // This operation can be done

$$\begin{pmatrix} 58 & 107 \\ 49 & 30 \\ 56 & 0 \end{pmatrix}$$

- F*B // This operation cannot be done
  Error: dimensions don't match
  [(Dom::Matrix(Dom::ExpressionField()))::_mult]

Multiplying F*C and C*F, you can see that multiplication of matrices is not commutative,
  ```

- C*F

$$\begin{pmatrix} 63 & 61 \\ 84 & 40 \end{pmatrix}$$

- F*C

$$\begin{pmatrix} 7 & 65 & 39 \\ 49 & 47 & 69 \\ 49 & 7 & 49 \end{pmatrix}$$

You can also produce an inverse matrix using the command 1/A or A^(-1). However, you need to remember when you can and cannot produce the inverse matrix. Just a quick reminder—matrix A should be a square matrix with a determinant not equal to zero.

- 1/B

$$\begin{pmatrix} \frac{1}{9} & -\frac{7}{54} & \frac{7}{432} \\ 0 & \frac{1}{6} & -\frac{1}{48} \\ 0 & 0 & \frac{1}{8} \end{pmatrix}$$

Expressions with matrices can be quite complicated. You might be surprised how many typical identities that are valid for numbers, are not valid for matrices. For example, do you think that $(A+B)^2 = A^2 + 2AB + B^2$? Here are two examples showing that this equality is not true in general.

- (A^2 + 2*A*B + B^2) - (A + B)^2 //we expect to get 0

$$\begin{pmatrix} -56 & 31 & -15 \\ 18 & 55 & -3 \\ 6 & 40 & 1 \end{pmatrix}$$

- (A^2 + 2*A*B + B^2)/(A + B)^2

$$\begin{pmatrix} -\dfrac{52089}{224} & -\dfrac{1749511}{3136} & \dfrac{104879}{784} \\[2mm] -\dfrac{13929}{112} & \dfrac{467911}{1568} & -\dfrac{27959}{392} \\[2mm] -\dfrac{6241}{56} & \dfrac{208875}{784} & -\dfrac{12323}{196} \end{pmatrix}$$

The last output was crowded with unpleasant fractions. You may apply the float procedure to it in order to get a floating point representation of the matrix.

- float(%)

$$\begin{pmatrix} -232.5401786 & 557.8797832 & -133.7742347 \\ -124.3660714 & 298.4126276 & -71.32397959 \\ -111.4464286 & 266.4221939 & -62.87244898 \end{pmatrix}$$

As you have seen above, you can use the float operation in order to convert all the entries of a matrix to floating point numbers. You can apply a number of other MuPAD procedures to all the entries of a given matrix in exactly the same way. Here are some of these operations:

diff(A, x) - *differentiate all the entries in respect to variable x*

expand(A) - *expand all the entries of A*

float(A) - *apply float to all the entries of A*

int(A, x) - *integrate all the entries of A*

subs(A, X=*expr*) - *substitute an expr in A in place of all the occurrences of X*

map(A, *function*) - *apply function to all entries of the matrix A*

zip(A, B, f) - *produces the matrix $C_{ij} = f(a_{ij}, b_{ij})$*

A number of other functions can be used in a similar manner. In the above list, there are two new procedures that had not been introduced in this book so far. These are the very useful procedures map and zip. Both can also be applied to lists, sequences and other structural MuPAD objects. Let us see what we can do with them.

The procedure map applies a given function to all the entries of the matrix A. Let us define a function of one variable:

- g := x -> x*exp(-x*3);

$$x \longrightarrow x \cdot e^{-3x}$$

Now we can use this function with the procedure map and the matrix A, where

$$A = \begin{pmatrix} 7 & 6 & 0 \\ 8 & 0 & 3 \\ 6 & 1 & 9 \end{pmatrix}$$

is still the same matrix defined for experimenting with operations on matrices.

- A := matrix([[7,6,0], [8,0,3], [6,1,9]]):
- map(A, g)

$$\begin{pmatrix} 7 \cdot e^{-21} & 6 \cdot e^{-18} & 0 \\ 8 \cdot e^{-24} & 0 & 3 \cdot e^{-9} \\ 6 \cdot e^{-18} & e^{-3} & 9 \cdot e^{-27} \end{pmatrix}$$

You will get an identical result by putting the declaration of the function directly inside the map procedure like here,

- map(A, x -> x*exp(-x*3))

The procedure zip requires a function of two variables $f(x,y)$. It will produce a new matrix C from the two given matrices A and B, so that $c_{ij} = f(a_{ij}, b_{ij})$.

Here is a typical use of this procedure for the matrices A and B defined earlier,

- f := (x,y) -> x*y:
  ZipMat := zip(A,B,f)

$$\begin{pmatrix} 63 & 42 & 0 \\ 0 & 0 & 3 \\ 0 & 0 & 72 \end{pmatrix}$$

You can easily check what happened by recalling both matrices here and comparing them with the obtained result. Thus, you have,

$$A = \begin{pmatrix} 7 & 6 & 0 \\ 8 & 0 & 3 \\ 6 & 1 & 9 \end{pmatrix}, B = \begin{pmatrix} 9 & 7 & 0 \\ 0 & 6 & 1 \\ 0 & 0 & 8 \end{pmatrix}, f(x,y) = x \cdot y \text{ and}$$

$$ZipMat = \begin{pmatrix} 7 \cdot 9 & 6 \cdot 7 & 0 \cdot 0 \\ 8 \cdot 0 & 0 \cdot 6 & 3 \cdot 1 \\ 6 \cdot 0 & 1 \cdot 0 & 9 \cdot 8 \end{pmatrix}$$

The `zip` procedure can be very useful when creating new operations on matrices.

Example 14.1 Solving systems of linear equations in matrix form

In linear algebra, we deal with systems of linear equations written either in expanded form or in matrix form. From time to time, we need to move from one form to the other. There is a number of tools in MuPAD that allow us to perform this transformation.

In this example, we need these two procedures.

linalg::expr2Matrix(*equations*, [*variables*]) - *transforms a system of linear equations into a matrix*

linalg::matlinsolve(*A, B*) - *solve a system of linear equations in matrix form*

Let us start from the system of equations,

• equations := [7*x+6*y = 1, 8*x+3*z = 2, 6*x+y+3*z = 3]
  $[7 \cdot x + 6 \cdot y = 1, 8 \cdot x + 3 \cdot z = 2, 6 \cdot x + y + 3 \cdot z = 3]$

Now you can transform this system of equations into matrix form:

• AM := linalg::expr2Matrix(equations,[x,y,z])

$$\begin{pmatrix} 7 & 6 & 0 & 1 \\ 8 & 0 & 3 & 2 \\ 6 & 1 & 3 & 3 \end{pmatrix}$$

Note that what we have obtained here is the augmented matrix of our system of linear equations. The terms from the right side of our equations were enclosed in this matrix as the last column. Now you can work with your equations on the matrix level. For example, you can extract a single row or column from such a matrix.

- B := linalg::col(AM,4)

$$\begin{pmatrix} 1 \\ 2 \\ 3 \end{pmatrix}$$

You can also delete a row or column from your matrix, like this:

- A := linalg::delCol(AM,4)

$$\begin{pmatrix} 7 & 6 & 0 \\ 8 & 0 & 3 \\ 6 & 1 & 3 \end{pmatrix}$$

We are thus able to reduce the system of equations to two matrices — the main matrix A of the system, and matrix B of the right-side terms of our system of equations.

Sometimes, you may need to solve a system of equations given in matrix form like the one that we just created. In order to do this, you need to use the command,

- linalg::matlinsolve(A,B)

$$\begin{pmatrix} -\frac{5}{19} \\ \frac{9}{19} \\ \frac{26}{19} \end{pmatrix}$$

You can easily verify that we have obtained exactly the same solution as the one that is produced by the command,

- solve(equations, {x,y,z})

$$\left\{\left[x = -\tfrac{5}{19},\ y = \tfrac{9}{19},\ z = \tfrac{26}{19} \right]\right\}$$

Note that the linalg::matlinsolve procedure can be applied in many different forms. Here, I have demonstrated its the most obvious form. Check MuPAD's help for other ways of using this procedure.

Going from a system of equations in matrix form to a system in expanded form is a bit more complicated. You need to write a simple procedure that will take two matrices and develop a system of equations using these matrices, if possible. ■

The `linalg` library contains a large collection of procedures to perform special operations on matrices. I will not discuss them here in detail, but I do need to at least mention some of them. So, you can find procedures in the `linalg` library that operate on matrices and also produce matrices, as well as procedures that produce numbers or other objects when applied to matrices.

Here are some of the procedures that produce results in the form of a matrix. Note that vectors are also matrices.

linalg::concatMatrix($A,B,...,C$) - *joins matrices horizontally*

linalg::stackMatrix($A,B,...,C$) - *joins matrices vertically*

linalg::transpose(A) - *produces transposition of A, i.e. A^t,*

linalg::gaussElim(A) - *performs Gaussian elimination of a matrix.*

linalg::gaussJordan(A) - *performs Gauss-Jordan elimination*

linalg::crossProduct(A,B) - *produces cross product of a 3D vectors*

linalg::normalize(V) - *normalizes a vector*

linalg::charmat(A,x) - *produces characteristic matrix*

In the same library, you can also find procedures that, when applied to matrices, produce numbers or polynomials.

linalg::angle(A,B) - *calculates the angle between two vectors*

linalg::charpoly(A, x) - *produces the char. polynomial of a matrix*

linalg::eigenvalues(A) - *produces eigenvalues of a matrix*

linalg::eigenvectors(A) - *produces eigenvectors of a matrix*

linalg::matdim(A) - *returns the dimensions of the matrix*

linalg::ncols(A) - *returns the number of columns in the matrix A*

linalg::nonZeros(A) - *returns the number of ≠0 elements in a matrix*

linalg::nrows(A) - *returns the number of rows in the matrix A*

linalg::scalarProduct(A, B) - *produces the scalar product of two vectors*

linalg::vecdim(V) - *returns the number of elements of a vector*

14.3.3 Local Matrix Operations

In example 14.1, we started using some of the local matrix operations, i.e. operations that can be applied to rows or columns. You can now perform with MuPAD the same operations that you had done by hand in a traditional course of linear algebra. You can swap the rows or columns of a matrix, remove rows or columns, insert new rows, etc. Let us list some of the most important procedures from the linalg library to perform local matrix operations.

linalg::addCol(*A*, *c*1, *c*2, *s*)- *replaces in A c2 by c2+s*c1*

linalg::addRow(*A*, *r1*, *r2*, *s*)- *produce a copy of matrix r2=r2+s*r1*

linalg::col(*A*, *c*) - *extracts column c from matrix A*

linalg::row(*A*, *r*) - *extracts row r from the matrix A*

linalg::delCol(*A*, *c1..c2*) - *deletes in A columns c1 through c2*

linalg::delRow(*A*, *r1..r2*) - *deletes in matrix A rows r1 through r2*

linalg::multCol(*A*, *c*, *s*) - *multiples in A column c by number s*

linalg::multRow(*A*, *r*, *s*) - *multiples in matrix A row r by number s*

linalg::setCol(*A*, *c*, *c1*) - *replaces in A column c by a new column*

linalg::setRow(*A*, *r*, *r1*) - *replaces in A row r by new row r1*

linalg::submatrix(*A*,*r1..r2*,*c1..c2*) - *extract submatrix defined by ranges*

linalg::substitute(*A*, *B*, *m*, *n*) - *substitutes matrix B inside of matrix A,*
$\qquad\qquad\qquad$ *element B*[1,1] *will be placed as A*[*m,n*]

linalg::swapCol(*A*, *c1*, *c2*) - *swaps columns c1 and c2 in matrix A*

linalg::swapRow(*A*, *r1*, *r2*) - *swaps rows r1 and r2 in matrix A*

The three large collections of procedures for linear algebra listed in this chapter will satisfy even the most demanding mathematician. However, you can still find more interesting procedures in the MuPAD libraries. I will leave this exploration to you. Now, let us practice with what we've learned so far.

Example 14.2 Gaussian elimination

Gaussian elimination is one of the most important topics in linear algebra. We can perform Gaussian elimination with MuPAD just by using a single command, or we can do it step by step on our own by applying some of the other commands that were mentioned in the last table. We will try both methods.

Let us consider the system of linear equations

$$3x + 2y + 5z = 4$$
$$-x + 5y + 3z = -1$$
$$x + 2y - 5z = 0$$

The augmented matrix of this system can be declared as follows:

- A := matrix(
      3,4,[[3, 2, 5, 4],[-1, 5, 3, -1],[1, 2, -5, 0]]
)

The above declaration does not assume anything about the terms of matrix A. However, you may request much more while declaring this matrix. For example, you may assume that its elements are from a specific domain. This will give you a chance to better shape your results. Thus, you may assume that the elements of A are integer numbers, or rational numbers, or floating point numbers. In each case, you will get a slightly different result with Gaussian elimination. Let us try out these options.

- A := Dom::Matrix(Dom::Integer)(
      [[3, 2, 5, 4], [-1, 5, 3, -1], [1, 2, -5, 0]]
)

$$\begin{pmatrix} 3 & 2 & 5 & 4 \\ -1 & 5 & 3 & -1 \\ 1 & 2 & -5 & 0 \end{pmatrix}$$

Let us declare the same matrix again as a matrix of rational and floating point numbers. The output from both declarations will look the same as it did for A, and I will omit it here. So, we declare,

- B := Dom::Matrix(Dom::Rational)(
      [[3, 2, 5, 4], [-1, 5, 3, -1], [1, 2, -5, 0]]
)
- C := Dom::Matrix(Dom::Float)(
      [[3, 2, 5, 4], [-1, 5, 3, -1], [1, 2, -5, 0]]
)

Now we can check how MuPAD will perform Gaussian elimination on these matrices. Of course, in each case MuPAD will try to follow the assumptions made in our declarations. Thus, our results will be slightly different each time. Here they are:

- linalg::gaussElim(A)

$$\begin{pmatrix} 3 & 2 & 5 & 4 \\ 0 & 17 & 14 & 1 \\ 0 & 0 & -132 & -24 \end{pmatrix}$$

- linalg::gaussElim(B)

$$\begin{pmatrix} 3 & 2 & 5 & 4 \\ 0 & \frac{17}{3} & \frac{14}{3} & \frac{1}{3} \\ 0 & 0 & -\frac{132}{17} & -\frac{24}{17} \end{pmatrix}$$

- linalg::gaussElim(C)

$$\begin{pmatrix} 3.0 & 2.0 & 5.0 & 4.0 \\ 0.0 & 5.666666667 & 4.666666667 & 0.3333333333 \\ 0.0 & 0.0 & -7.764705882 & -1.411764706 \end{pmatrix}$$

Now is the right time to think about how we could produce Gaussian elimination by hand. This is a good exercise to look inside the process of elimination of the terms of a matrix. The advantage of operating by hand is that you can produce an even better result than the one obtained by the gaussElim procedure.

Before we start our work, we should import some procedures that we will use in this exercise. This will save us a lot of unnecessary typing.

- export(linalg, addRow, swapRow, multRow):

Let us start again with matrix A in the simplest form:

- A := matrix(
      3, 4, [[3, 2, 5, 4],[-1, 5, 3, -1],[1, 2, -5, 0]]
)

$$\begin{pmatrix} 3 & 2 & 5 & 4 \\ -1 & 5 & 3 & -1 \\ 1 & 2 & -5 & 0 \end{pmatrix}$$

You will easily notice that it would not be difficult to eliminate the number 3 from the first column and the number -1 from the same column. For this, we will need the addRow procedure.

- addRow(A, 3, 2, 1)

$$\begin{pmatrix} 3 & 2 & 5 & 4 \\ 0 & 7 & -2 & -1 \\ 1 & 2 & -5 & 0 \end{pmatrix}$$

- addRow(%, 3, 1, -3)

$$\begin{pmatrix} 0 & -4 & 20 & 4 \\ 0 & 7 & -2 & -1 \\ 1 & 2 & -5 & 0 \end{pmatrix}$$

In the next step, we will swap the third and the first rows, in order to completely clear the first column of the matrix.

- swapRow(%, 1, 3)

$$\begin{pmatrix} 1 & 2 & -5 & 0 \\ 0 & 7 & -2 & -1 \\ 0 & -4 & 20 & 4 \end{pmatrix}$$

Eliminating the number 7 from the second column is again an easy task.

- multRow(%, 3, 1/4)

$$\begin{pmatrix} 1 & 2 & -5 & 0 \\ 0 & 7 & -2 & -1 \\ 0 & -1 & 5 & 1 \end{pmatrix}$$

- addRow(%, 3, 2, 7)

$$\begin{pmatrix} 1 & 2 & -5 & 0 \\ 0 & 0 & 33 & 6 \\ 0 & -1 & 5 & 1 \end{pmatrix}$$

Now we need to swap rows 2 and 3. This will end our calculations. However, if you wish to pursue this as far as possible, you could change −1 into 1 and obtain 1 on the intersection of the third row and third column.

- swapRow(%, 2, 3)

$$\begin{pmatrix} 1 & 2 & -5 & 0 \\ 0 & -1 & 5 & 1 \\ 0 & 0 & 33 & 6 \end{pmatrix}$$

- multRow(%,2, -1)   // remove -1

$$\begin{pmatrix} 1 & 2 & -5 & 0 \\ 0 & 1 & -5 & -1 \\ 0 & 0 & 33 & 6 \end{pmatrix}$$

- multRow(%, 3, 1/33)   // remove 33

$$\begin{pmatrix} 1 & 2 & -5 & 0 \\ 0 & 1 & -5 & -1 \\ 0 & 0 & 1 & \frac{2}{11} \end{pmatrix}$$

From here, it is easy to obtain the solution to our system of linear equations. You need to follow the back substitution path. Thus, you will obtain the solutions $z = \frac{2}{11}$, $y = -1 + 5 \cdot \frac{2}{11} = -\frac{1}{11}$ and $x = -2 \cdot \frac{-1}{11} + 5 \cdot \frac{2}{11} = \frac{12}{11}$

Finally, you can easily confirm the obtained results by asking MuPAD to solve the system of equations.

- linalg::matlinsolve(A)

$$\begin{pmatrix} \frac{12}{11} \\ -\frac{1}{11} \\ \frac{2}{11} \end{pmatrix}$$

As you have noticed, the results of Gaussian elimination can be quite different depending on the steps you perform while eliminating the terms of the matrix. However, the solution of the system of linear equations will always be the same.

A special case of Gaussian elimination is when the numbers on the diagonal are equal to 1, like what we obtained above. Such an elimination process is called Gauss-Jordan elimination. ∎

14.4 Determinants

Matrices, linear equations and determinants are the three main topics in linear algebra. It is now time to revise determinants.

We will start with two procedures to obtain the determinant of a matrix. One of them is `linalg::det`, and the other is `numeric::det`. The basic syntax of both procedures is identical, but `numeric::det` has a number of options specific to numerical calculations. For our purposes, the basic syntax is just enough.

> linalg::det(*A*)
>
> numeric::det(*A, options*)

Using both procedures is quite straightforward.

- A := matrix(3, 3, [[3, 2, 5], [-1, 5, 3], [1, 2, -5]])

$$\begin{pmatrix} 3 & 2 & 5 \\ -1 & 5 & 3 \\ 1 & 2 & -5 \end{pmatrix}$$

- linalg::det(A)

  -132

The `linalg::det` procedure and its numerical equivalent are enough for all undergraduate (and not only) examples. However, we may need a more sophisticated tool to demonstrate how we operate on determinants or, for example, to show how Cramer's rule works. This can be a very interesting example.

Example 14.3 Cramer's rule

Like in the previous example, we will begin by exporting all the procedures that we need here. I had developed this example before putting it in this book, so I know which procedures I need for it. However, if you wish to modify this example, do remember to also modify this declaration.

- `reset():`
- `export(linalg, nrows, ncols, delRow, delCol, det):`

Let A be a square matrix

$$A = \begin{pmatrix} 3 & 2 & 5 \\ -1 & 5 & 3 \\ 1 & 2 & -5 \end{pmatrix}$$

- `A := matrix([[3,2,5], [-1,5,3], [1,2,-5]])`

Our first goal is to produce a procedure to obtain the minor A_{ij} of the given element a_{ij} of the matrix A. We will start with a very simple procedure, and later we will improve it. So, here is our starting point:

```
• minor := proc(A, i, j)
      local B;
  begin
      B := delRow(A,i);
      B := delCol(B,j);
      return(det(B)) //this is a temporary declaration
  end:
```

We can check how the procedure works in the case of matrix A.

- `minor(A, 2, 2)`

    -20

In the next step, we need to write the complete procedure Det to expand a given determinant along a single row or column. For the purpose of this example, I will use row one, but later on you can modify the code to use any row or any column.

As the first step in this procedure, it is necessary to check if the input matrix is a square matrix. Otherwise, we may run into a disaster later on. For this purpose, it would be a good idea to use the conditional instruction if .. then.

```
if (nrows(A) <> ncols(A)) then
    error("Wrong input matrix")
...
```

The remaining part of the procedure is quite obvious — go along the first row of the matrix and produce the sum:

$$\sum_{j=1}^{ncols(A)} (-1)^{1+j} minor(A, 1, j)$$

In the situation when the minor is reduced to a determinant 1×1, we assume that Det(A)=A[1,1]. Here is the procedure implementing all of these steps.

```
• Det := proc(A)
  local r, j;
  begin
      r := 0;
      if (nrows(A) <> ncols(A)) then
          error("Wrong input matrix")
      else
          if nrows(A)=1 then
              return(A[1,1])
          else
              for j from 1 to ncols(A) do
                  r:=r+(-1)^(1+j)*A[1,j]*minor(A,1,j)
              end;
          end;
      end;
      return(r)
  end:
```

Now it is time to test the procedure and see if it works as we expected.

```
• Det(A)
  -132
```

We are definitely not satisfied with the Det procedure. It produces the determinant of the square matrix, but it does not really show us how Cramer's rule works. It would be good to see how Cramer's rule is applied step by step. Well, let us introduce some changes into the procedure minor. First, we will exchange the det procedure in it for

Det. This will allow us to produce the output fully using the Det procedure. However, this still won't give us the expected result. We need more changes. I hope you still remember the procedure hold — we use it to stop evaluating the function for a while. Let us use it here, in the procedure minor. This is the improved version of the procedure:

- ```
  minor := proc(A, i, j)
    local B;
    begin
       B := delRow(A,i);
       B := delCol(B,j);
       return(hold(Det)(B))
    end:
  ```

This will completely fulfill our requirements. Now we can trace, step by step, how the determinant of any matrix is calculated. Here I will show you how it works with the matrix A. Reading line after line, we can analyze what we got. This is quite an interesting example. By the way, notice that the procedure Det now calls the procedure minor, and the procedure minor calls Det. This is a kind of recursion that we call *mutual recursion*.

- Det(A)

$$5Det\left(\left(\begin{pmatrix} -1 & 5 \\ 1 & 2 \end{pmatrix}\right)\right) - 2Det\left(\left(\begin{pmatrix} -1 & 3 \\ 1 & -5 \end{pmatrix}\right)\right) +$$

$$3Det\left(\left(\begin{pmatrix} 5 & 3 \\ 2 & -5 \end{pmatrix}\right)\right)$$

- eval(%)

  $17Det((-5)) - 14Det((2)) - 19Det((1))$

- eval(%)

  -132

The above matrix was very simple. We should try using the Det procedure on a larger matrix.

- ```
  B := Dom::Matrix(Dom::Integer)(
       [[3,2,5,4], [2,5,3,8], [1,2,3,0], [2,3,4,5]]
     )
  ```

$$\begin{pmatrix} 3 & 2 & 5 & 4 \\ 2 & 5 & 3 & 8 \\ 1 & 2 & 3 & 0 \\ 2 & 3 & 4 & 5 \end{pmatrix}$$

Now, we can once again trace Cramer's rule. Note that MuPAD reduces the repeating terms, so we cannot really see all the terms that were produced in each step, but this is only a minor disadvantage in our procedure.

- Det(B)

$$5Det\left(\begin{pmatrix} 2 & 5 & 8 \\ 1 & 2 & 0 \\ 2 & 3 & 5 \end{pmatrix}\right) - 2Det\left(\begin{pmatrix} 2 & 3 & 8 \\ 1 & 3 & 0 \\ 2 & 4 & 5 \end{pmatrix}\right) -$$

$$4Det\left(\begin{pmatrix} 2 & 5 & 3 \\ 1 & 2 & 3 \\ 2 & 3 & 4 \end{pmatrix}\right) + 3Det\left(\begin{pmatrix} 5 & 3 & 8 \\ 2 & 3 & 0 \\ 3 & 4 & 5 \end{pmatrix}\right)$$

- eval(%)

$$28Det\left(\begin{pmatrix} 1 & 2 \\ 2 & 3 \end{pmatrix}\right) - 19Det\left(\begin{pmatrix} 1 & 0 \\ 2 & 5 \end{pmatrix}\right) +$$

$$Det\begin{pmatrix} 2 & 0 \\ 3 & 5 \end{pmatrix} + 4Det\left(\begin{pmatrix} 1 & 3 \\ 2 & 4 \end{pmatrix}\right) +$$

$$11Det\left(\begin{pmatrix} 3 & 0 \\ 4 & 5 \end{pmatrix}\right) + 16Det\left(\begin{pmatrix} 2 & 3 \\ 3 & 4 \end{pmatrix}\right)$$

- eval(%)

$$36Det((4)) - 20Det((3)) - 68Det((2)) + 16Det((5))$$

- eval(%)

28

And here is the final test if everything went right.

- `linalg::det(B)`

  28

Now it is your turn. You can try the `Det` procedure on your examples. You can also introduce a number of improvements in it. For example, it would be good to have some kind of type checking for input variables, and a few other forms of protection. ■

I cannot continue forever exploring linear algebra with MuPAD. For any book, there must come a time to stop writing and let the readers continue exploring the topic on their own. So, I will stop writing my book just here. If you liked this book, let me know about it. You can find my e-mail address on the book's web site.

14.5 Chapter Summary

In this chapter, I have introduced a large collection of MuPAD procedures. I have shown you how to use them to create matrices, solve systems of linear equations and to calculate determinants. Let me summarize all the procedures that were introduced in this chapter.

14.5.1 Procedures to Solve Systems of Linear Equations

linsolve([*lin. equations*], [*variables or expressions*], *domain declaration*)

linalg::matlinsolve(*A,B*) - *solve system of equations in matrix form*

14.5.2 Procedures to Create Matrices

matrix(*n, m,* [[*a1,...,am*], [*b1,...,bm*], ..., [*d1,...,dm*]]) - *create matrix*

linalg::expr2Matrix(*equations,* [*x,y,z*]) - *transforms system of lin.eq. into a matrix*

linalg::randomMatrix(*n, m, domain, bound, options*) - *creates a random matrix*

linalg::hilbert(n)- *produces nxn Hilbert matrix generated by the function*

   $h(i,j) = (i + j - 1)^{-1}$

linalg::invhilbert(n)- *produces inverse of the nxn Hilbert matrix generated by*

   the function $h(i,j) = (i + j - 1)^{-1}$

14.5.3 Global operations on Matrices from stdlib

diff(A, x) - *differentiate all the entries in respect to variable x*

expand(A) - *expand all the entries of A*

float(A) - *apply float to all the entries of A*

int(A, x) - *integrate all the entries of A*

subs(A, $X=expr$) - *substitute an expr in A in place of all the occurrences of X*

map(A, *function*) - *apply function to all entries of the matrix A*

zip(A, B, f) - *produces the matrix $C_{ij} = f(a_{ij}, b_{ij})$*

14.5.4 Global Operations on Matrices from linalg

linalg::concatMatrix($A,B,...,C$) - *joins matrices horizontally*

linalg::stackMatrix($A,B,...,C$) - *joins matrices vertically*

linalg::transpose(A) - *produces transposition of A, i.e. A^t,*

linalg::gaussElim(A) - *performs Gaussian elimination of a matrix,*

linalg::gaussJordan(A) - *performs Gauss-Jordan elimination*

linalg::crossProduct(A,B) - *produces cross product of a 3D vectors*

linalg::normalize(V) - *normalizes a vector*

linalg::charmat(A,x) - *produces characteristic matrix*

linalg::angle(A,B) - *calculates the angle between two vectors*

linalg::charpoly(A, x) - *produces the char. polynomial of a matrix*

linalg::eigenvalues(A) - *produces eigenvalues of a matrix*

linalg::eigenvectors(A) - *produces eigenvectors of a matrix*

linalg::matdim(A) - *returns the dimensions of the matrix*

linalg::ncols(A) - *returns the number of columns in the matrix A*

linalg::nonZeros(A) - *returns the number of $\neq 0$ elements in a matrix*

linalg::nrows(A) - *returns the number of rows in the matrix A*

linalg::scalarProduct(A, B) - *produces the scalar product of two vectors*

linalg::vecdim(V) - *returns the number of elements of a vector*

14.5.5 Local Matrix Operations

linalg::addCol(A, $c1$, $c2$, s)- *replaces in A c2 by c2+s*c1*

linalg::addRow(A, $r1$, $r2$, s)- *produce a copy of matrix r2=r2+s*r1*

linalg::col(A, c) - *extracts column c from matrix A*

linalg::row(A, r) - *extracts row r from the matrix A*

linalg::delCol(A, $c1..c2$) - *deletes in A columns c1 through c2*

linalg::delRow(A, $r1..r2$) - *deletes in matrix A rows r1 through r2*

linalg::multCol(A, c, s) - *multiples in A column c by number s*

linalg::multRow(A, r, s) - *multiples in matrix A row r by number s*

linalg::setCol(A, c, $c1$) - *replaces in A column c by a new column*

linalg::setRow(A, r, $r1$) - *replaces in A row r by new row r1*

linalg::submatrix(A,$r1..r2$,$c1..c2$) - *extract submatrix defined by ranges*

linalg::substitute(A, B, m, n) - *substitutes matrix B inside of matrix A,*
  element B[1,1] will be placed as A[m,n]

linalg::swapCol(A, $c1$, $c2$) - *swaps columns c1 and c2 in matrix A*

linalg::swapRow(A, $r1$, $r2$) - *swaps rows r1 and r2 in matrix A*

14.5.6 Procedures to Calculate Determinants

linalg::det(A) - *determinant of matrix A*

numeric::det(A) - *determinant of matrix A, uses numerical proc.*

14.6 Programming Exercises

1. Write the MuPAD declarations for the following systems of
   linear equations and use the `linsolve` procedure to obtain their
   solutions. Analyze the output you get. Some of these systems
   might not have solutions — why?

   a. $3x - 2y = -1$, $4x + 5y = 3$, $7x + 3y = 2$

   b. $2x + 2z = 1$, $3x - y + 4z = 7$, $6x + y - z = 0$

   c. $3x + 2z = 2$, $x - y + 4z = 1$, $6x + y - z = 0$

d. $2x + 2z = 1$, $3x - y + 4z = 7$, $2x + 2z = 0$

e. $2x + z = 1$, $x - y + 4z = 1$, $6x + y - z = 0$, $3x + y = 1$

f. $x + 2y - t + u = 1$, $3y + z - u = 2$, $z + 7t = 1$

g. $x + y = 1$, $y + z = 2$, $z + t = 3$, $t + u = 4$, $u + v = 5$

h. $y + z + t = 1$, $x + y + z = 1$, $x + y + t = 1$, $x + z + t = 1$,

2. Find a way to solve the following systems of equations using the methods learned in this chapter.

a. $x^2 + y^2 + z^2 = 6$

$2x^2 + y^2 - z^2 = 3$

$x^2 - y^2 - 2z^2 = 2$

b. $2x^2 + y^2 + z^2 = 0$

$x^2 + y^2 - 3z^2 = 3$

$x^2 - y^2 + 2z^2 = 2$

c. $x^2 + y^2 + z^2 = 3$

$x^2 + y^2 - z^2 = 3$

$x^2 - y^2 + 2z^2 = 2$

d. $\sin x + \sin y + \sin z = 2$, for $x, y, z \in [0, 2\pi]$

$2 \sin x - 3 \sin y - \sin z = 1$

$5 \sin z = 1$

e. $\sin x + \sin y + \sin z = 3$, for $x, y, z \in [0, 2\pi]$

$\sin x + \sin y - \sin z = 1$

$\sin x - \sin y - \sin z = -1$

3. For each of the systems of linear equations from exercise 1, use the procedure `linalg::expr2Matrix` to develop an augmented matrix and solve it using the `linalg::matlinsolve` procedure.

4. Use the local matrix operations available in MuPAD to perform

step by step a Gauss-Jordan elimination of matrices. Use the procedure `linalg::gaussJordan` to check if your results are similar. Supposing that the matrices below are augmented matrices of a system of linear equations, use back substitution to obtain the solutions for each system, if such solutions exist.

a.
$$\begin{pmatrix} 0 & 1 & 1 & 1 \\ 1 & 0 & 1 & 1 \\ 1 & 1 & 0 & 1 \end{pmatrix}$$

b.
$$\begin{pmatrix} 0 & 1 & 0 & 1 \\ 1 & 0 & 1 & 1 \\ 0 & 1 & 0 & 1 \end{pmatrix}$$

c.
$$\begin{pmatrix} 1 & 2 & 3 & 1 \\ 2 & 3 & 1 & 1 \\ 3 & 1 & 2 & 1 \end{pmatrix}$$

d.
$$\begin{pmatrix} 0 & a & b & 1 \\ a & 0 & a & 1 \\ b & a & 0 & 1 \end{pmatrix}$$

5. For this exercise, consider the matrices
$$A = \begin{pmatrix} 1 & 2 & 3 \\ 2 & 3 & 1 \\ 3 & 2 & 1 \end{pmatrix}, \quad B = \begin{pmatrix} 0 & 2 & 3 \\ 2 & 0 & 1 \\ 3 & 2 & 0 \end{pmatrix}, \quad C = \begin{pmatrix} 0 & 2 & 3 \\ 1 & 0 & 3 \\ 1 & 2 & 0 \end{pmatrix}$$

Compute, where possible:

a. $A \cdot B + A \cdot C$

b. $(A + B) \cdot (A + C)$

c. $(A \cdot B + A \cdot C)^{-1}$

d. $(A * B * C)^{-1}$

e. `map(A, f)`, where $f(x) = x^2$,

f. `map(B, f)`, where $f(x) = \sin(x\pi)$,

g. $zip(A,B,g)$, where $g(x,y) = x^2 y^2$

6. Use the Det procedure introduced in this chapter to trace how the following determinants are calculated.

a.
$$\begin{vmatrix} 1 & 2 & 3 \\ 2 & 1 & 3 \\ 3 & 3 & 1 \end{vmatrix}$$

b.
$$\begin{vmatrix} a & b & c \\ b & c & a \\ c & a & b \end{vmatrix}$$

c.
$$\begin{vmatrix} (a-b) & b & b & b \\ b & (a-b) & b & b \\ b & b & (a-b) & b \\ b & b & b & (a-b) \end{vmatrix}$$

d.
$$\begin{vmatrix} a & 0 & 0 & 0 \\ b & a & 0 & 0 \\ b & b & a & 0 \\ b & b & b & a \end{vmatrix}$$

7. Modify the procedure Det so that it will:

a. calculate the determinant using the first column of the matrix,

b. calculate the determinant using any row of the matrix selected by the user,

c. calculate the determinant using any column of the matrix selected by the user.

Appendix 1 ─────────────────────

MuPAD Libraries and Procedures

In this appendix, I have included all the libraries available in MuPAD, and almost all the procedures that are available in the most recent version of MuPAD at the time of writing. I have tried to make this list as complete as possible. Still, I could have missed a few things, as there are many procedures that are considered as internal. However, the border between internal procedures and those designated to be used by MuPAD users can be a bit dim. Therefore, it may happen that some of the procedures described here might be internal and some others might become internal in the future. In fact, from the user's point of view there is little difference between internal and official procedures. You can use them, as long as you know their syntax and where to find them.

For many procedures, I have provided a short syntax and description that may help the readers to identify the role of the procedure or find a procedure for a specific goal. You need to keep in mind that most of these procedures can be used in a wide variety of ways and with a number of parameters. It is therefore not possible to describe in a small chapter all the forms of syntax and parameters that can be applied to them. My intention when developing this appendix was to provide you with the most basic information about what you can find in MuPAD's libraries and in the whole MuPAD system.

There is a huge number of procedures and functions that I have grouped under the title "MuPAD Standard Collection". In fact, these procedures are not always included in any particular library. They are placed in various parts of the MuPAD system including kernel functions and those in standard library. The common feature for all of them is that you can access them directly without using the slot operator regardless of where they really are. Thus, if you can use a procedure like `float(1+exp(1))` rather than like `combinat::bell(15)`, then I will consider this procedure as part of the so-called standard collection.

Many of the procedures listed here were not even mentioned in my book. You can find information about them using MuPAD's help files. The order of libraries and procedures is alphabetic.

A1.1 MuPAD Libraries (version 3.0, 21/02/2004)

adt - *basic collection of abstract data types*

Ax - *basic axiom constructors*

Cat - *basic category constructors*

combinat - *functions for combinatorics*

detools - *tools for differential equations*

Dom - *domain constructors*

fp - *utilities for functional programming*

generate - *utilities to generate foreign formats from expressions*

Graph - *utilities for handling directed and undirected graphs*

groebner - *utilities to calculate Gröbner-bases for polynomial ideals*

import - *utilities for reading data in different formats*

intlib - *utilities for symbolic integration*

linalg - *the linear algebra package*

linopt - *the package for linear optimization*

listlib - *utilities for list operations*

matchlib - *tools for pattern matching*

module - *utilities for module management*

numeric - *functions for numerical mathematics*

numlib - *the package for elementary number theory*

orthpoly - *tools for orthogonal polynomials*

ode - *the library for ordinary differential equations*

output - *utilities for the output of data*

plot - *graphical primitives and functions for two- and three-dimensional plots*

polylib - *utilities for polynomials*

prog - *programming utilities*

property - *properties of identifiers*

RGB - *color functions and definitions*

Series - *tools and data structures for working with series*

solvelib - *methods for solving equations, systems of equations and inequalities*

stats - *statistical functions*
stringlib - *utilities for working with strings*
student - *the student package*
transform - *the library for integral transformations*
Type - *utilities for type expressions and properties*

A1.2 Operators Represented by Symbols

:= - *assign a value to a variable*
+, -, /, *, ^ - *arithmetical operations*
! - *factorial function*
!! - *double factorial function*
' - *derivative of a function*
=, <, <=, >=, > - *equality and inequality relations*
-> - *declaration of a function*
==>, <=> - *Boolean operators representing implication and equivalence*
. - *the dot operator to concatenate two lists or strings*
.. - *the range operator, 2..5*
... - *range operator, PI...5.1, returns a floating point interval including its ends*
@ - *composition of functions*
@@ - *the operator to iterate a function given number of times*
$ - *the operator to create a sequence*
?word - *displays help for a given word*
:: - *the slot operator, access to object slots*

A1.3 MuPAD Standard Collection

A1.3.1 Procedures in Standard Collection

This section lists the procedures that can be applied to objects other than numbers. All the arithmetic functions are listed in section A1.3.2

alias(x = object) - *defines x as an alias of a given object*
anames(All) - *returns identifiers that have values or properties in the current session of MuPAD*
and, or, not, xor - *Boolean operators*
args(x) - *function accessing procedure parameters*

array($k1..n1, k2..n2, ...$) - *creates an array*

assert(*condition*) - *declares a condition to be true at the moment when the statement is evaluated*

assign(*List*) - *assigns values given in the form of list of equation(s)*

assignElements($L, i = v, ..$) - *assigns values to entries of a list, array*

assume(x, *property*) - *assigns a mathematical property to a MuPAD object*

asympt(f, x) - *computes asymptotic series expansion*

bool(*expr*) - *produces Boolean value of the given expression*

break - *the procedure terminating execution of a loop or a case structure*

bytes() - *returns the current memory use*

C_ - *the set of complex numbers*

card(*set*) - *produces cardinality for a given set*

coeff(p) - *returns sequence of non-zero coefficients of a polynomial*

coerce(*object, U*) - *tries to convert object into an object of a domain U*

collect(p, x) - *collects coefficients of a given polynomial expression*

combine(*expr*) - *combines the terms of a given expression into a single power*

complexInfinity - *the constant representing infinity in complex numbers*

conjugate(z) - *produces the conjugate of a complex number*

contains(A, *object*) - *checks if a given element A is contained inside of a container object*

content(p) - *computes the content of the polynomial, i.e. gcd of its coefficients*

context(*object*) - *evaluates object in the given context of the calling procedure*

contfrac(x) - *produces continued fraction of a given number*

D(f) - *differential operator, equivalent to f, also partial derivative*

debug() - *starts MuPAD debugger*

degree(p, x) - *returns degree of the polynomial p with respect to x*

degreevec(p) - *returns a list of exponents of the leading term of a polynomial*

delete $x1, x2, ..xn$ - *deletes values of the given identifiers*

denom(*expr*) - *produces denominator of a given rational expression*

diff(f, x) - *produces the derivative of a function with respect to a given variable*

discont(f, x) - *produces all the discontinuities of a function f(x)*

divide($p(x), q(x)$) - *divides an univariate polynomial p(x) by q(x), returns quotient and reminder from division*

domtype(*object*) - *returns domain type of the given object*

doprint(*object*) - *prints large matrices on the screen*

error("*message*") - *breaks running procedure and produces error message*

eval(*object*) - *evaluates the given object*

evalassign(*x, value, depth*) - *evaluates x with the given depth and assigns value to the result*

evalp(*p, x = x0*) - *evaluates polynomial p for x = x0*

expand(*expression*) - *expands an arithmetical expression*

export(*library, procedures*) - *exports procedures from a given library*

expose(*procedure*) - *displays the source code of a given procedure or domain*

expr(*object*) - *converts object into an element of a basic domain*

expr2text(*object*) - *converts object into a string of characters*

external(*..*) - *returns the module function environment*

extnops(*object*) - *returns the number of operands of the given object in internal representation*

extop(*object*) - *returns all the operands of a domain element*

extsubsop(*d, i = newel*) - *produces a copy of the domain element with replaced i-th operand*

factor(*p*) - *factors polynomial into irreducible polynomials*

Factored(*f*) - *the domain of objects in factored form*

fclose(*n*) - *closes the file with descriptor n*

finput(*filename*) - *reads MuPAD objects from the given binary or ASCII file*

fname(*n*) - *returns the name of the file with specified descriptor*

fopen(*filename*) - *opens the file with the given name*

fprint(*filename, objects*) - *writes MuPAD objects into a file*

frandom() - *the floating point number random function*

fread(*filename*) - *reads and executes the specified MuPAD file*

freeze(*f*) - *creates an inactive copy of the function f*

ftextinput(*filename, x*) - *reads a line from a text file and assigns it to the identifier x*

funcenv(*f*) - *creates a function environment*

gcd(*p,q,..*) - *produces the greatest common divisor of polynomials*

gcdex(*p, q, x*) - *the extended Euclidean algorithm for polynomials*

genident() - *create a new identifier that was not used before in the current session*

genpoly(*n, b, x*) - *creates a polynomial p with variable x such that p(b) = n*

getpid() - *returns the ID of the running MuPAD process under the UNIX operating system*

getprop(*expr*) - *returns mathematical property of a given expression* ground(p) - *returns the constant coefficient of the polynomial p*

has(*object1, object2*) - *procedure checks syntax of objects to determine if object1 is part of the object2*

hastype(*object, type*) - *checks if a given object has a specified type*

help("*word*") - *the procedure to display help about a given word*

history(n), history() - *returns the n-th entry (or last one) in the history table*

hold(*object*) - *prevents evaluation of the given object*

hull(*object*) - *produces a floating point interval enclosing object*

icontent(p) - *computes the content of the polynomial with integer or rational coefficients*

id(x) - *evaluates x and returns result of evaluation*

in - *checks if an object is a an element of a given set, syntax: x in set*

indets(*expr*) - *produces indeterminates of the given expression*

indexval(x, i) - *accesses to entries of arrays and tables without evaluation*

infinity - *the constant representing infinity*

info(), info(*name*) - *display short information about given object*

input(..) - *interactive input of data to a MuPAD program*

int(f), int($f, x = a..b$) - *produces the definite/indefinite integral of a function*

int2text(n) - *converts integer n into a string of characters*

interpolate(*xList, yList*) - *computes an interpolating polynomial*

irreducible(p) - *tests if the given polynomial is irreducible*

is($x, property$) - *checks if the object has the given mathematical property*

isprime(n) - *checks if the given number is a prime number*

iszero(*object*) - *checking if the given object is zero element in its domain*

ithprime(i) - *produces the i-th prime number*

last(n) - *accesing the previously computed object, equivalent to n*

lasterror() - *reproduces the last error in the current session*

lcm(p, q) - *produces the least common multiplier of polynomials*

lcoeff(p) - *returns the leading coefficients of the polynomial*

ldegree(p,x) - *lowest degree of the terms of the polynomial p in respect to variable x*

length(*object*) - *returns an integer number representing the complexity of an object*

level(*object, n*) - *evaluates object until level of substitution n*

lhs(*equation*) - *returns left side of a given equation*

limit($f, x = x0$) - *computes the limit of a given expression* linsolve([*equations*],[*variables*]) - *solves a system of linear equations*

lllint(*A*) - *procedure applying LLL algorithm to the columns of a matrix*

lmonomial(*p*) - *returns the leading term of a polynomial*

loadmod("*name*") - *loads the specified dynamic module*

loadproc(*object, path, file*) - *loads MuPAD object from a specified file*

lterm(*p*) - *returns the leading term of a given polynomial*

map(*object, f*) - *applies a function to all operands of a given object*

mapcoeffs(*p, f, a1,a2,..*) - *procedure applies function f to the polynomial p replacing coefficients by f(c,a1,a2,..)*

maprat(*object, f*) - *applies a function to the rationalized object*

matrix(*m, n, [elements]*) - *procedure produces matrix with specified dimensions and elements*

monomials(*p*) - *sorted list of monomials of a polynomial p*

multcoeffs(*p, c*) - *multiplies all coefficients of the polynomial p by c*

new(*T, object1, object2...*) - *creates a new elements of the domain T with a given internal representation*

newDomain(*k*) - *creates a new domain with the key k*

nextprime(*n*) - *produces the smallest prime p such that n ≤ p*

NIL - *the nil object*

nops(*object*) - *number of operands of the given object*

norm(*M*) - *a norm of the matrix M*

normal(*x*) - *returns the normal form of the rational expression*

nterms(*p*) - *returns the number of terms of a polynomial*

nthcoeff(*p*) - *returns the n-th non zero coefficient of a polynomial*

nthmonomial(*p, n*) - *returns the n-th non-trivial monomial of the polynomial*

nthterm(*p, n*) - *returns the n-th non-zero term of the polynomial*

null() - *returns empty sequence of MuPAD expressions*

numer(*expression*) - *returns the numerator of the given rational expression*

O - *domain of Landau symbols*

ode(*equation, y(x)*) - *declares equation as a differential equation*

op(*object*) - *operands of the given object*

operator(*symb, f, type, priority*) - *declare a new operator defined by the given function f*

package(*directory*) - *loads a new library package*

pade(*f, x*) - *computes a Pade approximation of the given expression f* partfrac(*expr, x*) - *produces partial fractions of given rational expression*

patchlevel() - *returns the number of patch of the current MuPAD library*

pathname(*directory*) - *produces path name valid in the current o.s.*

pdivide(*p, q*) - *the pseudo-division operation for univariate polynomials p and q*

piecewise([*cond, f*], [*cond1, f1*],..) - *returns the piecewise function*

plot(objects) - *the procedure to plot graphical objects*

plotfunc2d(*f, x = a..b*) - *plots a graph of function of one variable*

plotfunc3d(*f, x = a..b, y = c..d*) - *plots a graph of function of two variables*

poly(*p, [x]*) - *declares polynomial p*

poly2list(*p*) - *converts polynomial p into a list of coefficients and exponents*

powermod(*p, a, m*) - *produces a modular power of a polynomial $p^a \bmod m$*

print(*object*) - *prints object on the screen*

product(*f(i), i = a..b*) - *produces definite and indefinite products*

protect(*identifier*) - *protects the given identifier*

protocol(*filename*) - *creates a protocol of the MuPAD session*

Q_ - *the set of rational numbers*

R_ - *the set of real numbers*

radsimp(*expr*) - *simplifies radicals in an arithmetical expression*

rationalize(*object*) - *transforms an expression into a rational expression*

read(*filename*) - *searches, reads and executes a file*

readbytes(*filename*) - *reads data from a binary file*

rec(*equation, y(n)*) - *represents a recurrence equation for the sequence y(n)*

rectform(*z*) - *produces the rectangular form of a complex expression*

reset() - *reinitializes the current MuPAD session*

return(*X*) - *terminates execution of the procedure and returns object X*

revert(*list*) - *reverses the order of elements in a given list, string or series*

rewrite(*expr, newfun*) - *transforms expression into an equivalent form using specified components*

rhs(*equality*) - *the right side of an equality*

RootOf(*p, x*) - *the roots of a polynomial*

rtime() - *returns the total real time of the current session, or executing a command(s)*

Rule(*pattern, replacement*) - *defines the equivalence rules for mathematical expressions*

save *X* - *saves the state of the identifier X*

select(*object, f*) - *selects operand of a given object*

series(*f, x*) - *produces a series expansion of a given expression*

setuserinfo(*f, n*) - *sets the information level*

simplify(*expr*) - *simplifies a given expression*

Simplify(*expr*) - *more powerful version of* simplify

slot(*d, "name"*) - *returns the value of the named slot of the object d*

slot(*d, "name", v*) - *creates the value of the named slot of the object d*

solve([*equations*], [*var*]) - *solves an equation or system of equations*

sort(*list*) - *sorts a given list*

split(*object, f*) - *splits object into a list of three objects*

strmatch(*text, pattern*) - *checks if the given patterns occurs in the string*

subs(*f, prev = new*) - *substitutes a new value into f*

subsex(*f,prev = new*) - *substitute expression*

subsop(*object,i = new*) - *substitutes the i-th operand*

substring(*str, i*) - *returns a substring of a given string*

sum(*f, n*) - *produces definite or indefinite sum*

system("*command*") - *executes a command of the operating system*

table(), table(*ind = obj,...*) - *creates a new empty table or table with specified elements*

taylor(*f, x = x_0*) - *produces the Taylor series of a function around a given point*

tbl2text(*strings*) - *concatenates strings in a table*

tcoeff(*p*) - *produces the trailing coefficient of the polynomial*

testeq(*exp1, exp2*) - *checks the mathematical equivalence of two given expressions*

testtype(*object, T*) - *checks if the given object has a syntactical type T*

text2expr(*text*) - *converts a given string into a MuPAD object*

text2int(*text*) - *converts a string of characters into an integer number*

text2list(*text, separators*) - *splits text into a list of substrings*

text2tbl(*text, separators*) - *splits text into a table of substrings*

textinput(*x*) - *the procedure allowing interactive input of text*

time() - *returns the total execution time of the current session*

traperror(*object*) - *traps errors produced while evaluating given object*

TRUE, FALSE, UNKNOWN - *Boolean constants*

type(*object*) - *returns the type of the given object*

unalias(*x*) - *deletes the alias x*

unassume(x) - *removes properties of the given variable*

undefined - *the constant representing an undefined object*

unexport(*library*) - *undoes the export of a library*

unfreeze(*object*) - *creates an active copy of the object with frozen function (see freeze)*

universe - *the constant representing the universe set*

unloadmod("*module*") - *unload the dynamic module*

unprotect(x) - *removes protection of x*

val(*object*) - *replaces every identifier in the given object by its value*

version() - *returns the version number of the MuPAD library*

warning("*message*") - *prints the specified warning message*

write(*file, values*) - *writes given values to the specified file*

writebytes(*filename*) - *writes data to a binary file*

Z_ - *the set of integer numbers*

zip(*l1, l2, ..*) - *combine listsor matrices*

A1.3.2 Functions in Standard Collection

abs(x) - *the absolute value function*

airyAi(z) - *the Airy function*

airyBi(z) - *the Airy function*

arccos(x) - *the inverse of the cosine function*

arccosh(x) - *the inverse of the hyperbolic cosine function*

arccot(x) - *the inverse trigonometric cotangent function*

arccoth(x) - *the inverse hyperbolic cotangent function*

arccsc(x) - *the inverse of the cosecant function*

arccsch(x) - *the inverse of the hyperbolic cosecant function*

arccsc(x) - *the inverse of the cosecant function*

arcsec(x) - *the inverse of the secant function*

arcsech(x) - *the inverse of the hyperbolic secant function*

arcsin(x) - *the inverse of the sine function*

arcsinh(x) - *the inverse of the hyperbolic sine function*

arctan(x) - *the inverse of the tangent function*

arctanh(x) - *the inverse of the hyperbolic tangent function*

arg(x) - *the argument of the complex number*

bernoulli(n) - *produces n-th Bernoulli number*

bernoulli(n, x) - *produces n-th Bernoulli polynomial of x*

besselI(v, z) - *the modified Bessel function*

besselJ(v, z) - *the Bessel function of the first kind*

besselK(v, z) - *the modified Bessel function*

besselY(v ,z) - *the Bessel function of the second kind*

beta(x, y) - *the beta function*

binomial(n, k) - *binomial function n over k*

ceil(x) - *returns the smallest integer n such that $x \leq n$*

Ci(x) - *the cosine integral function*

cos(x), sin(x), tan(x), cot(x), sec(x), csc(x) - *trigonometric functions*

cosh(x), sinh(x), tanh(x), coth(x), csch(x), - *hyperbolic functions*

dilog(x) - *the dilogarithm function*

dirac(x) - *the Dirac delta distribution*

dirac(x, n) - *the n-th derivative of the delta distribution*

div - *produces results of integer division*

Ei(x) - *exponential integral function*

erf(x) - *the error function*

erfc(x) - *the complementary error function*

exp(x) - *the exponential function e^x*

fact(n) - *factorial of given integer, same as n!*

factor(n) - *factors an integer number*

fract(x) - *fractional part of the number x*

floor(x) - *returns the largest integer n such that $n \leq x$*

float(x) - *returns floating point version of the given number x*

frac(x) - *the fractional part of the number x*

gamma(x) - *the gamma function*

gcd($n,m,..$) - *produces the greatest common divisor of if given integer numbers*

heaviside(x) - *the Heaviside step function*

hypergeom([$a1,..,an$],[$b1,..,bk$], z) - *hypergeometric function*

ifactor(n) - *produces prime factorization of n*

igamma(a, x) - *the incomplete Gamma function*

igcd($n1$, $n2,..$, nk) - *produces the greatest common divisor of integers*

igcdex(x, y) - *produces the gcd of integers using Euclidean algorithm* ilcm(*n1, n2,.., nk*) - *produces the least common multiple of integers*

Im(*z*) - *the imaginary part of a complex number*

iquo(*m, n*) - *produces the result of integer division m by n*

irem(*m, n*) - *produces the reminder of integer division m by n*

isqrt(*n*) - *produces integer approximation of the square root of integer n*

lambertV(*x*) - *produces the lower real branch of the Lambert function*

lcm(*p, q*) - *produces the least common multiple of integers*

ln(*x*), log(*n, x*) - *logarithms* ln(*x*) *and* log$_n$(*x*)

max(*x1, x2,..*) - *produces the maximum of given numbers*

min(*x1, x2,..*) - *produces the minimum of given numbers*

mod - *modulo function, x mod m*

modp(*x, m*) - *the division of x modulo m*

mods(*x, m*) - *produces the least absolute value r such that x-r is divisible by m*

meijerG(*lists of numbers*) - *the Meijer G function*

polylog(*n, x*) - *the polylogarithm function of index n*

psi(*x*) - *the digamma function*

random(), random(*n1..n2*) - *the random number generator*

Re(*z*) - *produces the real part of a complex number*

round(*x*) - *rounds x to the closest integer number*

Si(*x*) - *the sine integral function*

sign(*x*) - *the sign function*

signIm(*z*) - *the sign of the imaginary part of z*

sqrt(*x*) - *the function* $\sqrt{x}$

surd(*z,n*) - *returns the n-th root of z whose argument is closest to the argument of z, n must be a non-zero integer*

trunc(*x*) - *returns the integer part of x*

whittakerM, whittakerW - *the Whittaker functions*

zeta(*z*) - *the Riemann zeta function*

A1.4 Library 'adt' — Abstract Data Types

adt::Heap() - *abstract data type heap*

adt::Queue(*queue elements*) - *abstract data type queue*

adt::Stack(*stack elements*) - *abstract data type stack*

adt::Tree(*tree*) - *abstract data type tree*

A1.5 Library 'Ax' — Basic Axiom Constructors

Ax::canonicalOrder - *the axiom of canonically ordered sets*

Ax::canonicalRep - *the axiom of canonically representation*

Ax::canonicalUnitNormal - *the axiom of canonically unit normals*

Ax::closedUnitNormals - *the axiom of closed unit normals*

Ax::efficientOperation - *the axiom of efficient operations*

Ax::indetElements - *the axiom saying that indeterminetes may be elements*

Ax::noZeroDivisors - *the axiom of rings with no zero divisors*

Ax::normalRep - *the axiom of normal representation*

Ax::systemRep - *the axiom of facade domains*

A1.6 Library 'Cat' — Category Constructors

Cat::AbelianGroup - *the category of abelian groups*

Cat::AbelianMonoid - *the category of abelian monoids*

Cat::AbelianSemiGroup - *the category of abelian semi-groups*

Cat::Algebra - *the category of associative algebras*

Cat::AlgebraWithBasis - *the category of associative algebras with distinguished basis*

Cat::BaseCategory - *the base category*

Cat::CancellationAbelianMonoid - *the category of abelian monoids with cancellation*

Cat::CombinatorialClass - *the category of combinatorial classes*

Cat::CommutativeRing - *the category of commutative rings*

Cat::DifferentialFunction - *the category of differential functions*

Cat::DifferentialRing - *the category of ordinary differential rings*

Cat::DifferentialVariable - *the category of differential variables*

Cat::EntireRing - *the category of entire rings*

Cat::EuclideanDomain - *the category of euclidean domains*

Cat::FactorialDomain - *the category of factorial domains*

Cat::Field - *the category of fields*

Cat::FiniteCollection - *the category of finite collections*

Cat::GcdDomain - *the category of integral domains with gcd*

Cat::Group - *the category of groups*

Cat::HomogeneousFiniteCollection - *the category of homogeneous finite collections*

Cat::HomogeneousFiniteProduct - *the category of homogeneous finite products*

Cat::IntegerListsLexClass - *the category of combinatorial classes*

Cat::IntegralDomain - *the category of integral domains*

Cat::LeftModule - *the category of left R-modules*

Cat::Matrix - *the category of matrices*

Cat::Module - *the category of R-modules*

Cat::ModuleWithBasis - *the category of modules over distinguished basis*

Cat::Monoid - *the category of monoids*

Cat::OrderedSet - *the category of ordered sets*

Cat::PartialDifferentialRing - *the category of partial differential rings*

Cat::Polynomial - *the category of multivariate polynomials*

Cat::PrincipalIdealDomain - *the category of principal ideal domains*

Cat::QuotientField - *the category of quotient fields*

Cat::RightModule - *the category of right R-modules*

Cat::Ring - *the category of rings*

Cat::Rng - *the category of rings without unit*

Cat::SemiGroup - *the category of semi-groups*

Cat::SemiRing - *the category of semi-rings*

Cat::SemiRng - *the category of semi-rings without unit*

Cat::Set - *the category of sets of complex numbers*

Cat::SkewField - *the category of skew fields*

Cat::SquareMatrix - *the category of square matrices*

Cat::UnivariatePolynomial - *the category of univariate polynomials*

Cat::VectorSpace - *the category of vector spaces*

A1.7 Library 'combinat' — Combinatorial Functions

The combinat library is based on an open source project run by MuPAD users in France and coordinated by Nicholas M.Thiéry. More information about the project is available at

http://mupad-combinat.sourceforge.net/Wiki/

Currently, the project library contains functions to deal with usual combinatorial structures, symmetric functions, Schubert polynomials, characters of the symmetric group, and tools for constructing your own combinatorial algebras.

The core part of this library was integrated in the official MuPAD combinat library from version 2.5.0 onwards.

combinat::bell(*n*) - *computes the n-th Bell number*

combinat::binaryTrees(*list*) - *the function for generating and manipulating binary trees*

combinat::cartesianProduct(*set1, ...setN*) - *produces Cartesian product of given sets, result is a list*

combinat::catalan(*n*) - *produces Catalan numbers*

combinat::choose(*set, k*) - *computes all k-subsets of a given set*

combinat::composition(*n, k*) - *computes k-composition of an integer n*

combinat::composition(*n*) - *computes all compositions of an integer*

combinat::compositions - *functions for generating, counting and manipulating compositions*

combinat::decomposableObjects (*spec*) - *functions for generating, counting and drawing decomposable combinatorial objects*

combinat::dyckWords(*n*) - *words of zeros and ones*

combinat::generators - *sources of infinite streams of objects*

combinat::integerListLexTools - *internal library with functions for lexicographic generation of list of integers*

combinat::integerMatrices - *functions for counting, generating and manipulating integer matrices*

combinat::integerVectors(*n, m*) - *integer vectors of length m build from elements 0,...,n*

combinat::integerVectorsWeighted(*n, list*) - *weighted integer vectors*

combinat::labelledBinaryTrees - *functions for creating binary search trees*

combinat::linearExtensions - *functions for generating, counting and manipulating linear extensions of directed acyclic graphs*

combinat::modStirling(*q, n, k*) - *computes modified Stirling numbers*

combinat::nonCrossingPartitions - *functions for non crossing set partitions*

combinat::partitions(*n*) - *computes number of partitions of a given integer*

combinat::permutations(*list*) - *functions for generating, counting and manipulating permutations*

combinat::skewPartitions - *functions for generating, counting and manipulating skew partitions of a fixed size* combinat::permute(*list*) - *produces all permutations of a list (command obsolete)*

combinat::powerset(*set*) - *produces powerset of a given set or a list, command obsolete*

combinat::stirling1(*n, k*) - *computes Stirling numbers of the first kind*

combinat::stirling2(*n, k*) - *computes Stirling numbers of the second kind*

combinat::subsets(*set*) - *produces all subsets of a given set*

combinat::subwords(*list*) - *produces all subwords of a given set or list*

combinat::tableaux(*set*) - *Young tableaux operator*

combinat::warnDeprecated(*TRUE or FALSE*) - *determines whether to use syntax from version 2.0*

combinat::words(*n, k*) - *lists of k elements using 0,...,n integers*

A1.8 Library 'detools' — Methods for Differential Equations

detools::arbFuns(*q, alpha*) - *number of arbitrary functions in the general solution of an involutive partial differential equation*

detools::autoreduce(*sys, indvar, depvar*) - *autoreduction of a system of differential equations*

detools::cartan(*n, m, q, beta*) - *Cartan characters of a differential equation*

detools::charODESystem(*ldf, s*) - *the characteristic system of partial differential equation*

detools::charSolve(*ldf, init, pars*) - *solves partial differential equation with the method of characteristics*

detools::characteristics(*ldf, s*) - *a characteristics of partial differential equation*

detools::derList2Tree(*derlist*) - *minimal tree with a given list of derivatives as leaves*

detools::detSys(*deq, indvar, depvar*) - *determining system for Lie point symmetries*

detools::euler(*L, t, z*) - *Euler operator of variational calculus*

detools::hasHamiltonian(*vectfield, q, p*) - *checks for Hamiltonian vector field*

detools::hasPotential(*vectfield, x*) - *checks for gradient vector field*

detools::hilbert(*alpha, r*) - *Hilbert polynomial of a differential equation*

detools::modode(*psi, depvar, indvar, step, order*) - *modified equation*

detools::ncDetSys(*difeq, indvar, depvar*) - *determining system for non-classical Lie symmetries*

detools::pdesolve(*pdifeq, indvar, depvar*) - *the solver for partial differential equations* detools::transform(*difeq, indvar, depvar, mode*) - *changes variables in differential equations*

A1.9 Library 'Dom' — Domain Constructors

Dom::AlgebraicExtension - *the domain of algebraic field extensions*

Dom::ArithmeticalExpression - *the domain of arithmetical extensions*

Dom::BaseDomain - *the base domain (is contained in all other domains)*

Dom::Complex - *the field of complex numbers*

Dom::DifferentialExpression - *the domain of differential expressions*

Dom::DifferentialFunction - *the domain of differential functions*

Dom::DifferentialPolynomial - *the domain of differential polynomials*

Dom::DihedralGroup - *the domain of dihedral groups*

Dom::DistributedPolynomial - *the domain of distributed polynomials*

Dom::Expression - *the domain of all MuPAD objects of basic type*

Dom::ExpressionField - *the domain of all expressions forming a field*

Dom::Float - *the domain of all real floating point numbers*

Dom::FloatIV - *the algebra of finite unions of rectangular intervals on the complex plane*

Dom::Fraction - *the field of all fractions with integer components*

Dom::FreeModule - *the domains of free modules*

Dom::GaloisField - *the domain of finite fields*

Dom::Ideal - *the domain of sets of ideals*

Dom::ImageSet - *the domain of images of sets*

Dom::Integer - *the ring of integer numbers*

Dom::IntegerMod - *the rings of integers modulo*

Dom::Interval - *the domain of all intervals of real numbers*

Dom::LinearDifferentialFunction - *the domain of linear differential functions*

Dom::LinearDifferentialOperator - *the domain of linear differential operators*

Dom::LinearOrdinaryDifferentialOperator - *the domain of linear differential operators*

Dom::Matrix - *the domain of all matrices*

Dom::MatrixGroup - *Abelian group of m × n matrices*

Dom::MonoidAlgebra - *the domain of monoid algebras*

Dom::MonoidOperatorAlgebra - *the domain of monoid operator algebras*

Dom::MonomOrdering - *the domain of monomial orderings*

Dom::Multiset - *the domain of multisets*

Dom::MultivariatePolynomial - *the domain of multivariate polynomials*

Dom::MultivariateSeries - *the domain of multivariate series*

Dom::Numerical - *the field of numbers*

Dom::PermutationGroup - *the domain of permutation groups*

Dom::Polynomial - *the domain of polynomials*

Dom::Product - *the domain of homogenous products*

Dom::Quaternion - *the domain of quaternions*

Dom::Rational - *the domain of rational numbers*

Dom::Real - *the field of real numbers*

Dom::SparseMatrix - *the domain of sparse matrices over the component ring R*

Dom::SparseMatrixF2 - *the domain of sparse matrices over the field with two elements*

Dom::SquareMatrix - *the rings of square matrices*

Dom::SymmetricGroup - *the domain of all permutations of* $\{1, \ldots, n\}$ *elements*

Dom::TensorAlgebra - *the domain of tensor algebras*

Dom::TensorProduct - *the domain of tensor products*

Dom::UnivariatePolynomial - *the domains of univariate polynomials*

Dom::UnivariateSkewPolynomial - *the domain of univariate skew polynomials*

Dom::VectorField - *the domain of vector fields*

A1.10 Library 'fp' — Utilities for Functional Programming

fp::apply(*f, args*) - *apply function to arguments*

fp::bottom() - *the function that never returns because it produces an error*

fp::curry(*f*) - *returns the higher order function* $x \to (y \to f(x,y))$

fp::expr_unapply(*expr, x*) - *creates a functional expression from an expression*

fp::fixargs(*f, n*) - *creates function by fixing all but n-th argument*

fp::fixedpt(*f*) - *returns fixed point of a function*

fp::fold(*f, expr*) - *creates function which iterates over sequence of arguments*

fp::nest(*f, n*) - *the repeated composition of function*

fp::nestvals(*f, n*) - *the repeated composition returning intermediate values*

fp::unapply(*expr, x*) - *creates a procedure from a given expression*

A1.11 Library 'generate' — Generate Foreign Formats

generate::C(*expr*) - *generates C code*

generate::fortran(*expr*) - *generates FORTRAN code*

generate::Macrofort - *FORTRAN code generator*

generate::MathML(*expr*) - *generates MathML code for the given expression*

generate::optimize(*expr*) - *generates optimized code*

generate::TeX(*expr*) - *generates TEX formatted string from expressions*

A1.12 Library 'Graph' - Utilities for Directed and Undirected Graphs

Graph::addEdges(*G*, [*e1, e2, ..., en*]) - *adds the specified edges to the graph G*

Graph::addVertices(*G*, [*v1, v2, .., vn*]) - *adds the specified vertices to the graph G*

Graph::admissibleFlow(*G, flow*) - *checks a flow for admissibility in the graph G*

Graph::bipartite(*G*) - *checks if the given graph G is bipartite*

Graph::breadthFirstSearch(*G*) - *traverses through a graph via breadth first search*

Graph::checkForVertices([*e1, .., en*], [*v1, .., vn*]) - *checks if all vertices out of given list of edges are in the specified list of vertices*

Graph::chromaticNumber(*G*) - *returns the chromatic number of the graph G*

Graph::chromaticPolynomial(*G, x*) - *produces the chromatic polynomial of the graph G*

Graph::contract(*G, VetexTable*) - *adds to the given graph a new vertex and edges connecting it with vertices specified in the VertexTable*

Graph::convertSSQ(*G, q, s*) - *converts the given graph G into a single source sink graph*

Graph::createCircleGraph(*L*) - *produces a circle graph for the given list of vertices L*

Graph::createCompleteGraph(*n*) - *produces a complete graph with n vertices*

Graph::createGraphFromMatrix(*M*) - *produces a graph representing the given matrix M*

Graph::createRandomEdgeCosts(*G, x1..x2*) - *produces random edge costs in the range x1..x2*

Graph::createRandomEdgeWeights(*G, x1..x2*) - *produces random edge weights in the range x1..x2*

Graph::createRandomGraph(*Vn, En*) - *produces a random graph with the given number of vertices Vn and given number of edges En*

Graph::createRandomVertexWeights(*G,x1...x2*) - *produces random vertex weights in the range x1...x2*

Graph::depthFirstSearch(*G*) - *traverses through the graph G via depth first search*

Graph::getAdjacentEdgesEntering(*G, vertex*) - *produces all incoming edges to the given vertex*

Graph::getAdjacentEdgesLeaving(*G, vertex*) - *produces all outgoing edges from the given vertex*

Graph::getBestAdjacentEdge(*G, vertex*) - *returns the best incident edge according to the specified attributes*

Graph::getEdgeCosts(*G*) - *produces a table containing the edge costs of the specified graph G*

Graph::getEdgeDescriptions(*G*) - *produces a table containing the edge descriptions of the specified graph G*

Graph::getEdgeNumber(*G*) - *produces a list containing all edges of the specified graph G*

Graph::getEdgeWeights(*G*) - *produces a list containing all edge weights of the specified graph G*

Graph::getEdges(*G*) - *produces a list of all edges of the given graph G*

Graph::getEdgesEntering(*G*) - *produces a list of direct predecessors of each vertex in the given graph G*

Graph::getEdgesLeaving(*G*) - *produces a list of direct successors of each vertex in the given graph G*

Graph::getSubGraph(*G, [vertices]*) - *produces a subgraph that contains only specified vertices and belonging to them edges*

Graph::getVertexNumber(*G*) - *returns number of vertices in the given graph G*

Graph::getVertexWeights(*G*) - *returns a table of the vertex weights for the given graph G*

Graph::getVertices(*G*) - *returns a list of vertices of the given graph G*

Graph::inDegree(*G*) - *returns the number of edges coming to each vertex of the graph G*

Graph::isConnected(*G*) - *checks if the given graph is connected*

Graph::isDirected(*G*) - *checks if the given graph is directed*

Graph::isEdge(*G, [edges]*) - *checks if the given edges exist in the graph G*

Graph::isVertex(*G, [vertices]*) - *checks if the given vertices exist in the graph G*

Graph::longestPath(*G, v*) - *finds in the given graph G the longest path starting from the vertex v*

Graph::maxFlow($G, v1, v2$) - *computes the max flow through the graph G from vertex v1 to the vertex v2*

Graph::minCost(G) - *computes the minimal cost flow for the graph G*

Graph::minCut($G, v1, v2$) - *computes the minimal cut in the graph G separating vertex v1 from the vertex v2*

Graph::minimumSpanningTree(G) - *produces the minimum spanning tree of the given graph G*

Graph::outDegree(G) - *produces the number of edges leaving each vertex of the given graph G*

Graph::plotBipartiteGraph(G) - *produces a plot scene with the given graph in a bipartite layout; use the* plot *command to plot the scene*

Graph::plotCircleGraph(G) - *produces a plot scene with the given graph in a circle layout; use the* plot *command to plot the scene*

Graph::plotGridGraph(G) - *produces a plot scene with the given graph in a grid layout; use the* plot *command to plot the scene*

Graph::printEdgeCostInformation(G) - *prints the edge cost for the given graph G*

Graph::printEdgeDescInformation(G) - *prints the edge description for the given graph G*

Graph::printEdgeInformation(G) - *prints the edges used the given graph G*

Graph::printEdgeWeightInformation(G) - *prints the edge weights for the given graph G*

Graph::printGraphInformation(G) - *prints a summary of information for the given graph G*

Graph::printVertexInformation(G) - *prints information about the vertices used the given graph G*

Graph::removeEdge($G, [e1, e2, \ldots, en]$) - *removes the specified edges from the graph G*

Graph::removeVertex($G, [v1, v2, \ldots, vn]$) - *removes the specified vertices from the graph G*

Graph::residualGraph(G, f) - *produces the residual of the given graph with respect to the flow f*

Graph::revert(G) - *reverts all edges of the given graph G*

Graph::setEdgeCosts($G, [e1, .., en], [c1, c2, ..cm]$) - *assigns the specified costs to the given edges*

Graph::setEdgeDescriptions($G, [e1, .., en], [d1, d2, ..dn]$) - *assigns the specified edge descriptions to the given edges*

Graph::setEdgeWeights(*G*, [*e1*,.., *en*], [*w1*, *w2*,.. *wn*]) - *assigns the specified edge weights to the given edges*

Graph::setVertexWeights(*G*, [*v1*,.., *vn*], [*w1*, *w2*,.. *wn*]) - *assigns the specified vertex weights to the given vertices*

Graph::shortestPathAllPairs(*G*) - *produces the length of all shortest paths in the given graph G*

Graph::shortestPathSingleSource(*G*, StartVertex= *v*) - *produces the length of the shortest path from a given vertex v to any other vertex in the graph G*

Graph::stronglyConnectedComponents(*G*) - *produces a list of strongly connected components in the given graph G*

Graph::topSort(*G*) - *a topological sort of the given graph G*

A1.13 Library 'groebner' — Utilities for Groebner Bases

groebner::dimension(*polynomials*) - *the dimension of the affine variety generated by polynomials*

groebner::gbasis(*polynomials*) - *the computation of a reduced Gröbner basis*

groebner::normalf(*p, polynomials*) - *the complete reduction modulo a polynomial ideal*

groebner::spoly(*p1, p2*) - *the S-polynomial of two polynomials*

groebner::stronglyIndependentSets(*G*) - *strongly independent set variables of a Groebner basis*

A1.14 Library 'import' — Utilities for Reading Data

import::readbitmap("*file*") - *reads in the given bitmap file*

import::readdata("*file*") - *reads ASCII data files*

import::readlisp(*string*) - *parse Lisp-formatted string*

A1.15 Library 'intlib' — Definite and Indefinite Integration

intlib::byparts(*integral, du*) - *transforms integral using by parts formula*

intlib::changevar(*integral, equ*) - *transforms integral by changing variables*

A1.16 Library 'linalg' — the Linear Algebra Package

linalg::addCol(*A, c1, c2, s*) - *produces a copy of matrix A with c2= c2+s * c1* linalg::addRow(A, $r1$, $r2$, s) - *produces a copy of matrix A with $r2 = r2 + s * r1$*

linalg::adjoint(A) - *produces the adjoint of a matrix*

linalg::angle(A, B) - *calculates the angle between two vectors*

linalg::basis(S) - *produces the basis for a vector space*

linalg::charmat(A, x) - *produces characteristic matrix*

linalg::charpoly(A, x) - *produces characteristic polynomial of the matrix A*

linalg::col(A, c) - *extracts column c from matrix A*

linalg::companion(p) - *produces the companion matrix of a univariate polynomial p*

linalg::concatMatrix(A, B,.., C) - *joins matrices horizontally*

linalg::crossProduct(A, B) - *produces the cross product of two 3D vectors*

linalg::curl(v, x) - *produces the curl of a vector field*

linalg::delCol(A, $c1$, $c2$) - *deletes in the matrix A columns c1...c2*

linalg::delRow(A, $r1$, $r2$) - *deletes in the matrix A rows r1...r2*

linalg::det(A) - *produces determinant of the matrix A*

linalg::divergence(v, x) - *produces the divergence of a vector field*

linalg::eigenvalues(A) - *produces eigenvalues of the matrix A*

linalg::eigenvectors(A) - *produces the eigenvectors of the matrix A*

linalg::expr2Matrix($equations$, $[x,y,z]$) - *transforms system of linear equations into a matrix*

linalg::factorCholesky(A) - *produces the Cholesky decomposition of a matrix*

linalg::factorLU(A) - *produces LU-decomposition of a matrix*

linalg::factorQR(A) - *produces QR-decomposition of a matrix*

linalg::frobeniusForm(A) - *produces Frobenius form of a matrix*

linalg::gaussElim(A) - *performs Gaussian elimination of a matrix*

linalg::gaussJordan(A) - *performs Gauss-Jordan elimination of a matrix*

linalg::grad(f, x) - *produces the vector gradient*

linalg::hermiteForm(A) - *produces Hermite normal form of a matrix*

linalg::hessenberg(A) - *produces Hessenberg matrix*

linalg::hessian(f,x) - *produces Hessian matrix of a scalar function*

linalg::hilbert(n) - *produces n×n Hilbert matrix generated by the function* $h(i,j) = (i+j-1)^{-1}$

linalg::intBasis($S1$,$S2$...) - *produces the basis for the intersection of vector spaces*

linalg::invpascal(n) - *produces the inverse of the Pascal matrix*

linalg::invvandermonde($v1$,$v2$,...,vn) - *produces the inverse of the Vandermonde matrix*

linalg::inverseLU(*A*) - *computing the inverse of a matrix using LUdecomposition*

linalg::invhilbert(*n*) - *produces an inverse of the n × n Hilbert matrix generated by*
$h(i,j) = (i + j - 1)^{-1}$

linalg::isHermitean(*A*) - *checks whether a matrix is Hermitean*

linalg::isPosDef(*A*) - *tests a matrix for positive definiteness*

linalg::isUnitary(*A*) - *tests whether a matrix is unitary*

linalg::jacobian(*v, x*) - *produces Jacobian matrix of a vector function*

linalg::jordanForm(*A*) - *produces Jordan normal form of a matrix*

linalg::kroneckerProduct(*A, B*) - *produces the Kronecker product of matrices A
and B*

linalg::laplacian(*f*, [*x1,x2,..*]) - *produces Laplacian of the function*

linalg::matdim(*A*) - *produces dimensions of the matrix A*

linalg::matlinsolve(*A*) - *solves system of equations in matrix form*

linalg::matlinsolveLU(*A*) - *solves system of equations in matrix form*

linalg::minpoly(*A, x*) - *produces the minimal polynomial of a matrix*

linalg::multCol(*A, c, s*) - *multiples in matrix A column c by number s*

linalg::multRow(*A, r, s*) - *multiples in matrix A row r by number s*

linalg::ncols(*A*) - *returns number of columns in the matrix A*

linalg::nonZeros(*A*) - *returns number of non-zero elements in the matrix*

linalg::normalize(*A*) - *normalizes a vector A*

linalg::nrows(*A*) - *returns number of rows in the matrix A* *

linalg::nullspace(*A*) - *produces the basis for the null space of a matrix*

linalg::pascal(*n*) - *returns the Pascal matrix*

linalg::potential(*f,x*) - *checks if the vector field f is a gradient field and computes
the potential (if it exists)*

linalg::ogCoordTab[*ogName*](*u1, u2, u3*) - *produces the table of orthogonal
coordinate transformations*

linalg::orthog(*S*) - *produces orthogonalization of vectors in S*

linalg::permanent(*A*) - *produces the permanent of a matrix*

linalg::pseudoInverse(*A*) - *produces the Moore-Penrose inverse of a matrix*

linalg::randomMatrix(*n, m, domain, bound, options*) - *creates a new matrix with
random elements*

linalg::rank(*A*) - *calculates the rank of a matrix*

linalg::row(*A, r*) - *extracts row r from the matrix A*

linalg::scalarProduct(A, B) - *produces scalar product of two vectors*

linalg::setCol(A, c, $c1$) - *replaces in A column c by new column vector $c1$*

linalg::setRow(A, r, $r1$) - *replaces in A row r by new row vector $r1$*

linalg::smithForm(A) - *produces Smith canonical form of a matrix*

linalg::stackMatrix(A, B,.., C) - *joins given matrices vertically*

linalg::submatrix(A, $r1..r2$, $c1..c2$) - *produces the submatrix defined by the given ranges*

linalg::substitute(A, B, m, n) - *substitutes the given matrix B inside of the matrix A*

linalg::sumBasis($S1$, $S2$,..) - *produces the basis for the sum of vector spaces*

linalg::swapCol(A, $c1$, $c2$) - *swaps columns $c1$ and $c2$ in matrix A*

linalg::swapRow(A, $r1$, $r2$) - *swaps rows $r1$ and $r2$ in matrix A*

linalg::sylvester(p, q) - *produces a Sylvester matrix of two polynomials p and q*

linalg::toeplitz($m, n, [t_k,\ldots,t_{-k}]$) - *produces the m×n Toeplitz matrix*

linalg::toeplitzSolve(t, y) - *produces the solution of the linear Toeplitz system*

linalg::tr(A) - *produces the trace of a matrix*

linalg::transpose(A) - *produces the transposition of A, i.e. A^t*

linalg::vandermonde($v1,\ldots, vn$) - *produces the $n \times n$ Vandermonde matrix*

linalg::vandermondeSolve(v, y) - *solves a linear Vandermonde system*

linalg::vecdim(V) - *returns number of elements of a given vector*

linalg::vectorOf(R, n) - *specifies the type of vectors over the ring R*

linalg::vectorPotential(j, $[x1, x2, x3]$) - *produces the vector potential of a three-dimensional vector field*

linalg::wiedemann(A, b) - *solves linear systems using the Wiedemann's algorithm*

A1.17 Library 'linopt' — Tools for Linear Optimization

linopt::corners($[constr, obj]$) - *returns the feasible corners of a linear program*

linopt::maximize($[constr, obj]$) - *maximizes a linear or mixed-integer program*

linopt::minimize($[constr, obj]$) - *minimizes a linear or mixed-integer program*

linopt::plot_data($[constr, obj]$) - *plots the feasible region of a linear program*

linopt::Transparent($[constr, obj]$) - *returns the ordinary simplex of a linear program*

A1.18 Library 'listlib' — Operations on Lists

listlib::insert(*list, element*) - *inserts an element into a list*
listlib::insertAt(*list, element, place*) - *inserts an element into a list*
listlib::merge(*list1, list2*) - *merges two ordered lists*
listlib::removeDupSorted(*list*) - *removes duplicate entries from an ordered list*
listlib::removeDuplicates(*list*) - *removes duplicate entries*
listlib::setDifference(*list1, list2*) - *removes elements from a list*
listlib::singleMerge(*list1, list2*) - *merges two ordered lists without duplicates*
listlib::sublist(*list1, list2*) - *searches for sublists*

A1.18.1 Related Functions in MuPAD Standard Collection

. - *the dot operator to concatenate two lists, e.g.* [a,b,c].[x,y] *produces*
      [a,b,c,x,y]
_concat(*list1.list2*) - *the kernel function to concatenate two lists, strings*
append(*list,object*) - *adds a given object at the end of the list*
revert(*list*) - *reverts the order of elements in the given list*

A1.19 Library 'matchlib' — Pattern Matching Tools

matchlib::analyze(*expr*) - *analyses the structure of any expression.*

A1.20 Library 'module' — Module Management Tools

module("*modname*") - *loads a given module*
module::age() - *produces the module age in computer memory*
module::displace(*modname*) - *unloads module*
module::func(*modname*) - *creates a module function environment*
module::help(*modname*) - *displays information about specified module*
module::load(*modname*) - *loads module*
module::max(*nr*) - *sets the max number of simultaneously loadable modules*
module::stat() - *produces the status of the module manager*
module::which(*modname*) - *returns installation path of a dynamic module*

A1.21 Library 'numeric' — Tools for Numerical Methods

numeric::butcher(*method*) - *returns the Butcher parameters of the Runge-Kutta scheme method*

numeric::complexRound(z) - *rounds a complex number towards the real or imaginary axis*

numeric::cubicSpline($[x0, y0], ..$) - *returns the cubic spline function interpolating sequence of points*

numeric::cubicSpline2d($[x0, x1 ...], [y0, y1, ..], z$) - *returns the bi-cubic spline function interpolating sequence of data*

numeric::det(A) - *produces determinant of the matrix*

numeric::eigenvalues(A) - *produces numerical eigenvalues of a matrix*

numeric::eigenvectors(A) - *produces numerical eigenvectors of a matrix*

numeric::expMatrix(A) - *returns exponential matrix exp(A)*

numeric::fMatrix(f, A) - *produces the matrix f(A) for a square matrix A*

numeric::factorCholesky(A) - *returns the factor L of the Cholesky factorization of A*

numeric::factorLU(A) - *returns LU factorization of a matrix*

numeric::factorQR(A) - *returns QR factorization of a matrix*

numeric::fft(*data*) - *returns discrete Fourier transformation of the given data*

numeric::fsolve(*equations*) - *returns a numerical approximation of a solution of the system of equations*

numeric::gldata(n, *digits*) - *returns the weights and the abscissae of the Gauss-Legendre quadrature rule*

numeric::gtdata(n) - *returns the weights and the abscissae of the Gauss-Tschebyscheff quadrature rule*

numeric::indets(*object*) - *returns a set of indeterminantes contained in a given object*

numeric::int(f) - *computes a numerical approximation of an definite integral*

numeric::inverse(A) - *produces the inverse of a matrix*

numeric::invfft (*data*) - *produces the inverse discrete Fourier transformation*

numeric::leastSquares(A, b) - *computes least-squares solution of linear system*

numeric::linsolve(*equations*) - *solves a system of linear equations*

numeric::matlinsolve(A, B) - *solves a system of linear equations*

numeric::ncdata(n) - *produces weights and abscissae of the Newton-Cotes quadrature*

numeric::odesolve(*f, t0..t1, Y0*) - *produces a numerical solution of an ordinary differential equation*

numeric::odesolve2(*f, t0, Y0*) - *produces a numerical solution of an ordinary differential equation*

numeric::odesolveGeometric(*f, t0..t1, Y0*) - *produces a numerical solution of an ordinary differential equation on a homogenous manifold*

numeric::ode2vectorfield( ) - *converts a system of ordinary differential equations to vector field representation*

numeric::polyroots(*p*) - *produces numerical roots of a univariate polynomial*

numeric::polysysroots(*equations, var*) - *produces a numerical roots of a system of polynomial equations*

numeric::quadrature(*f(x), x = a..b*) - *produces a numerical integral*

numeric::rationalize(*object*) - *approximates floating point number by a rational*

numeric::realroot(*f(x), x = a..b*) - *numerical search for a real root of a real function f*

numeric::realroots(*f(x), x = a..b*) - *isolates intervals containing real roots of the function f*

numeric::singularvalues(*A*) - *produces numerical singular values of a given matrix*

numeric::singularvectors(*A*) - *produces numerical singular vectors of a matrix*

numeric::solve(*equations*) - *produces numerical solution of equations*

numeric::sort(*list*) - *sorts a list, output is given in floating point notation*

numeric::spectralradius(*A, x0, n*) - *produces the eigenvalue of the matrix A that has the largest absolute value*

numeric::sum(*f[i], i = a..b*) - *computes numerical approximation of the sum* $\sum_{i=a}^{b} f(i)$

numeric::startHardwareFloats() - *enables hardware floating point arithmetic*

A1.22 Library 'numlib' — Elementary Number Theory

numlib::contfrac(*x*) - *creates the continued fraction approximation for a real number x*

numlib::cornacchia(*a, b, m*) - *produces all pairs of positive and relatively prime integers x,y such that* $ax^2 + b^2 = m$

numlib::decimal(*x*) - *produces the decimal expansion of the rational number x*

numlib::divisors(*n*) - *produces the list of positive divisors of n* numlib::ecm(*n*) - *factorizes an integer using elliptic curve method*

numlib::factorGaussInt(*n*) - *produces the factorization of the Gaussian integer into Gaussian primes*

numlib::fibonacci(*n*) - *produces the n-th Fibonacci number*

numlib::fromAscii(*list*) - *converts a list of ASCII code to string*

numlib::g_adic(*n*) - *g-adic representation of a nonnegative integer*

numlib::ichrem(*a*, *m*) - *produces the least nonnegative integer representing Chinese remainder theorem*

numlib::igcdmult(*n1*, *n2*,.., *nk*) - *extends the* igcdex *function to more than two arguments*

numlib::invphi(*n*) - *produces the inverse of the Euler function for a given integer n*

numlib::ispower(*n*) - *checks if n has a form m^k for some positive integers m and k*

numlib::isquadres(*n*, *m*) - *tests is a given number is a quadratic residue modulo m*

numlib::issqr(*n*) - *tests if n is a square of an integer*

numlib::jacobi(*n*, *m*) - *produces the value of the Jacobi symbol (n | m)*

numlib::Lambda(*n*) - *returns the value of Mangoldt's function*

numlib::lambda(*n*) - *produces the value of Carmichael's function of n*

numlib::legendre(*n*, *p*) - *produces the Legendre symbol (n | p)*

numlib::lincongruence(*a*,*b*,*m*) - *produces the list of all solutions of the linear congruence $a * x \equiv b \pmod{m}$*

numlib::mersenne() - *produces the list of all Mersenne primes known to date*

numlib::moebius(*n*) - *produces the value of the Möbius function of n*

numlib::mpqs(*n*) - *applies multi-polynomial quadratic sieve to integer n*

numlib::mroots(*p*, *m*) - *produces the list of all integers x such that $p(x) \equiv 0 \pmod{m}$*

numlib::msqrts(*n*, *m*) - *produces the list of all integers such that $n^2 \equiv n \pmod{m}$*

numlib::numdivisors(*n*) - *produces the number of positive divisors of n*

numlib::numprimedivisors(*n*) - *produces the number of prime divisors of n*

numlib::Omega(*n*) - *produces the number of the prime divisors of a natural number n and their multiplicity*

numlib::omega(*n*) - *produces the number of prime divisors of n*

numlib::order(*n*, *m*) - *produces the order of the residue class of n modulo m*

numlib::phi(*n*) - *produces the Euler function of n*

numlib::pollard(*n*, *m*) - *tries to find a factor of n using Pollard's rho algorithm*

numlib::prevprime(*n*) - *produces the largest prime number p≤ n* numlib::primedivisors(*n*) - *produces the list of prime divisors of an integer n*

numlib::primroot(*n*) - *produces the last positive primitive root modulo n*

numlib::proveprime(*n*) - *tests if n is a prime number using elliptic curves*

numlib::sigma(*n*) - *produces the sum of all positive divisors of n*

numlib::sqrt2cfrac(*n*) - *produces the continued fraction expansion of $\sqrt{n}$*

numlib::sqrtmodp(*n*, *p*) - *produces a solution x of the congruence $x^2 \equiv n(\bmod p)$*

numlib::sumdivisors(*n*) - *produces the sum of all divisors of an integer n*

numlib::sumOfDigits(*n*) - *produces the sum of digits of the given number n*

numlib::tau(*n*) - *produces numbers of positive divisors of n*

numlib::toAscii(*string*) - *converts a given string to ASCII*

A1.23 Library 'orthpoly' — Orthogonal Polynomials

orthpoly::chebyshev1(*n*, *x*) - *produces n-th degree Chebyshev polynomial of the first kind*

orthpoly::chebyshev2(*n*, *x*) - *produces n-th degree Chebyshev polynomial of the second kind*

orthpoly::curtz(*n*, *x*) - *produces n-th degree Curtz polynomial*

orthpoly::gegenbauer(*n*, *a*, *x*) - *produces n-th degree Gegenbauer polynomial*

orthpoly::hermite(*n*, *x*) - *produces n-th degree Hermite polynomial*

orthpoly::jacobi(*n*, *a*, *b*, *x*) - *produces n-th degree Jacobi polynomial*

orthpoly::laguerre(*n*, *a*, *x*) - *produces n-th degree generalized Laguerre polynomial*

orthpoly::legendre(*n*, *x*) - *produces n-th degree Legendre polynomial*

A1.24 Library 'output' — Tools for the Output of Data

output::asciiArt("*string*") - *prints the given string using text terminal mode*

output::fence("*c*", "*d*", "*string*") - *prints the given string between two given delimiters c and d, where c and d can be (,),[,],{,},|,|*

output::ordinal(*n*) - *converts an integer into the English ordinal number*

output::tableForm(*object*) - *prints the given object in the table form*

output::tree(*tree*) - *formats internally represented trees to display them in graphical form*

A1.25 Library 'plot' — 2D and 3D Graphical Objects

plot::AmbientLight(r) - *the ambient light object of a given intensity* $0 < r < 1$

plot::Arc2d($r,p,a..b$) - *the circular arc object with radius r, center at* $p = [x,y]$ *and angle from b to a*

plot::Arrow2d(p,q) - *the arrow object in 2D from the point* $p = [x,y]$ *to the point* $q = [x_1,y_1]$

plot::Arrow3d(p,q) - *the arrow object in 3D from the point* $p = [x,y,z]$ *to the point* $q = [x_1,y_1,z_1]$

plot::Bars2d([$data1,data2,...$]) - *the bar chart object representing given lists of data*

plot::Bars3d([$data1,data2,...$]) - *the 3D bar chart object representing given lists of data, data can be organized in the matrix or list structure*

plot::Box(p,q) - *the 3D box object with two opposite vertices* $p = [x,y,z]$ *and* $q = [x_1,y_1,z_1]$ *and edges parallel to the coordinate axes*

plot::Boxplot(L) - *the statistical box plot of L, L can be organized as a list, list of lists, matrix, or statistical data samples*

plot::Camera(p,f,a) - *the camera object places at the position* $p = [x,y,z]$ *pointed at the point* $f = [x_1,y_1,z_1]$ *with camera angle equal a (in radians)*

plot::Canvas($objects$) - *the canvas object - a container for all objects to be plotted on the same picture*

plot::Circle2d(r,p) - *the circle in 2D object with radius r and given center* $p = [x,y]$

plot::Circle3d(r,p,n) - *the circle in 3D object with given radius r, center p and normal vector n*

plot::ClippingBox($x1..x2,y1..y2,x1..z2$) - *clipping box for the scene*

plot::Cone(r,p,q) - *the cone object with given radius r of the base, center of the base p and top vertex q*

plot::Conformal($f(z),z = z1..z2$) - *the conformal plot object of the function f(z) over the rectangle* $z1..z2$

plot::CoordinateSystem2d($objects$) - *the 2D coordinate system object*

plot::CoordinateSystem3d($objects$) - *the 3D coordinate system object*

plot::Curve2d([$x(t),y(t)$]$,t = t1..t2$) - *the planar curve object given by parametric equation* $[x(t),y(t)]$

plot::Curve3d([$x(t),y(t),z(t)$]$,t = t1..t2$) - *the 3D curve object given by parametric equation* $[x(t),y(t),z(t)]$

plot::Cylinder(r,p,q) - *the cylinder object with given radius r and centers p and q of the bottom and top base respectively* plot::Cylindrical($[r,phi,z], u = a..b, v = c..d$) - *the surface object in the cylindrical coordinates*

plot::Density($f(x,y), x = x1..x2, y = y1..y2$) - *the density plot object of the function* $z = f(x,y)$

plot::DistantLight(p,t,q) - *the distant light object with light rays parallel to the vector pt and given intensity* $0 < q < 1$

plot::Dodecahedron() - *the dodecahedron object*

plot::Ellipse2d(R,r,p) - *the 2D ellipse object with given two radii R and r and center* $p = [x,y]$

plot::Ellipsoid($rx,ry,rz,center$) - *the ellipsoid object with given three radii and center*

plot::Function2d($f(x), x = x1..x2$) - *the function 2d object*

plot::Function3d($f(x,y), x = x1..x2, y = y1..y2$) - *the function 3d object*

plot::Group2d(*objects*) - *the object that is a group of 2d objects*

plot::Group3d(*objects*) - *the object that is a group of 3d objects*

plot::HOrbital(n,l,m) - *the surface representing the electron orbital of a hydrogen atom with quantum numbers n,l,m*

plot::Hatch(f) - *the hatch object for the given function* $f(x)$

plot::Hexahedron() - *the hexahedron object*

plot::Histogram2d(*data*) - *the histogram object of the given data*

plot::Icosahedron() - *the icosahedron object*

plot::Implicit2d($f, x = x1..x2, y = y1..y2$) - *the 2D plot object of the given implict equation* $f(x,y) = 0$

plot::Implicit3d($f, x = x1..x2, y = y1..y2, z = z1..z2$) - *the 3D plot object of the given implict equation* $f(x,y,z) = 0$

plot::Inequality(*inequality*, $x = x1..x2, y = y1..y2$) - *plot of an inequality or list of inequalities with two variables*

plot::Iteration($f(x), x0, n, x = x1..x2$) - *the plot object to visualize the iteration* $x_i = f(x_{i-1})$ *for* $i = 1,...,n$.

plot::Line2d($[x,y],[x1,y1]$) - *the 2D segment object*

plot::Line3d($[x,y,z],[x1,y1,z1]$) - *the 3D segment object*

plot::Lsys(*angle,rules*) - *the Lindenmayer system object*

plot::Matrixplot(A) - *the plot object representing the given matrix A*

plot::MuPADCube() - *the MuPAD cube object*

plot::Octahedron() - *the octahedron object*

plot::Ode2d($f,..$) - *the 2D plot object representing solutions of a differential* equation *f*

plot::Ode3d(*f,...*) - *the 3D plot object representing solutions of a differential equation f*

plot::Parallelogram2d(*p, v1, v2*) - *the parallelogram object with the center p and sides parallel to vectors v1 and v2*

plot::Parallelogram3d(*p, v1, v2*) - *the parallelogram object in 3D space with the center p and sides parallel to vectors v1 and v2*

plot::Piechart2d([*data1,...,dataN*]) - *the piechart object with pieces representing the given data*

plot::Piechart3d([*data1,...,dataN*]) - *the 3D piechart object with pieces representing the given data*

plot::Point2d([*x, y*]) - *the 2D point object with coordinates x,y*

plot::Point3d([*x, y, z*]) - *the 3D point object with coordinates x,y,z*

plot::PointLight(*p, q*) - *the point light object located at the point p = [x,y,z] and with the given intensity* $0 < q < 1$

plot::PointList2d([*p1, p2,..., pn*]) - *the object representing the list of points on the plane*

plot::PointList3d([*p1, p2,..., pn*]) - *the object representing the list of 3D points on the space*

plot::Polar([*r(u), phi(u), u = u1..u2*]) - *the object representing the curve r(t) in polar coordinates*

plot::Polygon2d([*p1, p2,.., pn*]) - *the polygon object connecting points p1, p2,.., pn*

plot::Polygon3d([*p1, p2,..., pn*]) - *the 3D polygon object connecting points p1, p2,.., pn*

plot::Raster(*M, x = x1..x2, y = y1..y2*) - *the raster plot object representing the matrix M of RGB values*

plot::Rectangle(*x1..x2, y1..y2*) - *the rectangle object with given ranges of two sides*

plot::Rotate2d(*a, p, objects*) - *the plot object obtained by rotating given objects around point p and angle a (in radians)*

plot::Rotate3d(*a, p, q, objects*) - *the plot object obtained by rotating given objects around the vector $\overrightarrow{pq}$ and an angle a (in radians)*

plot::Scale2d([*r, s*], *objects*) - *the object obtained by scaling given objects by the vector* [*r, s*]

plot::Scale3d([*r, s, t*], *objects*) - *the object obtained by scaling given objects by the vector* [*r, s, t*]

plot::Scene2d(*objects*) - *the 2D scene object containing the given 2D objects* plot::Scene3d($objects$) - *the 3D scene object containing the given 3D objects*

plot::Sphere(r, c) - *the sphere object with given radius r and center c*

plot::Spherical($[r(u, v), phi(u, v), theta(u, v)]$) - *the plot object representing function r(u, v) in spherical coordinates*

plot::SpotLight($p, t, alpha, int$) - *the spot light object located at the point $p = [x, y, z]$, pointing at the point $t = [x_1, y_1, z_1]$ with the given angle alpha and intensity $0 < int < 1$*

plot::Sum($a(n), n = n1..n2$) - *the object representing partial sums of the given sequence a(n)*

plot::Surface($[x(u, v), y(u, v), z(u, v)], u = u1..u2, v = v1..v2$) - *the object representing the given parametric equation in 3D*

plot::SurfaceSTL($filename$) - *the 3D surface representing STL data (STL is a scripting format similar to VRML used for describing complex objects in 3D)*

plot::SurfaceSet($[p1, p2, .., pn]$) - *the surface object representing list of triangle or quad coordinates*

plot::Tetrahedron() - *the tetrahedron object*

plot::Text2d($string, p$) - *the plot object representing the specified text and located in the given point $p = [x, y]$*

plot::Text3d($string, p$) - *the plot object representing the specified text located in the given point $p = [x, y, z]$*

plot::Transform2d($b, A, objects$) - *the 2D object obtained by transforming given objects; A is matrix of transformation, b is the shift vector*

plot::Transform3d($b, A, objects$) - *the 3D object obtained by transforming given objects; here A is the matrix of transformation, b is the shift vector*

plot::Translate2d($[x, y], objects$) - *the object obtained by translating given objects along the vector $[x, y]$*

plot::Translate3d$[x, y, z], objects$) - *the object obtained by translating given objects along the vector $[x, y, z]$*

plot::Tube($[x(t), y(t), z(t)], r(t), t = t1..t2$) - *the object representing a 3D curve as a tube with given radius r(t)*

plot::Turtle($commands$) - *the object representing a turtle path*

plot::VectorField2d($[v(x, y), u(x, y], x = x1..x2, y = y1..y2$) - *the vector field object*

plot::XRotate($f(x), x = x1..x2$) - *the plot object representing surface of revolution of f(x) around x-axis*

plot::ZRotate($f(x), x = x1..x2$) - *the plot object representing surface of revolution of f(x) around z-axis* plot::copy(*object*) - *makes a copy of an existing graphical object*

plot::getDefault(*plot::object::attribute*) - *returns current value of the specified attribute*

plot::modify(*object*) - *the procedure to modify attributes of the given object*

plot::setDefault(*plot::object::attribute=value*) - *the procedure to change value of the specified attribute*

A1.25.1 Related Functions in MuPAD Standard Collection

plot(*objects*) - *plots all the given graphical objects*

plotfunc2d(*f1, f2,...*) - *quick plot of 2D functions*

plotfunc3d(*f1, f2...*) - *quick plot of 3D functions*

A1.26 Library 'polylib' — Tools for Polynomials

polylib::coeffRing(*p*) - *produces the coefficient ring of the specified polynomial p*

polylib::cyclotomic(*n, x*) - *produces the cyclotomic polynomials*

polylib::decompose(*p, x*) - *produces the functional decomposition of polynomials*

polylib::discrim(*p, x*) - *produces the discriminant of the polynomial with respect to variable x*

polylib::divisors(*p*) - *produces all divisors of a polynomial*

polylib::Dpoly(*f*) - *the differential operator for polynomials*

polylib::elemSym([*x1, .., xn*], *k*) - *produces k-th elementary symmetric polynomial*

polylib::makerat(*expr*) - *converts the given expression into a rational function*

polylib::minpoly(*a, n, x*) - *produces the minimal polynomial*

polylib::Poly([*x1,...,xn*], *R*) - *the domain of polynomials over the ring R*

polylib::primitiveElement(*f, g*) - *produces the primitive element*

polylib::primpart(*p*) - *produces the primitive part of the given polynomial*

polylib::randpoly() - *produces a random polynomial*

polylib::realroots(*p, epsilon*) - *produces intervals containing real roots of p*

polylib::representByElemSym(*p, [x1,...,xn]*) - *represents a given symmetric polynomial by elementary symmetric polynomials*

polylib::resultant(*p, q*) - *produces the resultant of p and q with respect to their first variable*

polylib::sortMonomials(*p*) - *sorts monomial with respect to the order of terms*

polylib::splitfield(*p*) - *produces the splitting field of a polynomial*

polylib::sqrfree(*p*) - *produces square-free factorizations of a polynomial*

A1.26.1 Related Functions in MuPAD Standard Collection

coeff(*p*) - *returns sequence of non-zero coefficients of a polynomial*

content(*p*) - *computes the content of the polynomial, i.e. gcd of its coefficients*

degreevec(*p*) - *returns a list of exponents of the leading term of a poly*

divide(*p, q*) - *divides two given polynomials*

expr(*object*) - *converts a given object into an element of a basic domain*

factor(*p*) - *factors a given polynomial into irreducible polynomials*

ground(*p*) - *returns the constant coefficient* $p(0, 0, ..0)$

lcoeff(*p*) - *returns the leading coefficients of the polynomial*

lmonomial(*p*) - *returns the leading term of a polynomial*

lterm(*p*) - *returns the leading term of a given polynomial*

nthcoeff(*p, n*) - *returns the n-th non zero coefficient of a polynomial*

poly(*f*) - *converts a polynomial expression into a polynomial*

tcoeff(*p*) - *produces the trailing coefficient of a polynomial*

A1.27 Library 'prog' — Programming Utilities

prog::allFunctions() - *examines and prints all functions and libraries*

prog::bless(*f, domain*) - *approves given the function f as a function of the specified domain*

prog::calltree(*statement*) - *visualize the call structure of nested function calls*

prog::changes(*object*) - *prints information about changes of the object*

prog::check(*object*) - *checks MuPAD objects, use to find errors in user defined objects*

prog::error(*number*) - *converts an internal error number into an error message*

prog::exprtree(*expression*) - *visualizes the given expression as a tree*

prog::find(*expression, operand*) - *produces all paths to the operand in an expression*

prog::getname(*object*) - *produces the name of the given MuPAD object*

prog::getOptions - *while called inside of a procedure gets and verifies all options from the parameters*

prog::init(*object*) - *initializes the given object*

prog::isGlobal(*identifier*) - *checks if the given identifier is used in the system*

prog::memuse(*statement*) - *shows the memory usage for executing the given statement*

prog::ntime() - *returns the speed of the current machine for typical library programs*

prog::profile(*statement*) - *displays timing data of nested function calls*

prog::remember - *the procedure extending the standard remember procedure*

prog::sort(*list, f*) - *the procedure to sort a given list according to the specified function f*

prog::tcov(*statement*) - *executes statement and for each program line counts the number of executions*

prog::test(*statement, res*) - *compares the calculation results*

prog::testerrors() - *returns a table of errors after running a test*

prog::testexit() - *closes automatic tests from test files*

prog::testfunc(*function*) - *initialize tests for the given MuPAD function*

prog::testinit(*protocol file*) - *initialize tests*

prog::testnum() - *returns the current test number*

prog::trace(*object*) - *prints result of an observation of a given MuPAD object*

prog::traced() - *lists all traced functions*

prog::untrace(*object*) - *terminates observation of a given object*

A1.28 Library 'property' — Properties of Identifiers

property::hasprop(*object*) - *checks is the given object has properties*

property::implies(*prop1, prop2*) - *tries to check if property 1 implies property 2*

property::Null - *the empty property*

property::simpex(*expr*) - *simplifies the Boolean expression*

A1.28.1 Related Functions in MuPAD Standard Collection

assume(*property*) - *assigns a property to a MuPAD object*

getprop(), getprop(*f*) - *returns mathematical property of a given expression*

is(*x, property*) - *checks if the given object has the given mathematical property*

unassume(*var*) - *removes properties of a given variable*

A1.29 Library 'RGB' - Color Names and color functions

RGB::ColorNames() - *produces list of predefined color names*

RGB::ColorNames(*string*) - *produces a list of color names that contain the specified string*

RGB::plotColorPalette(*string*) - *produces a picture with all colors such that their names contain the given string*

RGB::fromHSV(*HSVcolor*) - *converts the specified HSV color to RGB*

RGB::toHSV(*RGBcolor*) - *converts the specified RGB color to HSV*

RGB::*color*.[*a*] - *adds transparency to the specified* color, *here* $0 \leq a \leq 1$

Colors specified in the RGB library

The following list contains all colors predefined in the RGB library. In order to use any of them you have to specify it with the RGB:: prefix, for example RGB::AliceBlue, RGB::BLack30.

```
AliceBlue, AlizarinCrimson, Antique, Aqua, Aquamarine,
       AquamarineMedium,AureolineYellow, Azure,

Banana,Beige, Bisque, Black, Black0, Black10,Black100, Black15,
       Black20, Black25, Black30, Black35, Black40, Black45,
       Black5, Black50, Black55, Black60, Black65, Black70,
       Black75, Black80, Black85, Black90, Black95,
       BlanchedAlmond, Blue, BlueGrey, BlueLight, BlueMedium,
       BluePale, BlueViolet, Brick, Brown, BrownOadder,
       BrownOchre, Burlywood, BurntSienna, BurntUmber,

Cadet, CadmiumLemon, CadmiumOrange, CadmiumRedDeep,
       CadmiumRedLight, CadmiumYellow, CadmiumYellowLight,
       Carrot, Cerulean, Chartreuse, Chocolate,
       ChromeOxideGreen, CinnabarGreen, Cobalt, CobaltGreen,
       CobaltVioletDeep, ColdGray, ColdGrey, Coral, CoralLight,
       CornflowerBlue, Cornsilk, Cyan, CyanWhite,

DarkBlue, DarkGray, DarkGreen, DarkGrey, DarkOrange, DarkRed,
       DarkTeal, DeepOchre, DeepPink, DimGray,DimGrey,
       DodgerBlue,

Eggshell, EmeraldGreen, EnglishRed,

Firebrick, Flesh, FleshOchre, Floral, ForestGreen,

Gainsboro, GeraniumLake, Ghost, Gold, GoldOchre, Goldenrod,
       GoldenrodDark, GoldenrodLight, GoldenrodPale, Gray,
       Gray0, Gray10, Gray100, Gray15, Gray20, Gray25, Gray30,
       Gray35, Gray40, Gray45, Gray5, Gray50, Gray55, Gray60,
       Gray65, Gray70, Gray75, Gray80, Gray85, Gray90, Gray95,
       Green, GreenDark, GreenLight, GreenPale, GreenYellow,
       GreenishUmber, Grey, Grey0, Grey10, Grey100, Grey15,
       Grey20, Grey25, Grey30, Grey35, Grey40, Grey45, Grey5,
       Grey50, Grey55, Grey60, Grey65, Grey70, Grey75, Grey80,
       Grey85, Grey90, Grey95,
```

Honeydew, HotPink,

IndianRed, Indigo, Ivory, IvoryBlack, Khaki, KhakiDark,

LampBlack, Lavender, LavenderBlush, LawnGreen, LemonChiffon,
LightBeige, LightBlue, LightGoldenrod, LightGray,
LightGreen, LightGrey, LightOrange, LightSalmon,
LightTurquoise, LightYellow, Lime, LimeGreen, Linen,

MadderLakeDeep, Magenta, ManganeseBlue, Maroon, MarsOrange,
MarsYellow, Melon, MidnightBlue, Mint, MintCream,
MistyRose, Moccasin, MuPADGold,

NaplesYellowDeep, Navajo, Navy, NavyBlue,

Ochre, OldLace, Olive, OliveDrab, OliveGreen, OliveGreenDark,
Orange, OrangeRed, Orchid, OrchidDark, OrchidMedium,

PaleBlue, PapayaWhip, Peach, PeachPuff, Peacock, PermanentGreen,
PermanentRedViolet, Peru, Pink, PinkLight, Plum,
PowderBlue, PrussianBlue, Purple, PurpleMedium,

Raspberry, RawSienna, RawUmber, Red, Rose, RoseMadder,
RosyBrown, RoyalBlue,

SaddleBrown, Salmon, SandyBrown, SapGreen, SeaGreen,
SeaGreenDark, SeaGreenLight, SeaGreenMedium, Seashell,
Sepia, Sienna, SkyBlue, SkyBlueDeep, SkyBlueLight,
SlateBlue, SlateBlueDark, SlateBlueLight,
SlateBlueMedium, SlateGray, SlateGrayDark,
SlateGrayLight, SlateGrey, SlateGreyDark, SlateGreyLight,
Smoke, Snow, SpringGreen, SpringGreenMedium, SteelBlue,
SteelBlueLight,

Tan, Teal, TerreVerte, Thistle, Titanium, Tomato, Turquoise,
TurquoiseDark, TurquoiseLight, TurquoisePale,

Ultramarine, UltramarineViolet,

VanDykeBrown, VenetianRed, Violet, VioletDark, VioletRed,
VioletRedMedium, VioletRedPale, ViridianLight,

WarmGray, WarmGrey, Wheat, White,

Yellow, YellowBrown, YellowGreen, YellowLight, YellowOchre,

Zinc

A1.30 Library 'Series' — Tools for Series Expansions

Series::gseries(*f*(*x*), *x*) - *produces generalized series expansions*

Series::Puiseux(*f*(*x*), *x*) - *produces initial segment of the truncated Puiseux series*

A1.30.1 Related Functions in MuPAD Standard Collection

asympt($f(x)$, x) - *computes asymptotic series expansion*

series($f(x)$, x) - *produces a series expansion of a given expression*

A1.31 Library 'solvelib' — Tools for Solving Equations

solvelib::BasicSet - *represents the four infinite sets - integers, rationals, complex and real numbers*

solvelib::conditionalSort(*list*) - *sorts the list depending on parameters*

solvelib::getElement(*set*) - *returns an element of the given set*

solvelib::isFinite(*set*) - *checks if the given set is finite*

solvelib::pdioe(*p1, p2, p3, x*) - *solves polynomial Diophantine equations*

solvelib::preImage($f(x)$, x, *set*) - *produces the preimage of a set under mapping*

solvelib::Union(*set, t, parset*) - *produces the union of sets for all values of t from parset*

A1.32 Library 'stats' — Statistical Functions

stats::betaCDF(a, b) - *the cumulative distribution function of the beta distribution*

stats::betaPDF(a, b) - *the probability density function of the beta distribution*

stats::betaQuantile(a, b) - *the quantile function of the beta distribution*

stats::betaRandom(a, b) - *produces a procedure that returns beta deviates with shape parameters $a > 0$ and $b > 0$*

stats::binomialCDF - *the distribution function*

stats::binomialPF(n, p) - *produces a procedure representing the probability function*

stats::binomialQuantile(n, p) - *the binomial quantile function*

stats::binomialRandom(n, p) - *the binomial deviates*

stats::calc(*sample, [c1, c2,..], f1, f2, ...*) - *applies functions to the sample*

stats::cauchyCDF(n, p) - *the cumulative distribution function of the Cauchy distribution*

stats::cauchyPDF(a, b) - *the probability density function of the Cauchy distribution*

stats::cauchyQuantile(a, b) - *the quantile function of the Cauchy distribution*

stats::cauchyRandom(a, b) - *the random number generator function for Cauchy deviates* stats::chisquareCDF(*mean*) - *the cumulative distribution function of the chi-square distribution*

stats::chisquarePDF(*mean*) - *the probability density function of the chi-square distribution*

stats::chisquareQuantile(*mean*) - *the quantile function of the chi-square distribution*

stats::chisquareRandom(*mean*) - *the random number generator function for chi-square deviates*

stats::col(*sample, c1, c2, ..*) - *creates a new sample from selected columns of the given sample*

stats::concatCol(*s1, s2, ..*) - *creates a new sample containing columns of the given samples*

stats::concatRow(*r1, r2, ..*) - *creates a new sample containing rows of the given samples*

stats::correlation([*x1, x2, ..*],[*y1, y2, ..*]) - *produces the linear Bravais-Pearson correlation coefficient*

stats::correlationMatrix(*cov*) - *produces the correlation matrix for given covariance matrix*

stats::covariance([*x1, x2, ..*],[*y1, y2, ..*]) - *the covariance of data samples*

stats::csGOFT(*data, cells, CDF=f*) - *the classical chi-square goodness-of-fit test for the null hypothesis*

stats::empiricalCDF(*x1, x2, ..*) - *the empirical cumulative distribution function of a finite data sample*

stats::empiricalQuantile(*x1, x2, ..*) - *the quantile function of the empirical distribution*

stats::equiprobableCells(*k, q*) - *procedure to divide the real line into equiprobable intervals*

stats::erlangCDF(*a, b*) - *the cumulative distribution function of the Erlang distribution*

stats::erlangPDF(*a, b*) - *the probability density function of the Erlang distribution*

stats::erlangQuantile(*a, b*) - *the quantile function of the Erlang distribution*

stats::erlangRandom(*a, b*) - *random number generator function for Erlang deviates*

stats::exponentialCDF(*a, b*) - *the cumulative distribution function of the exponential distribution*

stats::exponentialPDF(*a, b*) - *the probability density function of the exponential distribution*

stats::exponentialQuantile(*a, b*) - *the quantile function of the exponential distribution*

stats::exponentialRandom(*a, b*) - *the random number generator function for exponential deviates*

stats::fCDF(*a, b*) - *the cumulative distribution function of the Fisher's f-distribution*

stats::fPDF(*a, b*) - *the probability density function of the Fisher's f-distribution*

stats::fQuantile(*a, b*) - *the quantile function of the Fisher's f-distribution*

stats::fRandom(*a, b*) - *the random number generator function for Fisher's f-deviates*

stats::gammaCDF(*a, b*) - *the cumulative distribution function of the gamma distribution*

stats::gammaPDF(*a, b*) - *the probability density function of the gamma distribution*

stats::gammaQuantile(*a, b*) - *the quantile function of the gamma distribution*

stats::gammaRandom(*a, b*) - *the random number generator function for gamma deviates*

stats::geometricCDF(*p*) - *the cumulative distribution function of the geometric distribution*

stats::geometricMean(*x1, x2, ..*) - *the geometric mean of a data samples*

stats::geometricPF(*p*) - *the probability function of the geometric distribution*

stats::geometricQuantile(*p*) - *the quantile function of the geometric distribution*

stats::geometricRandom(*p*) - *the random number generator for geometric deviates*

stats::harmonicMean(*x1, x2, ...*) - *produces the harmonic mean of a data sample*

stats::hypergeometricCDF(*N, X, n*) - *the cumulative probability function of the hypergeometric distribution*

stats::hypergeometricPF(*N, X, n*) - *the probability function of the hypergeometric distribution*

stats::hypergeometricQuantile(*N, X, n*) - *the quantile function of the hypergeometric distribution*

stats::hypergeometricRandom(*N, X, n*) - *the random number generator for the hypergeometric distribution*

stats::ksGOFT([*x1, x2, ...*], *CDF=f*) - *the Kolmogorov-Smirnov goodness-of-fit test*

stats::kurtosis(*x1, x2, ...*) - *the kurtosis of a data sample*

stats::linReg([*x1, x2, ...*], [*y1, y2, ...*]) - *linear regression*

stats::logisticCDF(*m, s*) - *the cumulative distribution function of the logistic distribution*

stats::logisticPDF(*m, s*) - *the probability density function of the logistic distribution*

stats::logisticQuantile(*m, s*) - *the quantile function of the logistic distribution* stats::logisticRandom(*m, s*) - *the random number generator for logistic deviates*

stats::mean(*x1, x2, ...*) - *produces the arithmetic mean of a data sample*

stats::meandev(*x1, x2, ...*) - *produces the mean deviation of a data sample*

stats::median(*x1, x2, ...*) - *produces the median value of a data sample*

stats::modal(*x1, x2, ...*) - *produces the most frequent value of a data sample*

stats::moment(*k, X, [x1, x2, ...]*) - *produces the k-th moment of a data sample*

stats::normalCDF(*m, v*) - *produces the cumulative distribution function of the normal distribution*

stats::normalPDF(*m, v*) - *produces the probability density function of the normal distribution*

stats::normalQuantile(*m, v*) - *produces the quantile function of the normal distribution*

stats::normalRandom(*m, v*) - *the random number generator for normal deviates*

stats::obliquity(*x1, x2, ...*) - *produces the obliquity of a data sample*

stats::poissonCDF(*m*) - *the cumulative distribution function of the Poisson distribution*

stats::poissonPF(*m*) - *the probability function of the Poisson distribution*

stats::poissonQuantile(*m*) - *the quantile function of the Poisson distribution*

stats::poissonRandom(*m*) - *the random number generator for the Poisson distribution*

stats::quadraticMean(*x1, x2, ...*) - *produces the quadratic mean of a data sample*

stats::reg(*data samples*) - *produces the general linear and nonlinear last squares fit*

stats::row(*sample, r1, r2, ..*) - *selects and rearranges rows of a given sample*

stats::sample([[*a11, .., a1n*], .. [*am1..., amn*]]) - *produces sample with m rows and n columns*

stats::sample2list(*sample*) - *converts a sample into a list of lists*

stats::selectRow(*s, c, x*) - *selects rows of a sample*

stats::sortSample(*sample, c1, c2, ..*) - *sorts rows of a given sample*

stats::stdev(*x1, x2, ..*) - *produces the standard deviation of a data sample*

stats::swGOFT([*x1, x2, ..*]) - *the Shapiro-Wilk goodness-of-fit for normality*

stats::tCDF(*a*) - *produces the cumulative distribution of the Student's t-distribution*

stats::tPDF(*a*) - *produces the probability density function of the Student's t-distribution*

stats::tQuantile(*a*) - *the quantile function of the Student's t-distribution*

stats::tRandom(*a*) - *the random generator for the Student's t-distribution* stats::tTest([*x1, x2, ..*]) - *the t-test for a mean*

stats::tabulate(*sample*) - *produces the statistics of duplicate rows in a sample*

stats::uniformCDF(*a, b*) - *the cumulative distribution function of the uniform distribution*

stats::uniformPDF(*a, b*) - *the probability density function of the uniform distribution*

stats::uniformQuantile(*a, b*) - *the quantile function of the uniform distribution*

stats::uniformRandom(*a, b*) - *the random number generator for uniformly continuos deviates*

stats::unzipCol(*list*) - *extracts columns from a list of lists*

stats::variance(*x1, x2, ..*) - *produces the variance of a data sample*

stats::weibullCDF(*a, b*) - *the cumulative distribution function of the Weibull distribution*

stats::weibullPDF(*a, b*) - *the probability density function of the Weibull distribution*

stats::weibullQuantile(*a, b*) - *the quantile function of the Weibull distribution*

stats::weibullRandom(*a, b*) - *the random number generator of the Weibull deviates*

stats::zipCol(*col1, col2, ..*) - *converts a sequence of columns into a list of lists*

A1.33 Library 'stringlib' — Tools for String Manipulation

stringlib::collapseWhitespace(*string*) - *replaces space patches in the given string by a single space*

stringlib::contains(*string1, string2*) - *tests if string 1 contains string 2*

stringlib::format(*str1, width*) - *adjusts the length of the string*

stringlib::formatf(*x, d*) - *converts a floating point number to a string*

stringlib::lower(*string*) - *converts the given string to lowercase*

stringlib::pos(*string, sstr*) - *produces position of the substring sstr in the given string*

stringlib::random() - *produces a random string*

stringlib::readText(*file*) - *reads in the given file and returns its content as a list of strings*

stringlib::remove(*string, sstr*) - *removes the substring sstr from the given string*

stringlib::split(*string, separator*) - *splits the given string in places where the specified separator occured*

stringlib::subs(*string, sstr=nsstr*) - *replaces a substring sstr by nsstr in a given string* stringlib::subsop(*string, n=newchar*) - *replaces character in position n by the new character*

stringlib::upper(*string*) - *convert the given string to uppercase*

stringlib::validIdent(*string*) - *returns TRUE if the given string is a valid identifier and FALSE otherwise*

A1.34 Library 'student' — the Student Package

student::equateMatrix(*A, variables*) - *produces a matrix equation*

student::isFree(*list of vectors*) - *tests for linear independence of vectors*

student::Kn(*n, F*) - *produces the vectors space of n-tuples over the field F*

student::plotRiemann(*f, x = a..b, n*) - *produces a plot object representing numerical approximation of the integral using n rectangles; use the* plot *procedure to plot the object*

student::plotSimpson(*f, x = a..b, n*) - *produces a plot object representing numerical approximation of the integral using the Simpson's rule; use the* plot *procedure to plot the object*

student::plotTrapezoid(*f, x = a..b, n*) - *produces a plot object representing numerical approximation of the integral using trapezoids; use the* plot *procedure to plot the object*

student::riemann(*f, x = a..b, n*) - *produces numerical approximation of the integral using n rectangles*

student::simpson(*f, x = a..b, n*) - *produces numerical approximation of the integral using the Simpson's rule*

student::trapezoid(*f, x = a..b, n*) - *produces numerical approximation of the integral using trapezoids*

A1.35 Library 'transform' — Integral Transformations

transform::fourier(*f, t, s*) - *produces the Fourier transformation*

transform::invfourier(*F, S, T*) - *produces the inverse Fourier transformation*

transform::invlaplace - *produces the inverse Laplace transformation*

transform::laplace(*f, t, s*) - *produces the Laplace transformation*

A1.36 Library 'Type' — Predefined Types

Type::AlgebraicConstant - *the type representing algebraic constants*

Type::AnyType - *the type representing an arbitrary MuPAD object*

Type::Arithmetical - *the type representing arithmetical objects*

Type::Boolean - *the type representing logical objects*

Type::Complex - *the type representing complex numbers*

Type::Constant - *the type representing constant objects*

Type::ConstantIdents - *the type representing constant identifiers in MuPAD*

Type::Equation - *the type representing equations*

Type::Even - *the type representing even integers*

Type::Function - *the type representing functions*

Type::Imaginary - *the type representing complex numbers with real part equal 0*

Type::IndepOf - *the type representing object that do not contain given identifiers*

Type::Integer - *the type representing integer numbers*

Type::Interval - *the type representing intervals of real numbers*

Type::ListOf - *the type representing lists of objects of the same type*

Type::ListProduct - *the type for testing lists*

Type::NegInt - *the type representing negative integers*

Type::NegRat - *the type representing negative rational numbers*

Type::Negative - *the type representing negative real numbers*

Type::NonNegInt - *the type representing non-negative integers* (≥ 0)

Type::NonNegRat - *the type representing non-negative rational numbers* (≥ 0)

Type::NonNegative - *the type representing non-negative real numbers* (≥ 0)

Type::NonZero - *the type representing complex numbers without 0*

Type::Numeric - *the type representing numerical objects*

Type::Odd - *the type representing odd integers*

Type::PolyExpr - *the type representing polynomial expressions*

Type::PolyOf - *the type representing polynomials*

Type::PosInt - *the type representing positive integers* (>0)

Type::PosRat - *the type representing positive rational numbers* (>0)

Type::Positive - *the type representing positive real numbers* (>0)

Type::Prime - *the type representing prime numbers*

Type::Product - *the type representing sequences*

Type::Property - *the type representing properties*

Type::RatExpr - *the type representing rational expressions*

Type::Rational - *the type representing rational numbers*

Type::Real - *the type representing real numbers*

Type::Relation - *the type representing relations*

Type::Residue - *the type representing a residue class*

Type::SequenceOf - *the type representing sequences of a given type*

Type::Series - *the type representing truncated Puiseux, Laurent and Taylor series*

Type::Set - *the type representing set-theoretic expressions*

Type::SetOf - *the type representing sets with elements of a given type*

Type::Singleton - *the type representing exactly one object*

Type::TableOf - *the type representing tables with specified entries*

Type::TableOfEntry - *the type representing tables with entries of a given type*

Type::TableOfIndex - *the type representing tables with specified indexes*

Type::Union - *the type representing objects having at least one of the specified types*

Type::Unknown - *the type representing variables*

Type::Zero - *the type representing a single number 0*

A1.37 MuPAD Environmental Variables

DIGITS:=*n* - *determines the number of digits in floating point calculations*

FILEPATH - *the variable that contains the path to a file*

HISTORY, HISTORY:=*n* - *determines the maximal number of entries in the history table*

LEVEL, LEVEL:=*n* - *determines the maximal substitution depth of identifiers*

LIBPATH - *the variable representing the directory where are library files*

MAXDEPTH, MAXDEPTH:=*n* - *prevents infinite recursion while calling a procedure*

MAXLEVEL, MAXLEVEL:=*n* - *determines the maximal substitution depth of identifiers*

NOTEBOOKFILE - *the variable representing the notebook file name*

NOTEBOOKPATH - *the variable representing the notebook file path*

ORDER, ORDER:=*n* - *the variable representing default number to be returned while producing series expansion*

PACKAGEPATH - *the variable representing the directory where the function package searches for packages*

READPATH - *the variable representing the directory from which library files will be loaded*

SEED - *the variable used to specify the seed for the random function, as a seed value use any non-zero integer*

WRITEPATH - *the variable representing the directory to which library files will be saved*

PRETTYPRINT, PRETTYPRINT:=*value* - *the variable controlling how the output is formatted*

TESTPATH:=*path* - *the directory where function* `prog::test` *will write files*

TEXTWIDTH, TEXTWIDTH:=*n* - *the variable controlling the number of characters in the output line on the screen*

Appendix 2

MuPAD Resources

The purpose of this appendix is to point out a few MuPAD-related resources. Most of the information about MuPAD can be found on the web site of the *MuPAD Research Group* at:

* http://www.mupad.de/

This place is the starting point leading to a number of MuPAD-related web sites. Here are just some of them.

The MuPAD Research Group

The *MuPAD Research Group* at the University of Paderborn is the scientific board of *SciFace Software, Inc*. This group develops ideas for new features of MuPAD, proposes new solutions, and implements them. Their web site can be found at:

* http://www.mupad.de/

This is the best place to look for MuPAD documentation and a complete bibliography of books and articles about MuPAD.

The SciFace Web Site

SciFace Software GmbH & Co. KG, in Germany, is the company that manages the MuPAD project, develops new graphical user interfaces and tools for visualization, and is responsible for the production and distribution of MuPAD all over the world. Their web site has two alternative addresses:

* http://www.mupad.com/

and

* http://www.sciface.com/

This web site contains all the information related to the commercial aspects of MuPAD—how to buy it, how to get a free license, or how to download evaluation versions. It also includes links to *SciFace* partners all over the world.

MuPAD TAN Server

The *MuPAD TAN Server* is the place where researchers, teachers, school administrators, and students can find detailed licensing information, download MuPAD, buy it online, and obtain registration numbers.

- `http://www.mupad.org/muptan.html`

If you wish to get MuPAD for your school or university, this is the best place to learn about discounts for schools and craft your MuPAD license according to the school needs and the school funds.

MuPAD in Education

There are two web sites with resources about the use of MuPAD in teaching. The main web site, *Schule und Studium*, is in German and thus is an invaluable resource for German-speaking users of MuPAD. Here is the link to German version of *Schule und Studium*:

- `http://www.mupad.de/schule/`

The English web site is published at:

- `http://www.mupad.de/schule/en/`

This web site is a bit less developed than its German counterpart, but the amount of material published here is growing rapidly. Both web sites contain a number of free MuPAD tutorials in PDF format.

MathPAD Online

MathPAD Online is an online magazine of the *MuPAD Research Group*. This web site started in the beginning of year 2003 and is still quite new. The objective of this web site is to collect the various publications about MuPAD, technologies that are used in MuPAD, and applications of MuPAD in research and education. Some of the articles published here were written by students and school teachers. This is the place where the MuPAD community can present its experience in using MuPAD and share ideas and MuPAD code. *MathPAD Online* can be found at:

- `http://www.mathpad.org`

As with many other online magazines, the content of the web site is updated frequently and new documents are added as soon as they arrive. On *MathPAD Online*, you can also find a list of MuPAD distributors all over the world.

MacKichan Software Inc.

MuPAD is distributed and supported by a number of companies worldwide. A significant source of information about MuPAD is the web site of one of its distributors, *MacKichan Software, Inc.* in the USA. Their web site is at:

- `http://www.mackichan.com/`

MacKichan Software, Inc. is widely known for their products Scientific WorkPlace (SWP), Scientific Word (SW), and Scientific Notebook (SNB). Allow me to mention that these products contain the best scientific word processor ever produced. It uses LaTeX as the format for its documents. With these programs you can edit any scientific text including mathematical documents with a lot of formulae. With SWP and SNB, you can also perform all the calculations inside of your document, because the programs also contain a powerful computing engine. My book was, in fact developed with the help of SNB. I used SNB to type the text, develop styles for printing, and produce the camera-ready pages.

Since the year 2000, MuPAD has been used as a computing engine for SWP and SNB. Thus, SWP and SNB can be considered as interactive, natural interfaces to MuPAD.

Book Resources

Finally, you can find a number of resources on *MuPAD Essentials*, the web site for my book, at:

- `http://www.mupad.com/majewski/`

or:

- `http://majewski.mupad.com`

This is the place where you should look for the source code of the many examples that were used throughout the pages of this book. I also intend to publish here the solutions to some of the more complicated programming exercises and information about book updates and revisions.

The web site for the second edition of this book is somewhat experimental. While developing this web site, I used the JavaView technologies (www.javaview.de) developed by Dr. Konrad Polthier and his research group *Geometry and Visualization* at *Zuse Institute* in Berlin, the SVG (Scalable Vector Graphics) format, and MathML. The JavaView format and applet are used to display interactive 3D graphics produced in MuPAD. While the JavaView display is quite impressive and the technology behind it is very powerful, the SVG part is a bit below my expectations. It contains only some of the static 2D examples that were created in MuPAD. Unfortunately, SVG is still in development and many graphical features that already exist in MuPAD cannot yet be translated into SVG format. Finally, MathML is used to display mathematical formulae on the web page.

In order to see each of these components, you need a few plugins. For the SVG graphics, you need the Adobe SVG plugin (www.adobe.com/svg/). The MathML formulae can be displayed using the MathML viewer from *Design Science* (www.dessci.com). You also can find links to these plugins on the book web site.

The 3D graphics in JavaView format can be seen through JavaView applets and they will come to your computer as soon as you try to open any of the 3D examples from the web site.

The primary goal of the *MuPAD Essentials* web site is to serve as a resource for the book. However, I will use it also as a showcase of technologies for the creation of online mathematical materials.

Index _____